About the Authors

DON O. BRUSH is currently Professor of Engineering in the Department of Civil Engineering at the University of California at Davis. He served as Chairman of this department from 1965 to 1968. Prior to his academic affiliation, from 1957 to 1964 Dr. Brush was a Senior Staff Scientist and a Senior Member of the Research Laboratories at the Lockheed Aircraft Corporation's Missiles and Space Company. He received his Ph.D. in 1957 from the University of Illinois.

BO O. ALMROTH is presently associated with the Lockheed Aircraft Corporation's Missiles and Space Company where he is a Senior Staff Scientist and a Senior Member of the Research Laboratories. His duties have primarily been in research with emphasis on stability and nonlinear behavior of thin shells. After graduation from the Techn. College in Örebro, Sweden, he worked as a stress analyst at Saab for thirteen years.

BUCKLING OF BARS, PLATES, AND SHELLS

**McGRAW-HILL
BOOK COMPANY**

New York
St. Louis
San Francisco
Düsseldorf
Johannesburg
Kuala Lumpur
London
Mexico
Montreal
New Delhi
Panama
Paris
São Paulo
Singapore
Sydney
Tokyo
Toronto

DON O. BRUSH

*Professor of Engineering
University of California, Davis*

BO O. ALMROTH

Lockheed Missiles and Space Company

Buckling of Bars, Plates, and Shells

This book was set in Times New Roman.
The editors were B. J. Clark and M. E. Margolies;
the cover was designed by Pencils Portfolio, Inc.;
the production supervisor was Dennis J. Conroy.
The drawings were done by Vantage Art, Inc.
Kingsport Press, Inc., was printer and binder.

Library of Congress Cataloging in Publication Data

Brush, Don Orr.
    Buckling of bars, plates, and shells.

    1. Buckling (Mechanics) 2. Structures, Theory of.
I. Almroth, B. O., joint author. II. Title.
TA656.2.B78   624′.171   74–9932
ISBN 0–07–008593–5

**BUCKLING OF
BARS, PLATES,
AND SHELLS**

1234567890 KPKP 7987654

CONTENTS

PREFACE

The subject matter of this book is the buckling behavior of structural members subjected to compressive loading. The members considered are columns, beams, beam columns, flat plates, circular rings, cylindrical shells, and general shells. Behavior is examined in terms of load-displacement curves for particular cases. Emphasis is on understanding the behavior of the structure and on calculation of the critical load.

Buckling analysis is fundamentally a subtopic of nonlinear rather than linear mechanics. The relative intractability of the governing nonlinear equations has led to the evolvement over the years of practical linear analysis procedures based on sometimes unstated approximations. As a consequence, many practicing structural analysts have come to view the theory of elastic stability as an arbitrary set of rules that is only marginally related to ordinary stress and deformation analysis. The role of stability analysis as a particular aspect of general nonlinear structural analysis is clearly described in advanced works on structural stability, but not in elementary ones. We believe there is need for an introductory textbook in which the methods of structural stability analysis are presented within the context of elementary nonlinear bending analysis.

In this book equilibrium and stability equations for bars, plates, and shells are systematically developed from a unified point of view. Except for analyses involving the twisting of bars with thin-walled open cross sections, the equilibrium and stability equations presented for columns, plates, circular rings, and cylindrical shells may be obtained by direct specialization of the equations given for shells of general shape. Familiarity with general shell theory on the part of the reader is not assumed, however. In each of the chapters on rings, plates, and shells, the nonlinear equilibrium equations and associated linear stability equations are developed, and then the stability equations are applied in the determination of the critical load for particular examples. For clarity in the presentations, emphasis throughout the book is placed on the simplest forms of the equations that illustrate the underlying principles.

The fundamental ideas, including geometric nonlinearity, primary and secondary equilibrium paths, bifurcation and limit points, the adjacent-equilibrium and minimum potential energy criteria for loss of stability, and the influence of initial imperfections, are presented in an elementary context in the first chapter. Chapters 2 to 6 cover applications of classical stability theory to columns, beams, plates, circular rings, cylindrical shells, and general shells. Advanced readers who are familiar with general shell theory may prefer to turn to Chapter 6 after completion of Chapter 1, and to consider the applications in Chapters 2 to 5 as special cases of the general shell equations in Chapter 6. The influence of initial imperfections is considered throughout the book and is the principal subject matter of Chapter 7. Some of the methods of numerical analysis that have been useful in the solution of linear stability equations with variable coefficients are examined in Chapter 8. Finally, in Chapter 9, the direct numerical solution of the governing nonlinear equations is presented for selected examples to illustrate the fact that the classical approach sometimes fails to yield all the information that is needed by the structural analyst.

Only static, conservative structural systems are treated in the basic analyses. Comments on the failure of columns, plates, and shells are included, but the design of structural elements is beyond the scope of the book.

The literature on elastic stability analysis is extremely extensive, especially for thin-walled shells and thin-walled open cross section beams. No attempt has been made to present a representative sampling of available solutions for various boundary conditions and methods of loading. Instead, characteristic examples have been selected to illustrate the kinds of problems that arise in elastic stability analysis, and references to other works in the literature are cited at the ends of the chapters. In every case we have attempted to choose the simplest example that illustrates the point.

The reader is assumed to have a thorough foundation in elementary mechanics of deformable bodies and in calculus and to have had some exposure to ordinary and partial differential equations. The book is intended as a textbook for first-year graduate students in aerospace and civil engineering and related fields and as a reference work for practicing structural analysts.

The authors express their sincere appreciation to John W. Hutchinson and Thomas J. Lardner for reading the manuscript and offering many valuable suggestions, and to Mrs. Florence Rose for typing most of the manuscript.

Don O. Brush
Bo O. Almroth

A	area
A, B	Lamé geometric coefficients
C	extensional stiffness parameter: $C = Eh/(1 - v^2)$
C_W	warping constant, in.^{6}
D	bending stiffness parameter: $D = Eh^3/[12(1 - v^2)]$
E	Young's modulus, psi
F	integrand in potential energy expression
G	shear modulus, psi
I	moment of inertia, in.^{4}
J	torsional constant, in.^{4}
K	column effective length factor
L	length, in.
M, M_{yy}, M_{zz}	bending moments, in.-lb
M_x, M_y, M_{xy}	bending and twisting moment intensities, in.-lb/in.
N	normal force, lb
N_x, N_y, N_{xy}	normal and shearing force intensities, lb/in.
P	load, lb
P_L	limit load, lb

Q	transverse shearing force, lb
Q_x, Q_y	transverse shearing force intensities, lb/in.
R_x, R_y	principal radii of curvature
U	strain energy
U_m, U_b	membrane and bending strain energy
V	total potential energy
X, Y, Z	rectangular cartesian coordinates
Z	cylinder geometry parameter: $Z = (1 - v^2)^{1/2}L^2/ah$
a	radius of ring, cylinder, or spherical cap, in.
a, b	plate length and width, respectively, in.
e_{xx}, e_{yy}, e_{xy}	linear displacement parameters
f	stress function
h	thickness, in.
k	load parameter: $k^2 = P/EI$
m, n	positive integers
p_x, p_y, p_z	surface load components, psi
q	load intensity, lb/in.
r	shell geometry parameter: $r = r_\theta \sin \phi$
r_ϕ, r_θ	principal radii of curvature for shell of revolution
s	coordinate variable
u, v, w	displacement components
x, y, z	orthogonal coordinates
Δ	displacement
Ω	potential energy of external forces
$\beta, \beta_{xx}, \beta_{yy}$	rotations (slope) of beam element
β_x, β_y	rotations of plate or shell element relative to the y and x coordinate directions, respectively
δ	displacement; variation
$\varepsilon_x, \varepsilon_y, \gamma_{xy}$	strain components
θ	circumferential coordinate
$\kappa_x, \kappa_y, \kappa_{xy}$	curvature changes and twist
λ	load parameter: $\lambda \equiv P/P_{\text{CL}}$
μ	imperfection amplitude parameter
v	Poisson's ratio
$\sigma_x, \sigma_y, \tau_{xy}$	stress components
ϕ	angle of twist, rad; meridional coordinate
$\chi_{xx}, \chi_{yy}, \chi_{xy}$	linear displacement parameters

Subscripts

0, 1   initial (prebuckling) and incremental quantities, respectively

x, y preceded by commas denote differentiation with regard to the respective coordinate variables

CL   classical value of critical load

cr   critical value of applied load

**BUCKLING OF
BARS, PLATES,
AND SHELLS**

INTRODUCTION

1.1 GEOMETRIC NONLINEARITY IN MECHANICS OF DEFORMABLE BODIES

In linear mechanics of deformable bodies, displacements are proportional to loads. The essence of buckling, however, is a disproportionate increase in displacement resulting from a small increase in load. Consequently, buckling analysis is a subtopic of nonlinear rather than linear mechanics. Nonlinearity in mechanics of deformable bodies is either physical or geometrical; i.e., it enters the theory either in the stress-strain relations or in expressions representing the influence of rotations of structural elements on the behavior of the structure. Except where otherwise noted, stresses and strains in this work are assumed to obey Hooke's law. Consequently, the nonlinearity is purely geometrical.

A simple illustration of the influence of geometric nonlinearity is provided by a clamped-end beam subjected to a concentrated lateral load P at midspan, as shown in Fig. 1.1. From linear beam theory the midspan deflection Δ is

$$\Delta = \frac{PL^3}{192EI} \qquad (1.1)$$

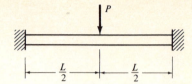

FIGURE 1.1
Beam clamped at both ends.

where E and I are Young's modulus and moment of inertia, respectively. A plot of P versus Δ for Eq. (1.1) is shown in Fig. 1.2*a*. The load-displacement relationship in Eq. (1.1) and Fig. 1.2*a*, however, is only an approximation of the actual P-Δ relationship. As the beam deflects, a tensile force is developed in the direction of the beam fibers. The vertical component of the tensile force becomes larger as the rotation or slope of off-center elements of the beam increases, and the deflection resistance of the beam increases in a nonlinear fashion. When the nonlinear interaction between applied load and resulting rotations is taken into account, the load-displacement plot is curved, as shown in Fig. 1.2*b*. The linear solution represented by Fig. 1.2*a* is satisfactory for sufficiently small values of rotation, but it is an approximation for all values of the applied load.

It is important to note that linear analysis in the mechanics of deformable bodies is an approximation except in cases of rotation-free deformation. We might also note that rotations are not assumed to equal zero in linear beam analysis; such an assumption would preclude deflection. They are assumed only to be so small that nonlinear interaction between rotations and loads may be neglected.

Load-displacement plots such as those in Fig. 1.2 are called *equilibrium paths*. Each point on a path represents an equilibrium configuration of the structure. In linear analysis all equilibrium paths are straight lines that pass through the origin. In this book all other equilibrium paths acquire their form

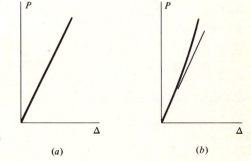

FIGURE 1.2
Load-displacement relations for beam clamped at both ends.

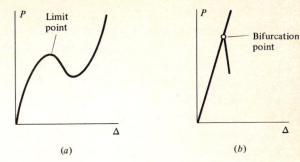

FIGURE 1.3
Equilibrium paths for Belleville spring and cylindrical panel.

as a result of rotations of structural elements as the structure deforms. The equilibrium path is one of the central concepts in the present treatment of the buckling behavior of structures.

Examples of more complex equilibrium paths are shown in Fig. 1.3a and b. Figure 1.3a illustrates a load-displacement curve for a Belleville spring, i.e., a shallow conical shell subjected to axial compression (Ref. 1.1). Here P denotes applied load and Δ the axial displacement of one end of the conical shell relative to the other. The point on the equilibrium path at which the load P is a relative maximum is called a *limit point*. Figure 1.3b illustrates the load-displacement relationship given in Ref. 1.2 for a thin-walled cylindrical panel subjected to axial compression. Here P denotes applied load and Δ the axial displacement of one end of the panel relative to the other. In Fig. 1.3b the *primary* (or *fundamental*) equilibrium path is intersected by a *secondary* path. At the point of intersection the displacement mode on the secondary branch is in some sense orthogonal to that on the primary branch. In the present example the primary branch represents cylindrical and the secondary branch noncylindrical (buckled) configurations of the panel.

Points at which equilibrium paths intersect are called *bifurcation points*. Such points are denoted by small circles, for emphasis, as shown in Fig. 1.3b. At such points the equilibrium equations have multiple solutions, one corresponding to each branch. The bifurcation point is another of the central concepts in the analysis of buckling.

The geometric nonlinearity in this book is limited to what may be called an intermediate class of deformations. For straight bars the kinematic relations used here are of the forms

$$\varepsilon = u' + \tfrac{1}{2}(w')^2 \qquad \kappa = -w'' \qquad (1.2)$$

where ε and κ are extensional strain and curvature, respectively, at points on the centroidal surface of the bar, u and w are axial and lateral displacement components at points on the centroidal surface, and primes denote differentiation with respect to the axial coordinate x. The term w' is the rotation or slope. These equations may be recognized as the counterpart for straight bars of the kinematic relations in the well-known von Kármán theory of flat plates (Ref. 1.3, p. 182). Kinematic relations of this general form are given for thin shells of arbitrary shape by, for example, Sanders (Ref. 1.4). They are derived by Sanders for deformations in which the strains are small, the rotations relative to the shell coordinate lines are moderately small, and rotations relative to normals to the shell middle surface are negligibly small. The Sanders kinematic relations are given here in terms of orthogonal curvilinear coordinates in Eqs. (6.6) and (6.7). All the kinematic relations for bars, plates, and shells in the present work may be obtained from the latter equations by appropriate specialization. Simplified kinematic relations of this general form underlie virtually all stability analyses and most large deflection analyses of specific problems in the literature.

To illustrate the sense in which these kinematic relations may be said to represent an intermediate class of deformations, let us insert at this point a brief derivation of Eqs. (1.2). Consider a line element (or fiber) of length dx that is initially parallel to the x axis, as shown in Fig. 1.4. After deformation in the xz plane, the length of the line element is dS^*, and the position of the element is described by new coordinate values x^*, z^*. If the x and z components of the displacement are denoted by $\bar{u}$ and $\bar{w}$, respectively (as distinguished from u and w, which denote the corresponding quantities at points on the centroidal surface only),

$$x^* = x + \bar{u} \qquad z^* = z + \bar{w}$$

Then differentiation with respect to x gives

$$\frac{dx^*}{dx} = 1 + \bar{u}' \qquad \frac{dz^*}{dx} = \bar{w}'$$

By the Pythagorean theorem,

$$(dS^*)^2 = (dx^*)^2 + (dz^*)^2$$

Consequently,

$$\left(\frac{dS^*}{dx}\right)^2 - 1 = 2\bar{u}' + (\bar{u}')^2 + (\bar{w}')^2 \qquad (1.3)$$

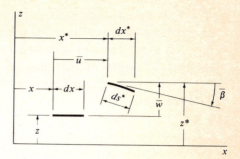

FIGURE 1.4
Line element before and after deformation.

Now let the extensional strain of the line element initially parallel to the x axis be denoted by $\bar{\varepsilon}$ and defined by the relation

$$\bar{\varepsilon} = \frac{dS^* - dx}{dx}$$

By suitable rearrangement this expression can be written in the form

$$\bar{\varepsilon} + \tfrac{1}{2}\bar{\varepsilon}^2 = \frac{1}{2}\left[\left(\frac{dS^*}{dx}\right)^2 - 1\right] \qquad (1.4)$$

Equations (1.3) and (1.4) are exact. If $\bar{\varepsilon}$ is much smaller than unity (a limitation that is satisfied if Hooke's law applies), Eq. (1.4) may be replaced by the approximate relation

$$\bar{\varepsilon} = \frac{1}{2}\left[\left(\frac{dS^*}{dx}\right)^2 - 1\right] \qquad (1.5)$$

Then Eqs. (1.3) and (1.5) give, for the strain-displacement relation, the expression

$$\bar{\varepsilon} = \bar{u}' + \tfrac{1}{2}[(\bar{u}')^2 + (\bar{w}')^2] \qquad (1.6)$$

This expression is limited only by the assumption that $\bar{\varepsilon}$ is small. The distinguishing characteristic of the present intermediate class of deformations is that the rotation $\bar{\beta}$ in Fig. 1.4 also is small (but not negligibly small). Then, from Fig. 1.4, it may be seen that $\bar{w}' = -\bar{\beta}$. If both $\bar{\varepsilon}$ and $(\bar{w}'^2)$ in Eq. (1.6) are small compared with unity, the $\bar{u}'$ also is small compared with unity. Then the square of $\bar{u}'$ may be neglected and, approximately,

$$\bar{\varepsilon} = \bar{u}' + \tfrac{1}{2}(\bar{w}')^2 \qquad (1.7)$$

This is the strain-displacement relationship for an intermediate class of deformations. In linear mechanics the corresponding strain-displacement relation is

$$\bar{\varepsilon} = \bar{u}' \qquad (1.8)$$

By comparison with Eq. (1.7), this expression is seen to be limited to deformations in which the square of the rotation is small compared with the strain component. The strain-displacement relationship in Eq. (1.7) is intermediate between those in Eqs. (1.6) and (1.8) in range of applicability, and it is in this sense that the geometric nonlinearity in this book represents an intermediate class of deformations.

The kinematic relation in Eq. (1.7) may be expressed in terms of centroidal-surface displacement components as follows. Slender-beam theory is based on the simplifying approximation that lines normal to the undeformed centroidal surface remain straight, normal, and inextensional during deformation. Then

$$\bar{u} = u + z\beta \qquad \bar{w} = w$$

where $\beta = -w' = \bar{\beta}$. Introduction into the intermediate strain-displacement relation in Eq. (1.7) and rearrangement gives

$$\bar{\varepsilon} = u' + \tfrac{1}{2}(w')^2 - zw''$$

This equation may be rewritten in the form

$$\bar{\varepsilon} = \varepsilon + z\kappa \qquad (1.9)$$

where

$$\varepsilon = u' + \tfrac{1}{2}(w')^2 \qquad \kappa = -w''$$

These are the centroidal-surface kinematic relations given in Eqs. (1.2) for the intermediate class of deformations.

Geometric nonlinearity in the equilibrium equations is a consequence of the nonlinear terms in the kinematic relations. In the derivation of equilibrium equations the nonlinearity is introduced directly through Eqs. (1.2) when the equations are derived by the so-called *variational* (or *energy*) method. (A descriptive development of the variational method is given in the Appendix.) As an alternative, the equilibrium equations may be obtained through summation of forces and moments. In that case the nonlinearity is introduced by considering the structural element in its deformed configuration. The latter procedure is followed in the analysis of the column in Sec. 1.2.

1.2 COLUMN ANALYSIS

A straight column of length L subjected to a centrally applied compressive load P is illustrated in Fig. 1.5. The notation and sign conventions for positive forces and moments acting on an element of the column are illustrated in Fig. 1.6, where N and Q are longitudinal and transverse components of the force on a cross section and M is bending moment. To account for the influence of

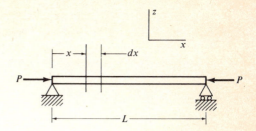

FIGURE 1.5
Centrally loaded straight column.

rotations on the behavior of the structure, we apply the equilibrium equations to the column element in a slightly deformed configuration, as shown in Fig. 1.7. (The resulting equations cover equilibrium in the compressed but straight form as a special case.) For the intermediate class of deformations the square of the rotation β is assumed to be small compared with unity, as noted. Consequently, $\sin \beta$ and $\cos \beta$ may be replaced by β and 1, respectively. The equations are obtained in their simplest form if forces are summed in the x and z directions. From summation of forces in the x direction we obtain

$$-N + (N + dN) - Q\beta + (Q + dQ)(\beta + d\beta) = 0$$

Therefore
$$N' + Q\beta' + \beta Q' = 0$$

where $N' \equiv dN/dx$, etc. Summation of forces in the z direction gives

$$-Q + (Q + dQ) + N\beta - (N + dN)(\beta + d\beta) = 0$$

or
$$-N\beta' - \beta N' + Q' = 0$$

From summation of moments,

$$M - (M + dM) + Q\,dx = 0$$

Consequently,
$$Q = M'$$

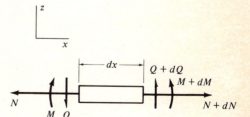

FIGURE 1.6
Sign conventions for positive forces and moments acting on a column element.

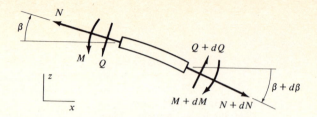

FIGURE 1.7
Forces and moments acting on column element in deformed configuration.

For the slender beams under consideration, transverse shearing stresses and forces are quite small. We adopt the usual assumption that quadratic terms representing nonlinear interaction between the small transverse shearing forces and the rotations may be neglected. In terms of this assumption the three equilibrium equations simplify to the form

$$N' = 0 \qquad (1.10a)$$

$$Q' - N\beta' = 0 \qquad (1.10b)$$

$$Q = M' \qquad (1.10c)$$

[The term $\beta N'$ does not appear in Eq. (1.10b), having been dropped as a consequence of Eq. (1.10a).] Elimination of Q from Eqs. (1.10b) and (1.10c) yields

$$N' = 0$$

$$M'' - N\beta' = 0$$

Thus the equilibrium analysis leads to two equations in the three unknowns N, M, and β.

As noted, however, $\beta = -w'$. From beam theory, furthermore, $M = -EIw''$, where I is the cross-sectional moment of inertia (see Prob. 1.4). Consequently, the equilibrium equations may be written in terms of two unknowns, N and w, as follows:

$$N' = 0$$

$$(EIw'')'' - Nw'' = 0$$

For constant EI these equations become

$$N' = 0 \qquad (1.11a)$$

$$EIw^{\mathrm{iv}} - Nw'' = 0 \qquad (1.11b)$$

Equation (1.11b) is quadratic in the dependent variables N and w; therefore the differential equation is nonlinear. From Eq. (1.11a), N is seen to be constant in x, however, and from the boundary conditions at $x = 0, L$, we see that $N = -P$. Consequently, Eqs. (1.11) simplify to the familiar expression

$$EIw^{iv} + Pw'' = 0 \qquad (1.12)$$

Equation (1.12) is the well-known expression governing equilibrium of the column in a slightly deflected form.

Although Eq. (1.12) takes into account the influence of geometric non-linearity (in the presence of the term Pw''), it is a linear differential equation (for fixed values of the parameter P). In this respect it is different from the corresponding expressions in Eqs. (3.18), (4.18), (5.8), and (6.10) for plates, circular rings, cylindrical shells, and shells of general shape, respectively, which are nonlinear. The peculiarity that distinguishes the column equation from the plate, ring, and shell equations is the fact that the N in Eq. (1.11b) is constant in x, as a result of Eq. (1.11a). In the general case the differential equations remain nonlinear. This characteristic difference is examined in greater detail in Sec. 1.3.

The reader is assumed to be familiar with Eq. (1.12) and its solution for columns with simply supported ends. Nevertheless, that solution is reviewed very briefly as background for discussion of the practical significance of the results. Solutions for other boundary conditions are examined in Chap. 2. Equation (1.12) is a homogeneous equation. It has a trivial solution $w = 0$, representing the primary equilibrium path. Nonzero solutions are found only for discrete values of the applied load P. For such values, called *eigenvalues*, the equation has multiple solutions corresponding to both straight and bent configurations of the bar. The solutions signify the existence of bifurcation points on the primary equilibrium path. The smallest eigenvalue is termed the *critical load* and is denoted by the symbol P_{cr}. The general solution of the constant-coefficient equation is readily found to be

$$w = C_1 \sin kx + C_2 \cos kx + C_3 x + C_4 \qquad (1.13)$$

where $k^2 \equiv P/EI$, and C_1, C_2, C_3, and C_4 are constants (see Prob. 1.5). The eigenvalues are determined by substitution of Eq. (1.13) into appropriate boundary-condition equations.

For simply supported ends the boundary-condition equations are

$$w = w'' = 0 \qquad \text{at } x = 0, L \qquad (1.14)$$

Introduction of Eq. (1.13) into Eqs. (1.14) gives

$$C_2 = C_3 = C_4 = C_1 \sin kL = 0 \qquad (1.15)$$

For the column in a deflected form, C_1 must be different from zero. For a non-zero value of C_1, Eqs. (1.15) are satisfied only if $\sin kL = 0$ and $k = n\pi/L$, where n is a positive integer. Equations (1.13) and (1.15) give, for the form of the deflected column, the equation

$$w = C_1 \sin \frac{n\pi x}{L} \qquad n = 1, 2, 3, \ldots \qquad (1.16)$$

and the relation $k = n\pi/L$ together with the definition $k^2 = P/EI$ yields, for the eigenvalues, the expression

$$P = n^2 \frac{\pi^2 EI}{L^2} \qquad (1.17)$$

The smallest eigenvalue in this case corresponds to $n = 1$. Therefore

$$P_{cr} = \frac{\pi^2 EI}{L^2} \qquad (1.18)$$

Equation (1.18) is the classical solution for the simply supported column. It is commonly called the *Euler column formula*, after Leonhard Euler (Ref. 1.5). (A brief biography of this remarkable man, the first to derive the equation for the critical load of a slender column, is given in Ref. 1.6, pp. 28–36.)

Equilibrium paths for the simply supported column may be obtained by plotting P versus midpoint deflection (i.e., versus w at $x = L/2$). Such a plot for $n = 1$ is shown in Fig. 1.8. Points along the primary (vertical) equilibrium path represent configurations of the column in the compressed but straight form; those along the secondary (horizontal) paths represent bent equilibrium configurations. (On the primary path the displacement component $u = -Px/EA$.) The symmetry of the plot indicates simply that the column may deflect in either direction. Equations (1.16) and (1.17) determine a distinct bifurcation point and secondary equilibrium path for each value of n. Because P is not a function of w, the secondary paths are horizontal. On the basis of Eqs. (1.16) and (1.17), the secondary paths extend indefinitely to the right and left, but the results lose significance when the deflection amplitude becomes so large that rotations can no longer be considered small.

A more accurate analysis suitable for larger rotations (Ref. 1.7, p. 80) shows that the secondary equilibrium paths for the column curve upward as rotations increase in size, as illustrated in Fig. 1.9. We emphasize that all the column, plate, ring, and shell analyses in this book are limited to moderately large rotations and, in this respect, are consistent in accuracy with the analysis underlying Fig. 1.8, not Fig. 1.9. This limitation to moderately large rotations applies to the nonlinear equilibrium analyses as well as to the linear stability analyses. When large rotations are taken into account as in Fig. 1.9, the column is called the *elastica*.

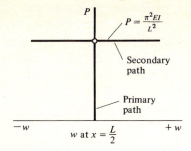

FIGURE 1.8
Equilibrium paths for initially straight column.

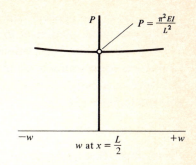

FIGURE 1.9
Column equilibrium paths when rotations are not assumed to be small.

1.3 ADJACENT EQUILIBRIUM

The differential equation governing equilibrium on the secondary path for the column [Eq. (1.12)] is linear because the factor N in Eqs. (1.11) happens to be independent of the axial coordinate. This results from our restriction of the analysis to an intermediate class of deformations, and it is a peculiarity of column analysis. For plates, rings, and shells, the corresponding equations for the same intermediate class of deformations are nonlinear differential equations. Even in the general case, however, points of intersection of primary and secondary paths can be obtained by solution of linear differential equations; i.e., once an equilibrium configuration has been established on the primary path, linear equations can be used to determine whether an adjacent equilibrium configuration exists under the same applied load. Existence of such a configuration signifies either a bifurcation point or a relative maximum or minimum point on the equilibrium path. If the primary path is, say, a straight line through the origin, as for a column, the adjacent equilibrium configurations occur at a bifurcation point. The linear equations necessary for this process may be derived from the nonlinear ones by use of a perturbation technique in which we replace $\bar{u}$ by $\bar{u}_0 + \bar{u}_1$, where $\bar{u}$ denotes the displacement field, $\bar{u}_0$ represents an equilibrium configuration on the primary path, and $\bar{u}_1$ is a small increment. All terms that do not contain the incremental displacement drop out of the resulting equations because $\bar{u}_0$ is an equilibrium configuration. If the incremental displacement is sufficiently small, furthermore, only first-order terms in the $\bar{u}_1$ need be retained in the equations. Then the resulting expressions are homogeneous linear equations that have nontrivial solutions only at discrete points along the primary path. At such points there exist two adjacent equilibrium configurations for a single value of the applied load, one corresponding to the displacement $\bar{u}_0$ and the other to the displacement $\bar{u}_0 + \bar{u}_1$.

To illustrate the derivation of the linearized equations, let us rederive the bifurcation equation for a column in a way that is applicable in the general case. The force component N in Eqs. (1.11) is related to the centroidal-surface strain ε by the familiar expression

$$N = EA\varepsilon \qquad (1.19)$$

where, from Eqs. (1.2), $\varepsilon = u' + \frac{1}{2}(w')^2$. Then, in terms of the displacement variables, the column equilibrium equations in Eqs. (1.11) may be written

$$[u' + \tfrac{1}{2}(w')^2]' = 0 \qquad (1.20a)$$

$$EIw^{iv} - EA[u' + \tfrac{1}{2}(w')^2]w'' = 0 \qquad (1.20b)$$

Now let

$$u \to u_0 + u_1$$
$$w \to w_0 + w_1 \qquad (1.21)$$

where the arrows are read "be replaced by," (u_0, w_0) represents a configuration on the primary equilibrium path, the incremental displacement (u_1, w_1) is infinitesimally small, and both (u_0, w_0) and (u, w) are equilibrium configurations. Then, in the equations obtained by introduction of Eqs. (1.21) into Eqs. (1.20), we see that

1 In each equation the sum of all terms containing u_0, w_0 alone is equal to zero because u_0, w_0 satisfy Eqs. (1.20).

2 Second- and higher-order terms in u_1, w_1 may be omitted because of the smallness of the incremental displacement.

Then the resulting equations simplify to the form

$$(u_1' + w_0' w_1')' = 0 \qquad (1.22)$$

$$EIw_1{}^{iv} - EA\{[u_0' + \tfrac{1}{2}(w_0')^2]w_1'' + w_0''[u_1' + w_0' w_1']\} = 0$$

These are seen to be linear equations in the unknowns u_1, w_1, as desired, the variables u_0, w_0 appearing as coefficients.

For the column, the primary equilibrium path represents undeflected configurations. Thus $w_0 = 0$ for all values of x, and Eqs. (1.22) reduce to the form

$$u_1'' = 0$$

$$EIw_1{}^{iv} - EAu_0' w_1'' = 0$$

For u_0 the equilibrium equation for the column in the undeflected form yields the relation

$$u_0 = -\frac{P}{EA}x$$

Introduction of this relationship gives the expression

$$EIw_1{}^{iv} + Pw_1'' = 0 \qquad (1.23)$$

Equation (1.23) is the final, linearized equation for equilibrium of the column in a slightly bent configuration.

As noted, the straight form of the column is an equilibrium form for all values of applied load. Nontrivial solutions to Eq. (1.23) represent equilibrium configurations adjacent to the straight form, and accordingly denote occurrence of a bifurcation point.

Equation (1.23) is identical with Eq. (1.12), except that w has been replaced by w_1. Its solution gives us the bifurcation-point load, but because w_1 is infinitesimally small, no information is obtained about the initial slope or shape of the secondary equilibrium path. This lack of information about the secondary equilibrium path is an invariable consequence of the linearization procedure. The implication of this loss of information is discussed in the following section.

1.4 THE SLIGHTLY CROOKED COLUMN

In the preceding sections the column has been considered to be perfectly straight before load application. Of course, no real column can be perfectly straight. In recent years a growing appreciation has developed of the occasionally profound influence of so-called initial imperfections (for the column, initial crookedness) on the behavior of structures under compressive loading. To illustrate the relation between the bifurcation phenomenon and the behavior of an initially imperfect structure, let us now consider a simply supported column that is slightly crooked from the outset.

Any initial shape can be represented by the infinite series

$$w^* = \sum_{m=1}^{\infty} C_m^* \sin\frac{m\pi x}{L} \qquad m = 1, 2, 3, \ldots \qquad (1.24)$$

where the C_m^* are known constants. We again let w denote displacement from the zero-load configuration. Summing forces and moments in Fig. 1.7 and noting that the small angle β now is augmented by a small angle $\beta^* = -(w^*)'$, we obtain the equilibrium equation [compare Eq. (1.12)]

$$EIw^{iv} + P(w + w^*)'' = 0 \qquad (1.25)$$

Introduction of Eq. (1.24) gives

$$EIw^{iv} + Pw'' = P \sum_{m=1}^{\infty} \left(\frac{m\pi}{L}\right)^2 C_m^* \sin \frac{m\pi x}{L} \qquad (1.26)$$

Unlike the corresponding equation for the initially straight column, Eq. (1.26) is inhomogeneous. The solution for simply supported ends is found to be (see Prob. 1.6)

$$w = \frac{PC_1^* \sin \dfrac{\pi x}{L}}{\dfrac{\pi^2 EI}{L^2} - P} + \frac{PC_2^* \sin \dfrac{2\pi x}{L}}{\dfrac{4\pi^2 EI}{L^2} - P} + \frac{PC_3^* \sin \dfrac{3\pi x}{L}}{\dfrac{9\pi^2 EI}{L^2} - P} + \cdots \qquad (1.27)$$

As the magnitude of the applied load P approaches the value $\pi^2 EI/L^2$ in Eq. (1.27), the magnitude of the lateral displacement w is seen to become arbitrarily large. This happens whatever the magnitude of C_1^* may be (provided that $C_1^* \neq 0$), and whatever the initial shape of the column. A plot of P versus midpoint displacement (for $C_1^* > 0$ and all other $C_m^* = 0$) is shown in Fig. 1.10. Comparison with the secondary equilibrium path for the straight column in Fig. 1.8 shows that the equilibrium paths for the two columns tend to converge as lateral displacements increase. The analyses for the straight and the crooked columns both lead to the result that large lateral displacements occur for $P = \pi^2 EI/L^2$.

Convergence of the equilibrium paths for perfect and for slightly imperfect structures for large deformations is typical of structures in general. A second example is given in Fig. 1.11, where the paths for a thin-walled cylindrical panel in Fig. 1.3b are shown again, along with the corresponding path for a slightly imperfect panel (Ref. 1.2). Although the paths for the cylindrical panels converge for large deformations, they are not close to one another in the region of the bifurcation point. The imperfect panel buckles at a limit-point load that is markedly smaller than the bifurcation-point load of the perfect panel. *In general, whether the bifurcation-point load for the perfect structure is close to the buckling load of the imperfect one depends on the shape of the secondary equilibrium path for the perfect structure.* For the column, the secondary path is horizontal near the bifurcation point and agreement is close. For the cylindrical panel in Fig. 1.11, the secondary path drops downward from the bifurcation point and agreement may not be close unless the imperfection is quite small. Lack of knowledge

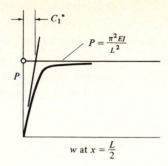

FIGURE 1.10
Equilibrium path for slightly crooked column.

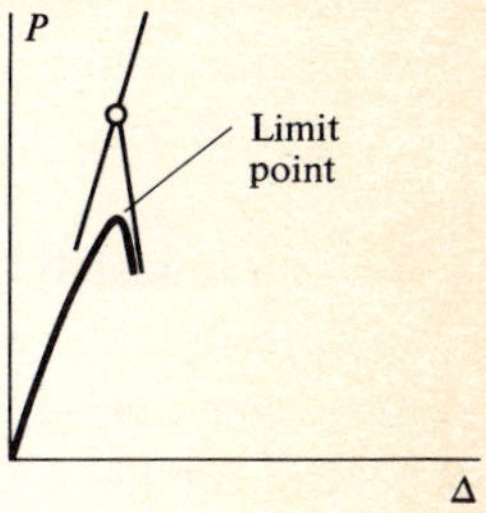

FIGURE 1.11
Equilibrium path for slightly imperfect cylindrical panel.

of the shape of the secondary equilibrium path introduces an element of uncertainty and is a major disadvantage of the use of linearized equations. This matter is examined in detail in later chapters.

1.5   STABILITY THEORY

Points on equilibrium paths represent configurations of equilibrium. At some of the points the equilibrium is stable and at others it is unstable. According to the dynamical criterion for loss of stability, an equilibrium configuration is stable if and only if, for a fixed value of the applied load, a small perturbation of the configuration results in oscillations that are confined to the immediate vicinity of the equilibrium configuration (Ref. 1.8). All the structural systems treated in this work are static and conservative unless otherwise noted. For such systems the stability determination may be based on a criterion known as the *minimum potential energy criterion*. According to this criterion, a structure is in a configuration of stable equilibrium if and only if the change in total potential energy corresponding to any sufficiently small, kinematically admissible displacement is positive. (A displacement is kinematically admissible if the displacement function satisfies certain continuity and boundary conditions as described in the Appendix.) For static, conservative systems this criterion is equivalent to the dynamical criterion (Ref. 1.8). The *critical load* is defined as the smallest load at which the equilibrium of the structure fails to be stable as the load is slowly increased from zero.

In Secs. 1.2 and 1.3 a bifurcation point, i.e., a point of intersection between equilibrium paths, was determined through consideration of possible multiple solutions to the column equilibrium equations. Such a procedure is found to lead to differential equations for the determination of bifurcation points that are identical with the equations for loss of stability given by the minimum potential energy criterion. Consequently, configurations on the primary path above the lowest bifurcation point given by this criterion are unstable. This procedure for the determination of loss of stability is known as the *adjacent-equilibrium criterion*. According to this criterion, a structure is in a configuration of neutral equilibrium if a second, infinitesimally adjacent equilibrium configuration exists for the same value of applied load. For static, conservative systems this criterion also is equivalent to the dynamical criterion (Ref. 1.8).

Both the minimum potential energy criterion and the adjacent-equilibrium criterion are used in the analysis of the static, conservative structural systems of this book. A brief descriptive development of the energy criterion is included in the Appendix. A detailed development of all three criteria is given in Ref. 1.8. An extensive discussion of the present status of the fundamentals of stability theory may be found in Ref. 1.9.

Both the minimum potential energy and the adjacent-equilibrium criteria determine the location of limit points as well as bifurcation points on primary paths. Examples of the loss of stability at a limit point and at a bifurcation point, respectively, are given by the equilibrium paths for the Belleville spring in Fig. 1.3a and the cylindrical panel in Fig. 1.3b. For the Belleville spring, points on the rising portions of the path represent stable configurations (provided that load is the controlled variable), and those on the falling portion represent unstable configurations. As the load applied to a Belleville spring is slowly increased, the structure follows the initial portion of the path up to the limit point, where the structure is set in motion on a nonequilibrium path. At a bifurcation point, on the other hand, the equilibrium equations have multiple solutions, one corresponding to each of the intersecting paths. As the applied load is slowly increased, the structure follows the primary path until the bifurcation point is reached. The portion of the primary path that lies above the bifurcation point is unstable. Thus the structure either follows a stable secondary path, or is set in motion. (In the absence of a lateral disturbance, an ideal structure could follow the unstable portion of the primary path. In reality, some disturbance or asymmetry is always present: a sharpened pencil cannot be balanced on its point.) For the cylindrical panel in Fig. 1.3b the secondary path is unstable, and motion ensues at the bifurcation point.

Both the adjacent-equilibrium and minimum potential energy criteria are applied in the determination of bifurcation-point loads of plates, rings, and shells in Chaps. 3 to 6. With minor exceptions, all the stability equations are derived on both bases. Loss of stability at a limit point is investigated in Chaps. 7 and 9. The influence of initial imperfections is considered throughout the book and is the principal subject matter of Chap. 7.

1.6 METRIC SYSTEM UNITS

Dimensions in the book are given in English units. Factors for conversion to the metric system are given in the following table. The unit of force in the International System of units is the newton (N). In European countries, however, the commonly used unit is the kilogram-force (kg). Both units are included in the table. One newton equals 10^5 dynes, and one kilogram-force equals 9.81 N.

$$
\begin{aligned}
1 \text{ in.} &= 2.54 \text{ cm} \\
1 \text{ lb-force} &= 4.45 \text{ N} &&\text{or } 0.454 \text{ kg-force} \\
1 \text{ psi} &= 0.690 \text{ N/cm}^2 &&\text{or } 0.0703 \text{ kg/cm}^2 \\
1 \text{ lb/in.} &= 1.751 \text{ N/cm} &&\text{or } 0.1788 \text{ kg/cm} \\
1 \text{ lb-in.} &= 11.31 \text{ N-cm} &&\text{or } 1.152 \text{ kg-cm}
\end{aligned}
$$

1.7 GENERAL REFERENCES

Some of the more general references on the buckling of structures are collected in this section for convenience. References cited in the text are listed at the ends of the respective chapters.

Relatively recent textbooks and monographs on buckling include those by Bleich (Ref. 1.10), Timoshenko and Gere (Ref. 1.7), Gerard (Ref. 1.11), Cox (Ref. 1.12), Ziegler (Ref. 1.8), Britvec (Ref. 1.13), Thompson and Hunt (Ref. 1.14), and Chajes (Ref. 1.15).

The authors' viewpoint on the subject has been strongly influenced by the works of Koiter, including Refs. 1.9, 1.16, 1.17, and 1.18.

The design of structural elements is beyond the scope of this work. For design criteria for columns and plates the Column Research Council's, "Guide to Design Criteria for Metal Compression Members" (Ref. 1.19) is highly recommended. An extensive collection of theoretical and experimental results for the buckling of plates and shells is available in the six-part Handbook for Structural Stability, published by the National Advisory Committee for Aeronautics (Ref. 1.20). A large amount of similar material has been made available in the recent book by Baker, Kovalevsky, and Rish (Ref. 1.21).

Reviews of recent papers and books on the buckling of structures may be found in the monthly journal *Applied Mechanics Reviews* under the section entitled Buckling. A general survey article by Langhaar entitled "General Theory of Buckling" appeared in volume 11 (Ref. 1.22), and one by Hutchinson and Koiter entitled "Postbuckling Theory" appeared in volume 23 (Ref. 1.23).

REFERENCES

1.1 WEMPNER, G. A.: The Conical Disk Spring, *Proc. Third U.S. Natl. Congr. Appl. Mech.*, 1958, pp. 473–478.

1.2 KOITER, W. T.: Buckling and Post-buckling Behavior of a Cylindrical Panel under Axial Compression, *Natl. Aeronaut. Res. Inst. Rep.* S. 476, Amsterdam, 1956.

1.3 NOVOZHILOV, V. V.: "Foundations of the Nonlinear Theory of Elasticity," Graylock Press, Rochester, N.Y., 1953.

1.4 SANDERS, J. L.: Nonlinear Theories for Thin Shells, *Q. Appl. Math.*, vol. 21, no. 1, pp. 21–36, 1963.

1.5 EULER, L.: On the Strength of Columns, *Acad. Roy. Sci. Belles Lettres Berlin Mém.*, vol. 13, p. 252, 1759. English translation by J. A. Van den Broek, *Am. J. Phys.*, . vol. 15, pp. 315–318, 1947.

1.6 TIMOSHENKO, S. P.: "History of Strength of Materials," McGraw-Hill, New York, 1953.

1.7 TIMOSHENKO, S. P., and J. M. GERE: "Theory of Elastic Stability," 2d ed., McGraw-Hill, New York, 1961.

1.8 ZIEGLER, H.: "Principles of Structural Stability," Blaisdell, Waltham, Mass., 1968.

1.9 KOITER, W. T.: Purpose and Achievements of Research in Elastic Stability, *Recent Advanc. Eng. Sci.*, vol. 3, pp. 197–218, 1966.

1.10 BLEICH, F.: "Buckling Strength of Metal Structures," McGraw-Hill, New York, 1952.

1.11 GERARD, G.: "Introduction to Structural Stability Theory," McGraw-Hill, New York, 1962.

1.12 COX, H. L.: "The Buckling of Plates and Shells," Macmillan, New York, 1963.

1.13 BRITVEC, S. J.: "The Stability of Elastic Systems," Pergamon, New York, 1973.

1.14 THOMPSON, J. M. T., and G. W. HUNT: "A General Theory of Elastic Stability," Wiley, New York, 1974.

1.15 CHAJES, A.: "Principles of Structural Stability Theory," Prentice-Hall, Englewood Cliffs, N. J., 1974.

1.16 KOITER, W. T.: On the Stability of Elastic Equilibrium (in Dutch), thesis, Delft, H. J. Paris, Amsterdam, 1945. English translation, *Air Force Flight Dyn. Lab. Tech. Rep.* AFFDL-TR-70-25, Wright Patterson Air Force Base, Ohio, 1970.

1.17 KOITER, W. T.: Elastic Stability and Postbuckling Behavior, *Proc. Symp. Nonlinear Probl.*, University of Wisconsin Press, Madison, 1963, pp. 257–275.

1.18 KOITER, W. T.: General Equations of Elastic Stability for Thin Shells, *Proc. Symp. Theory of Shells to Honor Lloyd Hamilton Donnell*, University of Houston, Houston, Tex., 1967, pp. 187–223.

1.19 B. G. JOHNSTON (ed.): "Guide to Design Criteria for Metal Compression Members," 2d ed., Column Research Council, Wiley, New York, 1966.

1.20 GERARD, G., et al.: Handbook of Structural Stability, Parts I–VI, *NACA* TN 3781–3786, 1957–1958.

1.21 BAKER, E. H., L. KOVALEVSKY, and F. L. RISH: "Structural Analysis of Shells," McGraw-Hill, New York, 1972.

1.22 LANGHAAR, H. L.: General Theory of Buckling, *Appl. Mech. Rev.*, vol. 11, pp. 585–589, 1958.

1.23 HUTCHINSON, J. W., and W. T. KOITER: Postbuckling Theory, *Appl. Mech. Rev.*, vol. 23, pp. 1353–1366, 1970.

PROBLEMS

1.1 A straight column 20 in. long has a rectangular cross section 1 in. wide and 0.5 in. thick. If the column ends are simply supported and $E = 29 \times 10^6$ psi, determine the elastic buckling load.   *Ans.* $P_{cr} = 7,450$ lb

1.2 A slightly crooked column 20 in. long has a rectangular cross section 1 in. wide and 0.5 in. thick (compare Prob. 1.1). The ends are simply supported, and $E = 29 \times 10^6$ psi. (*a*) Let $\Delta \equiv w$ at $x = L/2$, and use Eq. (1.27) to plot P versus Δ (with $C_m^* = 0$ for $m > 1$) for three respective values of C_1^*: 0.05, 0.005, and 0.0005 in. Let the range of abscissa values in the graph be $0 \leq \Delta \leq 1.0$ (compare Fig. 1.10). (*b*) For $P = 7,000$ lb, determine the magnitude of the maximum compressive stress on the cross section at $x = L/2$ for each of the three cases. Assume the stresses to remain elastic and use the equation $\sigma = P/A + Mc/I$, where $c = 0.25$ in. and $M = P(C_1^* + \Delta)$.

1.3 A slightly crooked column 20 in. long has a rectangular cross section 1 in. wide and 0.5 in. thick (compare Probs. 1.1 and 1.2). The ends are simply supported and $E = 29 \times 10^6$ psi. The values of C_m^* in the expression for the zero-load shape in Eqs. (1.24) and (1.27) are $C_1^* = 0.0005$ in., $C_2^* = -0.05$ in., and $C_m^* = 0$ for $m > 2$. Let $\Delta_1 \equiv w$ at $x = L/4$, and plot P versus Δ_1 on the basis of Eq. (1.27). Let the range of abscissa values in the graph be $-0.02 \leq \Delta_1 \leq 0.20$. For approximately what value of P on the graph does the sign of Δ_1 change from minus to plus? (The calculations for the graph are simplified if P is treated as the independent variable.)

1.4   The bending moment M in Eqs. (1.10) is defined by the relationship $M = \int \bar{\sigma} z \, dA$, where $\bar{\sigma}$ is normal stress at any point in the column cross section. Using Hooke's law, Eq. (1.9), and the second of Eqs. (1.2), derive the constitutive relationship $M = -EIw''$.

1.5   Show that Eq. (1.13) is the general solution of the differential equation for column equilibrium in Eq. (1.12).

1.6   Show that Eq. (1.27) is the solution for simply supported boundary conditions of the equilibrium equation for the slightly crooked column in Eq. (1.26).

1.7   Derive the two column equilibrium equations in Eqs. (1.11) from the potential energy expression in Eqs. (A.40) by the variational method, i.e., by application of the principle of stationary potential energy (Sec. A.4). Use Eqs. (1.19) and (A.34).

2

STRAIGHT BARS

2.1   INTRODUCTION

This chapter treats the flexural and torsional instability of initially straight columns, the lateral instability of beams, and the nonlinear bending and lateral-torsional instability of beam columns. For all these cases the governing differential equations are linear, and the linearization procedure discussed in Chap. 1 is unnecessary. The chapter is concluded with a discussion of the failure of columns, beams, and beam columns.

2.2   FLEXURAL INSTABILITY OF COLUMNS

For an initially straight column the straight form is an equilibrium form for all values of applied load. Consequently, any adjacent nonstraight equilibrium form represents a secondary equilibrium configuration for the corresponding load and signifies the occurrence of a bifurcation point. Thus the critical load for flexural instability is the smallest load at which the column can be maintained in equilibrium in a slightly bent form.

2.2a Influence of Boundary Conditions

The equation governing equilibrium of an initially straight column in a slightly bent form was found in Eq. (1.12) to be

$$EIw^{iv} + Pw'' = 0 \qquad (2.1)$$

The general solution of the fourth-order equation was given in Eq. (1.13) as

$$w = C_1 \sin kx + C_2 \cos kx + C_3 x + C_4 \qquad (2.2)$$

where $k^2 \equiv P/EI$ and C_1, C_2, C_3, and C_4 are constants. As noted, the eigenvalues are determined by substitution of this expression for w into appropriate sets of boundary-condition equations. Eigenvalues for simply supported ends were determined in Chap. 1. The critical load and deflection mode shape for that case are shown in Table 2.1. As additional examples of typical eigenvalue problems, let us next determine the critical load and deflection mode for a column that is clamped at both ends and for one that is simply supported at one end and clamped at the other, as illustrated in Table 2.1.

For a column clamped at both ends, the boundary-condition equations are

$$w = w' = 0 \quad \text{at} \quad x = 0, L \qquad (2.3)$$

Introduction of Eq. (2.2) into Eqs. (2.3) gives the four equations

$$\begin{aligned} C_2 + C_4 &= 0 \\ kC_1 + C_3 &= 0 \\ C_1 \sin kL + C_2 \cos kL + C_3 L + C_4 &= 0 \\ kC_1 \cos kL - kC_2 \sin kL + C_3 &= 0 \end{aligned} \qquad (2.4)$$

Table 2.1 COMPARISON OF BOUNDARY-CONDITION EFFECTS

Boundary conditions	Critical load P_{cr}	Deflection mode shape	Effective length KL
Simple support–simple support	$\dfrac{\pi^2 EI}{L^2}$		L
Clamped-clamped	$4\dfrac{\pi^2 EI}{L^2}$		$\frac{1}{2}L$
Clamped–simple support	$2.04\dfrac{\pi^2 EI}{L^2}$		$0.70L$
Clamped-free	$\dfrac{1}{4}\dfrac{\pi^2 EI}{L^2}$		$2L$

For a nontrivial solution of this homogeneous equation system to exist, the determinant of the coefficients of the C_1, C_2, C_3, C_4 must equal zero. After some rearrangement, the determinant is found to be the product of two factors, with the result that either

$$\sin \frac{kL}{2} = 0 \qquad \text{or} \qquad \tan \frac{kL}{2} = \frac{kL}{2} \qquad (2.5)$$

The smallest nonzero value of $kL/2$ for which $\sin (kL/2) = 0$ is π, whereas the smallest value for which $\tan (kL/2) = kL/2$ is 4.49. (The roots of the equations are easily determined by trial and error from a table of natural trigonometric functions of the angles in radians.) Consequently, the smallest eigenvalue is $kL/2 = \pi$, or since $k^2 = P/EI$,

$$P_{\text{cr}} = 4 \frac{\pi^2 EI}{L^2} \qquad (2.6)$$

This result for the critical load for a column that is clamped at both ends is seen to be four times that for a column that is simply supported at both ends. For the eigenvalue $kL = 2\pi$, the constants in Eq. (2.2) are found from Eqs. (2.4) to be $C_1 = C_3 = 0$ and $C_4 = -C_2$. Consequently, from the general solution,

$$w = C_4 \left(1 - \cos \frac{2\pi x}{L}\right) \qquad (2.7)$$

A plot of the deflection mode shape given by Eq. (2.7) is shown in Table 2.1.

For a column clamped at one end (say, at $x = 0$) and simply supported at the other, the boundary-condition equations are

$$\begin{aligned} w = w' = 0 & \qquad \text{at } x = 0 \\ w = w'' = 0 & \qquad \text{at } x = L \end{aligned} \qquad (2.8)$$

Equations (2.2) and (2.8) lead to the condition

$$\tan kL = kL \qquad (2.9)$$

The smallest root corresponds to $kL = 4.49$. Therefore $k^2 = 2.04\pi^2/L^2$ and

$$P_{\text{cr}} = 2.04 \frac{\pi^2 EI}{L^2} \qquad (2.10)$$

This value is seen to be intermediate between those in Eqs. (1.18) and (2.6), as would be expected on intuitive grounds. For this eigenvalue, $C_2 = -kLC_1$, $C_3 = -kC_1$, $C_4 = kLC_1$, and

$$w = C_1[\sin kx - kL \cos kx + k(L - x)] \qquad (2.11)$$

where $k = 4.49/L$. A plot of this deflection mode shape also is shown in Table 2.1.

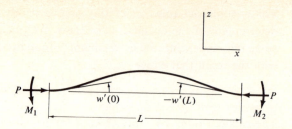

FIGURE 2.1
End moments on column.

The preceding examples are special cases of a column whose ends are elastically restrained against rotation and rigidly restrained against lateral displacement. General boundary conditions for such a column may be expressed as follows (compare Fig. 2.1):

$$w = 0 \quad \text{and} \quad w' = -\frac{M_1}{\alpha_1} \qquad \text{at } x = 0$$

$$w = 0 \quad \text{and} \quad w' = \frac{M_2}{\alpha_2} \qquad \text{at } x = L \tag{2.12}$$

where M_1, M_2 are end moments and α_1, α_2 are rotational stiffnesses. Introduction of Eq. (2.2) into (2.12) gives

$$C_2 + C_4 = 0$$

$$\sin kL\, C_1 + \cos kL\, C_2 + LC_3 + C_4 = 0$$

$$kC_1 + C_3 = -\frac{M_1}{\alpha_1} \tag{2.13}$$

$$k \cos kL\, C_1 - k \sin kL\, C_2 + C_3 = \frac{M_2}{\alpha_2}$$

But in general,

$$M = -EIw''$$

Therefore, from Eq. (2.2),

$$M_1 = EIk^2 C_2$$

$$M_2 = EIk^2(C_1 \sin kL + C_2 \cos kL)$$

Introduction into Eqs. (2.13) gives the homogeneous-equation system

$$C_2 + C_4 = 0$$

$$\sin \Phi\, C_1 + \cos \Phi\, C_2 + LC_3 + C_4 = 0 \tag{2.14}$$

$$\Phi C_1 + \lambda_1 \Phi^2 C_2 + LC_3 = 0$$

$$(\Phi \cos \Phi - \lambda_2 \Phi^2 \sin \Phi)C_1 - (\Phi \sin \Phi + \lambda_2 \Phi^2 \cos \Phi)C_2 + LC_3 = 0$$

where nondimensional parameters λ_1, λ_2, and Φ, defined by the relations

$$\lambda_1 \equiv \frac{EI}{\alpha_1 L} \qquad \lambda_2 \equiv \frac{EI}{\alpha_2 L} \qquad \Phi \equiv kL \qquad (2.15)$$

have been introduced. Equations (2.14) have nontrivial solutions only for discrete values of the load parameter Φ. For a nontrivial solution, the determinant of the coefficients must equal zero. Thus

$$\begin{vmatrix} 0 & 1 & 0 & 1 \\ \sin\Phi & \cos\Phi & L & 1 \\ \Phi & \lambda_1\Phi^2 & L & 0 \\ (\Phi\cos\Phi - \lambda_2\Phi^2\sin\Phi) & (-\Phi\sin\Phi - \lambda_2\Phi^2\cos\Phi) & L & 0 \end{vmatrix} = 0 \qquad (2.16)$$

After some rearrangement, Eq. (2.16) gives the equation

$$(1 - \lambda_1 - \lambda_2 - \lambda_1\lambda_2\Phi^2)\Phi\sin\Phi + (2 + \lambda_1\Phi^2 + \lambda_2\Phi^2)\cos\Phi - 2 = 0 \qquad (2.17)$$

Equation (2.17) is seen to be symmetrical in λ_1, λ_2, as it must be.

For equal elastic restraints at the two ends, $\lambda_1 = \lambda_2 = \lambda$, and Eq. (2.17) simplifies to the expression

$$\left(\tan\frac{\Phi}{2} + \lambda\Phi\right)\left[(2 + \lambda\Phi^2)\tan\frac{\Phi}{2} - \Phi\right] = 0$$

Then the critical load is seen to be determined by either of two factors. In terms of $\Phi/2$ the factors may be written

$$\tan\frac{\Phi}{2} = -2\lambda\frac{\Phi}{2}$$

and

$$\tan\frac{\Phi}{2} = \frac{\Phi/2}{1 + 2\lambda(\Phi/2)^2}$$

But $\lambda \geq 0$. Therefore $(-2\lambda\Phi/2) \leq 0$, and $0 \leq (\Phi/2)/[1 + 2\lambda(\Phi/2)^2] \leq (\Phi/2)$. Examination of these values in relation to a plot of $\tan(\Phi/2)$ versus $(\Phi/2)$ reveals that the lowest root corresponds to the equation

$$\tan\frac{\Phi}{2} = -\lambda\Phi \qquad (2.18)$$

Equation (2.18) for equal restraint at the two ends is the counterpart of Eq. (2.17) for the more general case.

Appropriate specialization of the general results in Eq. (2.17) or (2.18) gives Eqs. (1.18), (2.6), and (2.10), respectively, for the three special cases treated earlier, as may be seen, as follows:

1 For a column with simply supported ends, $\alpha_1 = \alpha_2 = 0$. Then $\lambda \to \infty$, and Eq. (2.18) gives $\Phi = \pi$. Then, from Eqs. (2.15), $kL = \pi$, and $P_{cr} = \pi^2 EI/L^2$, as before.

2 For a column clamped at both ends, $\lambda = 0$, and for a nontrivial solution, Eq. (2.18) gives $\Phi = 2\pi$. Therefore $kL = 2\pi$, and $P_{cr} = 4\pi^2 EI/L^2$, as before.

3 For a column clamped at $x = 0$ and simply supported at $x = L$, $\lambda_1 = 0$ and $\lambda_2 \to \infty$. For this case the terms in Eq. (2.17) may be divided by λ_2 to give

$$\left(\frac{1}{\lambda_2} - \frac{\lambda_1}{\lambda_2} - 1 - \lambda_1 \Phi^2\right)\Phi \sin \Phi + \left(\frac{2}{\lambda_2} + \frac{\lambda_1}{\lambda_2}\Phi^2 + \Phi^2 \cos \Phi\right) - \frac{2}{\lambda_2} = 0$$

Then, for $\lambda_1 = 0$ and $\lambda_2 \to \infty$,

$$\Phi(-\sin \Phi + \Phi \cos \Phi) = 0$$

Therefore either $\Phi = 0$ or $\tan \Phi = \Phi$. The first equation is the trivial solution. In terms of the parameter k, the second equation may be written $\tan kL = kL$. This is the equation obtained in the earlier analysis of this case, and again, $P_{cr} = 2.04\pi^2 EI/L^2$.

Examples in which $0 < \lambda < \infty$ will be considered in the section on frame stability. Equation (2.18) for equal elastic restraints is readily solved for any value of λ, but a numerical analysis usually is necessary for the solution of Eq. (2.17). A convenient approximate solution based on numerical calculations has been provided by Newmark (Ref. 2.1) in the equation

$$\Phi^2 = \pi^2 \frac{(\lambda_1 + 0.4)(\lambda_2 + 0.4)}{(\lambda_1 + 0.2)(\lambda_2 + 0.2)} \qquad (2.19)$$

Since $\Phi = kL = (P/EI)^{1/2}L$, Eq. (2.19) may be rewritten in the alternative form

$$P_{cr} = \frac{(\lambda_1 + 0.4)(\lambda_2 + 0.4)}{(\lambda_1 + 0.2)(\lambda_2 + 0.2)} \frac{\pi^2 EI}{L^2} \qquad (2.20)$$

For the three special cases considered at the outset, Eq. (2.20) is seen to give the values 1.0, 4.0, and 2.0, respectively, for the numerical coefficients 1.0, 4.0, and 2.04 in Eqs. (1.18), (2.6), and (2.10).

In all the preceding examples the ends of the column have been rigidly restrained against lateral displacement (that is, $w = 0$ at $x = 0, L$). In structural frames one end of a column sometimes may displace laterally relative to the other in a deformation mode called *sidesway*. As an example of such end conditions, let us consider a column that is clamped at one end (say, at $x = 0$) and free at the other, and is subjected to a compressive force P that remains parallel to its original direction as the column deflects, as shown in the fourth example in Table 2.1. For the slightly deflected column, the transverse shearing force at the free end is approximately $Q = Pw'$ [see Eq. (1.10b)]. But $Q = -EIw'''$ and $k^2 = P/EI$. Therefore the boundary-condition equations may be written

$$w = w' = 0 \qquad \text{at } x = 0$$

$$w'' = 0 \qquad \text{and} \qquad w''' = -k^2 w' \qquad \text{at } x = L \tag{2.21}$$

Introduction of Eq. (2.2) into Eq. (2.21) gives

$$C_1 = C_3 = 0 \qquad C_2 = -C_4 \, \cdot \quad C_2 \cos kL = 0$$

Therefore
$$w = C_4(1 - \cos kx) \tag{2.22}$$

where $k = (2n + 1)\pi/2L$ and n is a nonnegative integer. Then

$$P = \frac{(2n + 1)^2 \pi^2}{4L^2} EI \qquad n = 0, 1, 2, \ldots \tag{2.23}$$

The smallest eigenvalue corresponds to $n = 0$. Therefore

$$P_{\text{cr}} = \frac{1}{4} \frac{\pi^2 EI}{L^2} \tag{2.24}$$

and
$$w = C_4\left(1 - \cos \frac{\pi x}{2L}\right) \tag{2.25}$$

This value for the critical load is much lower than those for the first three cases, as would be expected on intuitive grounds. The deflection mode shape given by Eq. (2.25) is shown in Table 2.1.

It should be noted that if the load applied to the end of the clamped-free column remains tangent to the free end rather than parallel to its original direction as the column deflects, the system is unconservative and may not be analyzed by the methods presented in this book. An analysis of the tangential-load problem may be found in Ref. 2.2, p. 21.

Results for the four specific sets of boundary conditions treated in this section and summarized in Table 2.1 show that the critical load for the column is extremely sensitive to the influence of boundary conditions. In structural

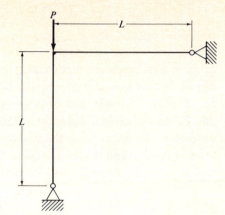

FIGURE 2.2
Two-bar frame.

applications it is convenient to consider the column with simply supported ends as the basic case, and to express the critical load for other sets of boundary conditions in terms of an "equivalent" simply supported column of *effective length KL*, as follows:

$$P_{cr} = \frac{\pi^2 EI}{(KL)^2} \qquad (2.26)$$

The factor $1/K^2$ is seen to represent the numerical coefficient in the expression for the critical load for the respective sets of boundary conditions. Thus, for the last three cases in Table 2.1, $1/K^2$ is equal to 4, 2.04, and 1/4, respectively, and K is equal to 1/2, 0.70, and 2, as shown in the table.

2.2*b*   Instability of Frames

Let us apply the results of the preceding section in the analysis of the instability of a planar frame, i.e., a two-dimensional frame that is constrained to deform only in the plane of the frame. Consider as an example the two-bar frame shown in Fig. 2.2. The column and beam are of equal lengths L and bending stiffness EI. They are rigidly joined to one another and are simply supported at their far ends. At the lower end of the column the rotational stiffness $\alpha_1 = 0$. Therefore $\lambda_1 \rightarrow \infty$ and Eq. (2.17) reduces to the form

$$(1 + \lambda_2 \Phi^2) \sin \Phi = \Phi \cos \Phi$$

Then, since $\Phi = kL$,

$$\tan kL = \frac{kL}{1 + \lambda_2 (kL)^2}$$

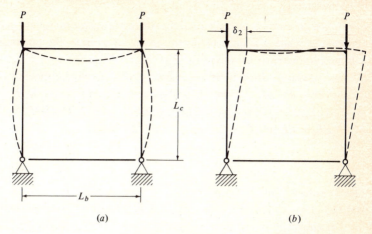

FIGURE 2.3
Rectangular frame.

The value of the parameter λ_2 for the upper end of the column may be determined from a deformation analysis of the beam. From beam theory the rotation β_2 of the left end of a simply supported beam that is subjected to a bending moment M_2 at that end is $\beta_2 = M_2 L/3EI$. (The influence of the vertical displacement of the left end of the beam can be shown to be small and is neglected.) Therefore $\alpha_2 = M_2/\beta_2 = 3EI/L$ and $\lambda_2 = EI/(\alpha_2 L) = 1/3$. Consequently, for this example,

$$\tan kL = \frac{kL}{1 + \frac{1}{3}(kL)^2} \qquad (2.27)$$

The smallest root of Eq. (2.27) is found to be $kL = 3.72$. But $k^2 \equiv P/EI$. Therefore the critical load is

$$P_{\text{cr}} = 1.406 \frac{\pi^2 EI}{L^2} \qquad (2.28)$$

This result corresponds to the critical load of an equivalent simply supported column whose effective length $KL = (1/1.406)^{1/2}L = 0.84L$.

As a second example, consider a frame consisting of two equal-sized columns that are rigidly attached to a beam at their upper ends and are simply supported at their lower ends, as shown in Fig. 2.3a. As in the preceding example, $\lambda_1 \to \infty$ for the lower end. Let us suppose that sidesway (in-plane horizontal displacement of the top of the frame) is prevented by bracing not shown. Then, when the columns buckle, the beam will be deformed into either an arc of a circle or an S shape, depending on whether the buckled configuration of the frame is symmetrical or unsymmetrical. Since the restoring moments supplied by the beam are smaller when it is deformed into an arc of a circle, the buckled configuration will be symmetrical as shown in Fig. 2.3a. Then the rotational stiffness factor α_2 may be determined from an analysis of a simply supported beam subjected to end couples M_2. From beam theory the end rotation β_2 is found to be $\beta_2 = M_2 L_b/2EI_b$. Then $\alpha_2 = 2EI_b/L_b$ and $\lambda_2 = (EI_c/EI_b)/(2L_b/L_c)$. For the case $EI_c = EI_b = EI$ and $L_c = L_b = L$, $\lambda_2 = 1/2$. Then

$$\tan kL = \frac{kL}{1 + \frac{1}{2}(kL)^2}$$

The smallest root is found to be $kL = 3.59$. Therefore $P_{cr} = 1.302\pi^2 EI/L^2$, and the effective length of the columns is $KL = (1/1.302)^{1/2}L = 0.87L$.

If sidesway is not prevented, the load P_{cr} is much smaller. The frame in a slightly deformed configuration with sidesway is shown in Fig. 2.3b. In this configuration the beam is subjected to clockwise moments M_2 at both ends. From a deformation analysis of the beam, $\alpha_2 = M_2/\beta_2 = 6EI_b/L_b$. For the columns the lateral displacement $w \neq 0$ at the upper end, and therefore Eq. (2.17) for elastic rotational restraint does not apply. For either of the columns the expression for equilibrium in the slightly deformed configuration of Fig. 2.3b is given by the second-order beam equation

$$EIw'' = -M = -Pw$$

and the boundary conditions are $w = 0$ at $x = 0$ and $w = -\delta_2$ at $x = L_c$. Integration of the differential equation and introduction into the boundary conditions gives

$$w = -\frac{\delta_2}{\sin kL_c} \sin kx$$

where $k^2 \equiv P/EI_c$. Therefore, at $x = L_c$, $\beta_2 = k\,\delta_2/\tan kL_c$ and $M_2 = EI_c k^2\,\delta_2$. Thus, for the column, $\alpha_2 = M_2/\beta_2 = EI_c k \tan kL_c$. Introduction of the corresponding expression for α_2 for the beam and rearrangement gives

$$kL_c \tan kL_c = 6\frac{EI_b}{EI_c}\frac{L_c}{L_b}$$

For $EI_b = EI_c = EI$ and $L_b = L_c = L$, for example,

$$kL \tan kL = 6$$

Here the smallest root is found to be $kL = 1.35$. Therefore $P_{cr} = 0.184\pi^2 EI/L^2$, and the effective-length factor is $K = (1/0.184)^2 = 2.33$. In general, $K \leq 1$ if sidesway is prevented and $K > 1$ if it is permitted.

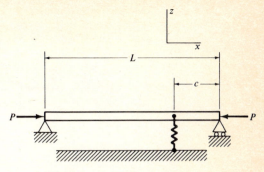

FIGURE 2.4
Column with a single elastic lateral support.

The procedure followed in these examples is convenient only for relatively simple frames. For more complicated examples a procedure that utilizes one of the methods of indeterminate structural analysis such as the slope-deflection method is more suitable. A more extensive treatment of frame stability is included in the excellent textbook on frame analysis by Galambos (Ref. 2.3).

2.2c   Elastic Lateral Support

Let us consider the stability of a straight bar whose lateral displacement is resisted by a single elastic spring, as shown in Fig. 2.4. For convenience, only simply supported ends are considered. Let

$$F_s = -k_s w_s \qquad (2.29)$$

where

F_s = force applied by spring to column, lb (Fig. 2.5)
w_s = lateral displacement at $x = L - c$, in.
k_s = spring stiffness, lb/in.

From beam theory,

$$EIw'' = -M$$

and for the column in a slightly deflected configuration, from Fig. 2.6,

$$M = Pw + \frac{F_s c}{L} x \qquad\qquad x \le (L - c)$$

$$M = Pw + \frac{F_s(L - c)(L - x)}{L} \qquad\qquad x \ge (L - c)$$

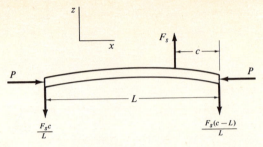

FIGURE 2.5
Column in deformed configuration.

Therefore

$$EIw'' + Pw = -\frac{F_s c}{L} x \qquad x \le (L - c) \qquad (2.30)$$

and

$$EIw'' + Pw = -\frac{F_s(L - c)(L - x)}{L} \qquad x \ge (L - c) \qquad (2.31)$$

The general solutions of these constant-coefficient inhomogeneous equations are

$$w = A \cos kx + B \sin kx - \frac{F_s c}{PL} x \qquad x \le (L - c) \qquad (2.32)$$

and

$$w = C \cos kx + D \sin kx - \frac{F_s(L - c)}{PL}(L - x) \qquad x \ge (L - c) \qquad (2.33)$$

where $k^2 \equiv P/EI$.

For the second-order equation, the simple-support boundary-condition equations are $w = 0$ at $x = 0, L$. Two additional relations are found from the

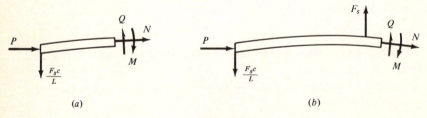

(a)   (b)

FIGURE 2.6
Forces and moments acting on column sections.

requirement that the displacement w and rotation w' be continuous at $x = L - c$. Evaluation of the four arbitrary constants gives

$$w = \frac{F_s \sin kc}{Pk \sin kL} \sin kx - \frac{F_s c}{PL} x \qquad x \le (L - c) \qquad (2.34)$$

and

$$w = \frac{F_s \sin k(L - c)}{Pk \sin kL} \sin k(L - x)$$
$$- \frac{F_s(L - c)}{PL}(L - x) \qquad x \ge (L - c) \qquad (2.35)$$

At $x = L - c$, from either equation,

$$w_s = \frac{F_s \sin kc \sin k(L - c)}{Pk \sin kL} - \frac{F_s c(L - c)}{PL} \qquad (2.36)$$

But $w_s = -F_s/k_s$. Therefore division by the common factor F_s gives, for $F_s \ne 0$,

$$-\frac{1}{k_s} = \frac{\sin kc \sin k(L - c)}{Pk \sin kL} - \frac{c(L - c)}{PL} \qquad (2.37)$$

Equation (2.37) may be written in somewhat more convenient form by introduction of nondimensional parameters defined by the relations

$$\bar{c} \equiv \frac{c}{L} \qquad \bar{k}_s \equiv \frac{L^3}{EI} k_s$$

where $\bar{c}$ and $\bar{k}_s$ are spring location and spring stiffness parameters, respectively. Then

$$\frac{(kL)^3}{\bar{k}_s} = \bar{c}(1 - \bar{c})kL - \frac{\sin \bar{c}kL \sin (1 - \bar{c})kL}{\sin kL} \qquad (2.38)$$

where, from Eqs. (2.15), $kL = (P/EI)^{1/2}L$. Equation (2.38) represents the condition for equilibrium of the bar in a slightly deflected configuration.

For given values of $\bar{c}$ and $\bar{k}_s$, Eq. (2.38) may be solved numerically to determine critical values of the load parameter kL. The procedure is simplified if values of $\bar{k}_s$ are calculated for selected values of $\bar{c}$ and kL and the results are plotted and then cross-plotted in terms of kL versus $\bar{c}$ and $\bar{k}_s$. Results of such a procedure are shown in Fig. 2.7, where the ordinates are expressed in terms of the ratio of critical loads for the column with and without a lateral support.

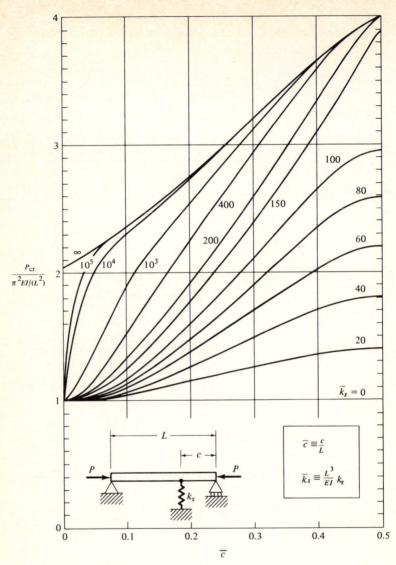

FIGURE 2.7
Buckling coefficients for column stabilized by a lateral spring.

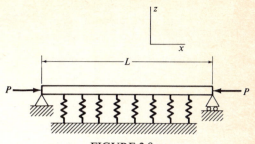

FIGURE 2.8
Column on an elastic foundation.

If the lateral spring is sufficiently stiff, it will enforce a node at the spring. Then higher values of spring stiffness will not further increase the critical load. For example, if a spring located at midlength ($\bar{c} = 0.5$) enforces a node, the column buckles into two half sine waves. Then $kL = 2\pi$ and, from Eq. (2.38), $\bar{k}_s = 16\pi^2 = 158$. For off-center springs, higher values of spring stiffness are required to enforce nodes, as may be seen from Fig. 2.7.

An infinitely stiff spring located infinitesimally close to one end of the column will prevent rotation of that end of the column. Then the numerical coefficient will be the same as that given in Table 2.1 for the clamped–simple-support column, namely, 2.04.

2.2d Column on an Elastic Foundation

Instability mode shape is an important factor in the stability analysis of plates and shells. The role of mode shape may be illustrated by the analysis of a simply supported column on a so-called *Winkler foundation*. Such a foundation is characterized by the force-displacement relationship:

$$q(x) = -k_f w(x) \qquad (2.39)$$

where

$q = $ force intensity between column and foundation, lb/in.

$w = $ lateral displacement, in.

$k_f = $ foundation modulus, psi

The coefficient k_f is a constant because the foundation reaction at any point depends only on the displacement of the beam at that point; the conceptual model for a Winkler foundation is an infinite set of parallel springs with no shear coupling between springs, as illustrated in Fig. 2.8. In a more general

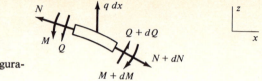

FIGURE 2.9
Beam element in deformed configuration.

case, the displacement at any one point affects the foundation reaction at others, and the foundation must be treated as an elastic continuum.

The differential equation for loss of stability of the column may be obtained by summation of forces and moments for a beam element in a slightly deformed configuration, as shown in Fig. 2.9. This figure differs from Fig. 1.7 only in the presence of the lateral force $q\,dx$, and the resulting equilibrium equation differs from Eq. (1.12) only in the addition of a term q on the right side, as follows:

$$EIw^{iv} + Pw'' = q \qquad (2.40)$$

Equations (2.39) and (2.40) may be combined to give the homogeneous equation

$$EIw^{iv} + Pw'' + k_f w = 0 \qquad (2.41)$$

For $P = 0$, Eq. (2.41) reduces to the homogeneous form of the well-known equation for a beam on an elastic foundation (Ref. 2.4, p. 2).

For convenience, we consider only simply supported ends. Then the boundary-condition equations are given by Eqs. (1.14). A solution of the form of Eq. (1.16), i.e.,

$$w = C_1 \sin \frac{n\pi x}{L} \qquad n = 1, 2, 3, \ldots$$

satisfies both Eqs. (1.14) and (2.41). Introduction into Eq. (2.41) gives, for the eigenvalues, the expression

$$P = \left(\frac{n\pi}{L}\right)^2 EI + \left(\frac{L}{n\pi}\right)^2 k_f \qquad (2.42)$$

For $k_f = 0$, Eq. (2.42) reduces to Eq. (1.17). Unlike that case, however, the smallest eigenvalue here does not necessarily correspond to $n = 1$. For given values of L, EI, and k_f, the n for which P is smallest may be determined by trial. For example, for $L = 31.4$ in., $EI = 100$ lb-in.^{2} and $k_f = 100$ psi, Eq. (2.42) becomes

$$P = n^2 + 10^4 n^{-2} \qquad (2.43)$$

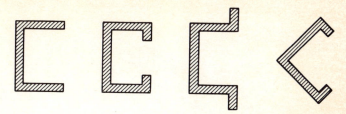

FIGURE 2.10
Typical thin-walled open cross sections.

The values of P for $n = 9$, 10, and 11 are found to be 204, 200, and 203 lb, respectively, and higher values of P are found for all other values of n. Consequently, the critical load for the column is 200 lb, and the instability mode shape is

$$w = C_1 \cos \frac{10\pi x}{L}$$

For a given column an increase in foundation stiffness causes a decrease in wavelength of the critical-load instability mode.

Although n is not a continuous variable, formal minimization of P with respect to n by application of the relation $dP/dn = 0$ to equations such as Eq. (2.43) obviously yields a close estimate of the critical load and corresponding value of n, provided n is much larger than unity (see Prob. 2.4). Such a procedure is widely used in the stability analysis of plates and shells.

2.3 TORSIONAL INSTABILITY OF COLUMNS

In the preceding sections only buckling by bending has been considered. If a column has a thin-walled open cross section such as those illustrated in Fig. 2.10, its twisting stiffness is relatively small and it may buckle by twisting about a longitudinal axis or by a combination of bending and twisting. The action of the compressive load in causing loss of stability by twisting is illustrated in Fig. 2.11 by a longitudinally compressed flat bar that is hinged both on the ends and along one edge and free along the other edge. If the free edge is bent as shown, the longitudinal forces acting on the element dx will have a net component in the yz plane, the component will have a moment with respect to the x axis, and the moment will cause the bar to twist. For torsional instability, the critical

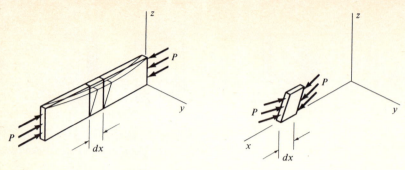

FIGURE 2.11
Twisting deformation caused by a compressive axial load.

value of the compressive load P is defined as the smallest load at which the bar can be maintained in equilibrium in a slightly twisted configuration.

An approximate analysis of torsional instability of slender columns with thin-walled open cross sections may be developed in terms of the assumption that the shape of the cross section remains unchanged during the deformation. (Columns with more flexible cross sections must be treated as thin-walled shells.) This assumption underlies the analyses in this section and in Secs. 2.4 and 2.5, and the equations in these sections may not be obtained by direct specialization of the general shell equations in Chap. 6. The literature based on the assumption that the shape of a thin-walled open cross section remains unchanged is extremely extensive, and the material in these three sections is only a brief introduction to one of the important fields of structural analysis. A more detailed treatment may be found in, for example, Ref. 2.3.

The present analysis of torsional-flexural instability follows in some respects the analysis in Ref. 2.5. A thin-walled column of length L subjected to a uniformly distributed compressive load P is shown in Fig. 2.12 in the undeformed configuration. The boundary conditions at the ends of the bar will be specified later. The x axis is the longitudinal axis through the centroid of the cross section, as before, and y and z are the principal centroidal axes in the cross section. The centroid and shear center of the cross section are denoted by C and S, respectively, as indicated in Fig. 2.13. (A discussion of the shear center for thin-walled open cross sections may be found in textbooks on mechanics of solids, such as Ref. 2.6.) The shear center derives its significance from the fact that a bar that is fixed at one end and subjected to a transverse load applied through the shear center at the other end undergoes bending without twisting. Conversely, a twisting moment M_{xx} applied to such a bar produces no transverse displacement of the shear center. Consequently, the shear-center axis remains

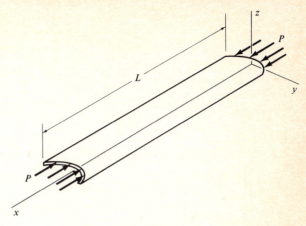

FIGURE 2.12
Thin-walled open-cross-section column subjected to compressive axial load.

straight during twisting, and the cross sections of the bar rotate about the shear center during the deformation.

In the present analysis we consider equilibrium of a column in a slightly bent and twisted configuration. We let v and w denote the y and z components, respectively, of the displacement of the shear center, and we let ϕ denote the small angle of rotation of the cross section about the shear center. Then, from Fig. 2.13, the displacement components of the centroid C are

$$v_C = v + z_o\phi \qquad w_C = w - y_o\phi \qquad (2.44)$$

where y_o and z_o represent the location of the shear center in relation to the centroid prior to deformation. Three equilibrium equations for the column in a

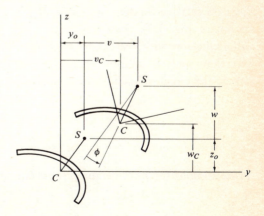

FIGURE 2.13
Column cross section before and after
deformation.

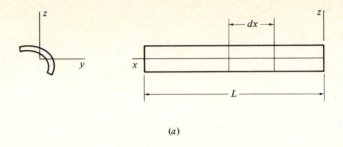

(a)

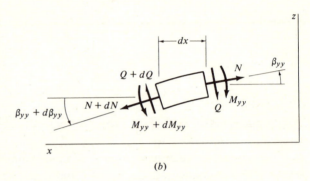

(b)

FIGURE 2.14
Column element bent in the xz plane.

slightly deformed configuration may be derived in terms of the three displacement variables v, w, and ϕ, as shown in the following paragraphs.

Let us first consider bending of the column. A column segment of length dx is shown in a deformed configuration in Fig. 2.14, where β_{yy} is the angle of rotation relative to the y axis and M_{yy} is the corresponding bending moment (compare Fig. 1.7). From the figure, $\beta_{yy} = -w'_C$, where the prime signifies differentiation with respect to x. From summation of forces in the x and z directions and summation of moments relative to the y axis, respectively, we obtain the equations [compare Eqs. (1.10)]

$$N' = 0 \qquad (2.45a)$$

$$Q' + Nw''_C = 0 \qquad (2.45b)$$

$$M'_{yy} - Q = 0 \qquad (2.45c)$$

Proceeding as for pure bending, these equations may be combined to give

$$M''_{yy} - Pw''_C = 0 \qquad (2.46)$$

But from beam theory, the elastic relationship for the shear center is

$$M_{yy} = -EI_{yy} w'' \qquad (2.47)$$

where $I_{yy} = \int z^2 \, dA$. Introduction into Eq. (2.46) gives, for constant EI_{yy},

$$EI_{yy} w^{iv} + P w''_C = 0 \qquad (2.48)$$

Therefore, from Eqs. (2.44),

$$EI_{yy} w^{iv} + P(w - y_o \phi)'' = 0 \qquad (2.49)$$

A similar analysis of bending in the xy plane leads to the equilibrium equation

$$EI_{zz} v^{iv} + P(v + z_o \phi)'' = 0 \qquad (2.50)$$

where $I_{zz} = \int y^2 \, dA$.

Now let us consider twisting of the column. The third equilibrium equation is obtained by summing moments about the shear center. This portion of the analysis is relatively long and is presented in two parts: (1) development of the equilibrium relationship between the applied load P and the resulting non-uniform twisting moment $M_{xx}(x)$ in the bar, and (2) development of the elastic relationship between $M_{xx}(x)$ and the resulting angle of twist $\phi(x)$ in terms of the torsional and the warping resistance of the bar.

For the column in the undeformed configuration, the axial load P causes no twisting moment relative to the shear center. In the deformed configuration, however, a column segment of length dx (compare Fig. 2.11) is subjected to a force component in the yz plane that does have a moment relative to the shear center. A moment exists even for cases for which the shear center S coincides with the centroid C. Therefore, to determine the moment, we must examine not merely a cross-sectional segment of length dx, as in Fig. 2.14, but a general element of that segment of area $dA = h \, ds$, as shown in Fig. 2.15. From the figure the displacements v_B, w_B of the element dA in terms of the displacements v, w of the shear center are [see Fig. 2.13 and Eqs. (2.44)]

$$v_B = v + (z_o - z)\phi \qquad w_B = w - (y_o - y)\phi \qquad (2.51)$$

In the deformed configuration the ends of the volume element of area dA and length dx are subjected to longitudinal forces of magnitude $\sigma_x \, dA$, where, approximately, $\sigma_x = -P/A$ (compare Fig. 2.15b with Fig. 2.14b). Thus, in the xz plane, the element of length dx is subjected to a lateral force intensity whose magnitude is $(\sigma_x \, dA)w''_B$ lb/in. [Compare Eq. (2.48) and the fourth-order beam equation, $EI_{yy} w^{iv} = q$.] Introduction of the expression for w_B from Eqs. (2.51)

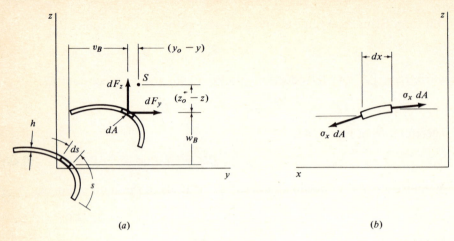

FIGURE 2.15
Element of cross section before and after deformation.

and multiplication by the length of the element gives, for the lateral force, the expression

$$dF_z = (\sigma_x \, dA)[w - (y_o - y)\phi]'' \, dx$$

Similarly, the component in the y direction is, approximately,

$$dF_y = (\sigma_x \, dA)[v + (z_o - z)\phi]'' \, dx$$

The forces dF_z and dF_y acting on the volume element are shown in Fig. 2.15a. Now let m_{xx} represent the twisting moment per unit length of column. Then the moment of the forces dF_y, dF_z with respect to the shear center S is, approximately

$$dm_{xx} \, dx = (z_o - z) \, dF_y - (y_o - y) \, dF_z \tag{2.52}$$

Introduction of the expressions for dF_y and dF_z and rearrangement gives

$$dm_{xx} = \{(z_o - z)[v + (z_o - z)\phi]'' - (y_o - y)[w - (y_o - y)\phi]''\} \, \sigma_x \, dA \tag{2.53}$$

Introduction of the relationship $\sigma_x = -P/A$, integration over the cross-sectional area A, and use of the relations $\int_A y \, dA = \int_A z \, dA = 0$ leads to the expression

$$m_{xx} = Py_o w'' - Pz_o v'' - P\frac{I_o}{A}\phi'' \tag{2.54}$$

where

$$I_o = I_{yy} + I_{zz} + A(y_o{}^2 + z_o{}^2)$$

$$I_{yy} = \int_A z^2 \, dA \qquad I_{zz} = \int_A y^2 \, dA \tag{2.55}$$

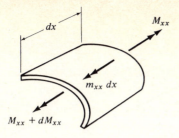

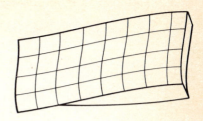

FIGURE 2.16
Twisting moments acting on column element.

FIGURE 2.17
Twisting and warping deformation.

But from Fig. 2.16,

$$M'_{xx} = -m_{xx}$$

Therefore

$$M'_{xx} = -P\left(y_o w'' - z_o v'' - \frac{I_o}{A}\phi''\right) \qquad (2.56)$$

Equation (2.56) is an expression for the rate of change of the twisting moment due to the compressive load P.

Now we turn to the development of the elastic relationship between the twisting moment M_{xx} and the angle of twist ϕ. When a twisting moment is applied to a bar, a cross section of the bar rotates through an angle ϕ relative to a fixed cross section. If the cross section is not axisymmetric, the cross section warps out of its plane, as well as rotates. The nature of warping may be visualized with the aid of a rectangular-cross-section eraser on which parallel lines have been drawn, as shown in Fig. 2.17. If some cross sections (for example, the end cross sections) are prevented from warping by physical constraints, the twisting resistance of the bar is increased. In such cases we may let

$$M_{xx} = M_{xx1} + M_{xx2} \qquad (2.57)$$

where M_{xx1} is the portion of the applied twisting moment resisted by torsion of the bar, and M_{xx2} is the portion associated with warping.

From elementary mechanics [for example, Ref. 2.6, eq. (6.5)], the relation between M_{xx1} and angle of twist ϕ is simply

$$M_{xx1} = GJ\phi' \qquad (2.58)$$

where J is the *torsional constant*, in.[4] For thin-walled open cross sections the

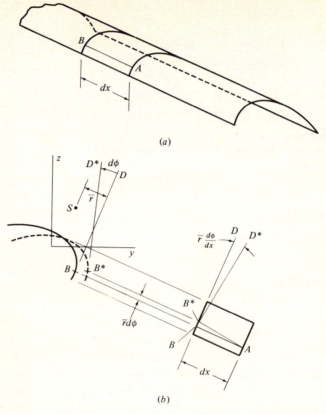

FIGURE 2.18
Segment of cross section showing warping deformation.

torsional constant is given approximately by the relation [Ref. 2.4, eq. (221) and art. 47]

$$J = \tfrac{1}{3} \sum m_i h_i^3 \qquad (2.59)$$

where $\sum m_i$ is the developed length of the middle line of the cross section and h_i is wall thickness.

Now let us develop a corresponding relationship between the angle of twist and the moment M_{xx2}. Warping represents a longitudinal displacement of points in the column cross section. An equation relating the longitudinal displacement component u to angle of twist ϕ may be derived as follows. The middle surface of a column element of length dx is shown in the undeformed configuration in Fig. 2.18a. The element is shown in orthographic projections in Fig. 2.18b. We refer to the orthographic projection on the right as the *lateral view* and that on the left as the *end view*. Line AB in the lateral view is a longitudinal line on the middle surface prior to deformation, and BD is tangent to the middle surface at B and perpendicular to line AB. When the bar is twisted, one end of the longitudinal element rotates about the shear center through a small angle $d\phi$, as shown. Then point B moves to B^*, and the angle BAB^* in the lateral view is $\bar{r}\,d\phi/dx$, where $\bar{r}$ is the perpendicular distance from the shear center to the tangent BD, as shown. The variable $\bar{r}$ is positive if a vector along the tangent in the direction of increasing s acts counterclockwise about the shear center. After the deformation, the tangent B^*D^* remains perpendicular to the longitudinal line AB^*. Thus, in the lateral view, the angle between the tangents before and after deformation is $\bar{r}\,d\phi/dx \equiv \bar{r}\phi'$. But that angle is the rate of change of the displacement u in the s direction. Thus the equation relating the displacement u to the angle of twist ϕ is

$$\frac{\partial u}{\partial s} = -\bar{r}\phi' \qquad (2.60)$$

Integration of Eq. (2.60) gives

$$u = u_0 - \phi' \int_0^s \bar{r}\,ds$$

where $u_0 = u$ at $s = 0$. Now let an average displacement be defined by

$$\bar{u} = \frac{1}{m} \int_0^m u\,ds$$

where, as noted, m is the length of the middle line of the cross section.

Then

$$\bar{u} = \frac{1}{m} \int_0^m \left(u_0 - \phi' \int_0^s \bar{r}\,ds \right) ds$$

and

$$u - \bar{u} = \frac{1}{m}\phi' \int_0^m \left(\int_0^s \bar{r}\,ds \right) ds - \phi' \int_0^s \bar{r}\,ds$$

For convenience, let us measure u from the plane for which $\bar{u} = 0$. Then

$$u = \frac{1}{m}\phi' \int_0^m \left(\int_0^s \bar{r}\,ds \right) ds - \phi' \int_0^s \bar{r}\,ds$$

Now let us define *warping functions* as follows:

$$\omega_s = \int_0^s \bar{r}\,ds \qquad \bar{\omega}_s = \frac{1}{m} \int_0^m \omega_s\,ds \qquad (2.61)$$

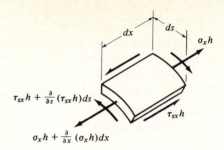

FIGURE 2.19
Element of column illustrating equilibrium of forces in the longitudinal direction.

Then the displacement u may be written in the simpler form

$$u = \phi'(\bar{\omega}_s - \omega_s) \qquad (2.62)$$

With this expression for the warping displacement u, we may relate the twisting moment M_{xx2} to the angle of twist ϕ as follows. From elementary mechanics, $\varepsilon_x = \partial u/\partial x$. Furthermore, for a uniaxial state of stress, $\sigma_x = E\varepsilon_x$. Therefore

$$\sigma_x = E(\bar{\omega}_s - \omega_s)\phi'' \qquad (2.63)$$

To establish a relationship between the normal stress σ_x and the shearing stress τ_{sx}, let us consider the longitudinal equilibrium of a rectangular element $ds\,dx$, shown in Fig. 2.19. From summation of forces in the x direction we obtain

$$\frac{\partial}{\partial s}(h\tau_{sx}) = -\frac{\partial}{\partial x}(h\sigma_x)$$

$$= -Eh(\bar{\omega}_s - \omega_s)\phi'''$$

Integration gives

$$h\tau_{sx} = -E\phi'''\int_0^s (\bar{\omega}_s - \omega_s)h\,ds$$

But the shearing stress τ_{sx} is related to the twisting moment M_{xx2} by the equation

$$M_{xx2} = \int_0^m \tau_{xs}\bar{r}h\,ds$$

Therefore

$$M_{xx2} = -E\phi'''\int_0^m \left[\int_0^s (\bar{\omega}_s - \omega_s)h\,ds\right]\bar{r}\,ds$$

From the first of Eqs. (2.61), $\bar{r} \, ds = d(\omega_s)$ or, since $\bar{\omega}_s$ is not a function of s,

$$\bar{r} \, ds = -\frac{d(\bar{\omega}_s - \omega_s)}{ds} ds$$

Introduction of this expression into the equation for M_{xx2} and integration by parts gives

$$M_{xx2} = E\phi''' \int_0^m (\bar{\omega}_s - \omega_s)h \, ds \, (\bar{\omega}_s - \omega_s) - \int_0^m (\bar{\omega}_s - \omega_s)^2 h \, ds$$

But the twisting moment M_{xx2} produces no net force in the longitudinal direction, that is, $\int_0^m \sigma_x h \, ds = 0$ (Ref. 2.5, p. 221). Therefore from Eq. (2.63),

$$\int_0^m (\bar{\omega}_s - \omega_s)h \, ds = 0$$

and the expression for M_{xx2} simplifies to the form

$$M_{xx2} = -E\phi''' \int_0^m (\bar{\omega}_s - \omega_s)^2 h \, ds$$

Now let us define the *warping constant* C_W (in.^{6}) by the relation

$$C_W = \int_0^m (\bar{\omega}_s - \omega_s)^2 h \, ds \qquad (2.64)$$

Then the expression for M_{xx2} may be written

$$M_{xx2} = -EC_W\phi''' \qquad (2.65)$$

Introduction of the expressions for M_{xx1} and M_{xx2} into Eq. (2.57) finally gives, for the elastic relationship between twisting moment and angle of twist, the equation

$$M_{xx} = -EC_W \phi''' + GJ\phi' \qquad (2.66)$$

The third equilibrium equation now is obtained by differentiation of Eq. (2.66) with respect to x and combination with Eq. (2.56) to give

$$EC_W \phi^{iv} - \left(GJ - \frac{I_o}{A}P\right)\phi'' - Py_o w'' + Pz_o v'' = 0 \qquad (2.67)$$

Equations (2.49), (2.50), and (2.67) are the three governing equations for combined torsional-flexural instability. They are seen to be coupled homogeneous constant-coefficient equations in the variables v, w, and ϕ.

For simplicity, we consider only simply supported ends, i.e., ends for which the boundary conditions are

$$v = w = \phi = v'' = w'' = \phi'' = 0 \qquad \text{at } x = 0, L \qquad (2.68)$$

A solution of the form

$$\phi = A_1 \sin \frac{\pi x}{L} \qquad v = A_2 \sin \frac{\pi x}{L} \qquad w = A_3 \sin \frac{\pi x}{L} \qquad (2.69)$$

where A_1, A_2, A_3 are constants, is seen to satisfy both the differential equations and boundary conditions. Introduction into Eqs. (2.49), (2.50), and (2.67) and rearrangement gives

$$Pz_o A_1 + (P - P_z)A_2 = 0$$

$$Py_o A_1 - (P - P_y)A_3 = 0 \qquad (2.70)$$

$$\frac{I_o}{A}(P - P_\phi)A_1 + Pz_o A_2 - Py_o A_3 = 0$$

where

$$P_\phi = \frac{A}{I_o}\left(\frac{\pi^2 EC_W}{L^2} + GJ\right) \qquad P_y = \frac{\pi^2 EI_{yy}}{L^2} \qquad P_z = \frac{\pi^2 EI_{zz}}{L^2} \qquad (2.71)$$

For a nontrivial solution, the determinant of the coefficients of the A_1, A_2, A_3 in Eqs. (2.70) must equal zero. Thus

$$\frac{I_o}{A}(P - P_\phi)(P - P_y)(P - P_z) - P^2 z_o^{\,2}(P - P_y) - P^2 y_o^{\,2}(P - P_z) = 0 \qquad (2.72)$$

where, as noted, y_o, z_o are the coordinates of the shear center and I_o is defined by Eq. (2.55). Equation (2.72) is the final expression for the simply supported column for the critical value of the load P in combined torsional-flexural instability. The smallest of the three roots of Eq. (2.72) is the critical load.

If the column bends without twisting, $\phi = A_1 = 0$ in Eqs. (2.69) and (2.70). Then nontrivial solutions are given by $P = P_y$ and $P = P_z$. Consequently, P_y and P_z in Eqs. (2.71) represent critical values of P for flexural instability. Similarly, if the column twists without bending, $v = w = A_2 = A_3 = 0$, and from Eq. (2.70), P_ϕ is seen to represent the critical value of P for purely torsional instability.

If the cross section does not have an axis of symmetry, the centroid and shear center will not coincide, and bending and twisting will always occur to-

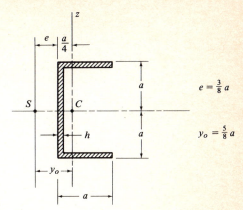

$$e = \tfrac{3}{8} a$$

$$y_o = \tfrac{5}{8} a$$

FIGURE 2.20
Channel section.

gether. In such cases the smallest root of Eq. (2.72) will always be smaller than any of the values P_y, P_z, or P_ϕ.

In most applications of thin-walled open sections, the cross section has at least one axis of symmetry, as illustrated in Fig. 2.10. If the y axis is the symmetry axis, $z_o = 0$, and Eq. (2.72) simplifies to the form

$$(P - P_z)\left[\frac{I_o}{A}(P - P_y)(P - P_\phi) - P^2 y_o{}^2\right] = 0 \qquad (2.73)$$

If P_z is the smallest of the three roots of Eq. (2.73), the column will buckle in pure bending. Otherwise, the buckling will be in combined bending and twisting.

If the cross section has two axes of symmetry, $y_o = z_o = 0$, and Eq. (2.72) simplifies to the form

$$(P - P_z)(P - P_y)(P - P_\phi) = 0 \qquad (2.74)$$

In this case the three roots are P_y, P_z, and P_ϕ, and the column buckles in pure bending or pure twisting, depending on which of the three roots is smallest.

As a numerical example, let us consider a column with the cross section shown in Fig. 2.20, where $a = 4$ in. and $h = 0.20$ in. Because of the symmetry with respect to the y axis, the load P_{cr} is the smallest root of Eq. (2.73), where P_y, P_z, and P_ϕ are given by Eqs. (2.71). The cross-sectional area is $A = 4ah$, and the torsional constant is $J = \sum m_i h_i{}^3/3 = 4ah^3/3$. The warping constant is $C_W = \int_0^m (\bar{\omega}_s - \omega_s)^2 h \, ds$, where $\bar{\omega}_s$ and ω_s are given by Eqs. (2.61). The coordinate s in Eqs. (2.61) is measured counterclockwise around the cross section, and $\bar{r}$ is positive when a vector along the tangent to the cross section in the direction of increasing s also acts counterclockwise about the shear center.

Accordingly, from Eqs. (2.61),

$$\omega_s = \int_0^s a\,ds = as \qquad\qquad\qquad 0 \le s \le a$$

$$\omega_s = a^2 - \int_a^s e\,ds = a(a+e) - es \qquad\qquad a \le s \le 3a$$

$$\omega_s = a(a-2e) + \int_{3a}^s a\,ds = -2a(a+e) + as \qquad 3a \le s \le 4a$$

Therefore

$$\bar{\omega}_s = \frac{1}{4a}\left[\int_0^a as\,ds + \int_a^{3a}(a^2 + ae - es)\,ds + \int_{3a}^{4a}(-2a^2 - 2ae + as)\,ds\right]$$

$$= a(a - e)$$

Introduction into the expression for C_W now gives

$$C_W = \int_0^a [a(a-e) - as]^2 h\,ds + \int_a^{3a}[a(a-e) - a(a+e) + es]^2 h\,ds$$

$$+ \int_{3a}^{4a}[a(a-e) + 2a(a+e) - as]^2 h\,ds$$

$$= \tfrac{2}{3}a^3 h(a^2 - 3ae + 4e^2)$$

From Fig. 2.20, $e = 3a/8$. Therefore $C_W = 7a^5h/24$. Introduction of the values $a = 4$ and $h = 0.20$ into the expressions for A, J, and C_W, respectively, now gives $A = 3.2$ in.^{2}, $J = 0.0427$ in.^{4}, and $C_W = 59.7$ in.^{6}. For these dimensions, furthermore, $I_{yy} = 34.1$ in.^{4} and $I_{zz} = 5.34$ in.^{4} But $I_o = I_{yy} + I_{zz} + Ay_o^2$ and, from Fig. 2.20, $y_o = 2.5$ in. Therefore $I_o = 59.4$ in.^{4} and $I_o/A = 18.57$ in.^{2}

Now suppose the column in the example has a length $L = 120$ in. and is made of steel with $E = 29 \times 10^6$ psi and $G = 11.2 \times 10^6$ psi. Then

$$P_y = \frac{\pi^2 E I_{yy}}{L^2} = \frac{\pi^2(29 \times 10^6)(34.1)}{(120)^2} = 678,000 \text{ lb}$$

$$P_z = \frac{\pi^2 E I_{zz}}{L^2} = \frac{\pi^2(29 \times 10^6)(5.34)}{(120)^2} = 106,100 \text{ lb}$$

$$P_\phi = \frac{A}{I_o}\left(\frac{\pi^2 E C_W}{L^2} + GJ\right) = \frac{1}{18.57}\left[\frac{\pi^2(29 \times 10^6)(59.6)}{(120)^2} + (11.2 \times 10^6)(0.0422)\right]$$

$$= 89,600 \text{ lb}$$

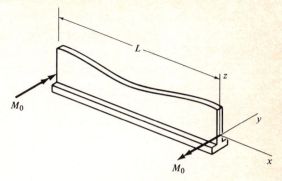

FIGURE 2.21
Lateral instability of beam.

We see that P_z is not smaller than both P_y and P_ϕ, and therefore the column will buckle in combined bending and twisting. Then, from Eq. (2.73), the critical load will be the smaller root of the quadratic equation

$$\frac{I_o}{A}(P - P_y)(P - P_\phi) - P^2 y_o{}^2 = 0$$

Introduction of the numerical values gives

$$18.57(P - 678,000)(P - 89,600) - (2.5)^2 P^2 = 0$$

From the quadratic formula the smaller root is found to be $P_{cr} = 85,500$ lb. The corresponding average stress is $P_{cr}/A = 85,500/3.2 = 26,700$ psi.

Solutions for other boundary conditions may be found in, for example, Refs. 2.5 and 2.7. Extensive numerical results based on Eq. (2.73) for cross sections with a single axis of symmetry are given in Ref. 2.8.

2.4 LATERAL INSTABILITY OF BEAMS

A slender beam loaded in a plane of symmetry may buckle laterally if its resistance to lateral bending is much smaller than its resistance to bending in the plane of symmetry. The action of in-plane bending loads in causing loss of stability through lateral displacement is illustrated by the example shown in Fig. 2.21, where double-headed vectors represent bending couples in accordance with the right-hand-screw convention. For lateral instability the critical value of the applied load may be defined as the smallest value at which the beam can be maintained in equilibrium with a small lateral displacement.

The following approximate analysis of lateral instability is developed in terms of the thin-walled elastic beam theory of Sec. 2.3. In that section the symbols M_{xx}, M_{yy}, M_{zz} were used to denote moments in the deformed configuration. In this section and in Sec. 2.5 it is convenient to reserve these symbols for the undeformed configuration and to let $M_{x^*x^*}$, $M_{y^*y^*}$, $M_{z^*z^*}$ represent the corresponding quantities in the deformed configuration. Then, from Eq. (2.66) and elementary beam theory, the elastic relations for twisting and bending are

$$EC_W \phi''' - GJ\phi' = -M_{x^*x^*} \qquad (2.75a)$$

$$EI_{yy} w'' = -M_{y^*y^*} \qquad (2.75b)$$

$$EI_{zz} v'' = M_{z^*z^*} \qquad (2.75c)$$

where C_W and J are given by Eqs. (2.64) and (2.59), respectively.

A slender prismatic beam is illustrated in Fig. 2.22a. Both the y and z axes are axes of symmetry, and $I_{yy} \gg I_{zz}$. The portion of the beam to the left of section mn is shown in a deformed configuration in Fig. 2.22b. The axes x^*, y^*, z^* represent coordinates for the deformed configuration, and the angle of twist ϕ and angles of rotation v' and w' are assumed to be small, as before. The beam is loaded only in the xz plane; i.e., it is subjected only to transverse forces acting in the z direction and to bending couples M_{yy}. The transverse forces cause an additional bending moment M_{yy}. When the beam is displaced laterally, the loading also causes a relatively small twisting moment M_{xx}. The sign conventions for positive moments are indicated in the sketch. For small angles of rotation, the bending and twisting moments on a cross section in the deformed configuration may be expressed in terms of M_{xx} and M_{yy} by the relations

$$M_{x^*x^*} = M_{xx} \cos(x, x^*) + M_{yy} \cos(y, x^*)$$

$$M_{y^*y^*} = M_{xx} \cos(x, y^*) + M_{yy} \cos(y, y^*) \qquad (2.76)$$

$$M_{z^*z^*} = M_{xx} \cos(x, z^*) + M_{yy} \cos(y, z^*)$$

where (x, x^*) denotes the angle between the x and x^* directions, etc. From the sketch, approximately,

$$\cos(x, x^*) = \cos(y, y^*) = 1$$

$$\cos(y, x^*) = -\cos(x, y^*) = v'$$

$$\cos(x, z^*) = -w'$$

$$\cos(y, z^*) = -\phi$$

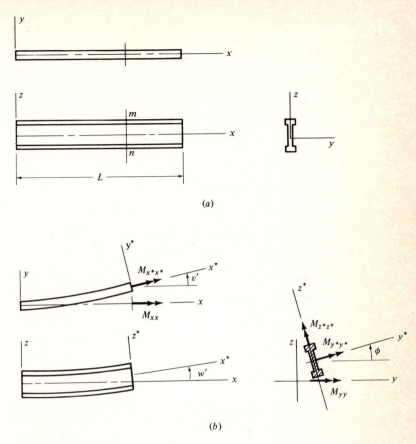

FIGURE 2.22
Beam in undeformed and deformed configurations—symbols and sign conventions.

If we neglect products of small angles and small twisting moments M_{xx}, Eqs. (2.75) simplify to the expressions

$$M_{x^*x^*} = M_{xx} + M_{yy}v' \qquad (2.77a)$$

$$M_{y^*y^*} = M_{yy} \qquad (2.77b)$$

$$M_{z^*z^*} = -M_{yy}\phi \qquad (2.77c)$$

Equations (2.75b) and (2.77b) give the usual expression for bending of the beam in the xz plane:

$$EI_{yy}w'' = -M_{yy}$$

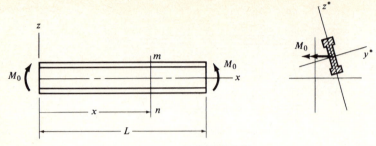

FIGURE 2.23
Beam subjected to bending couples M_0.

Equations (2.75a), (2.75c), (2.77a), and (2.77c) are the equations for lateral instability of doubly symmetrical beams. They lead to two coupled equations in the lateral displacement v and the angle of twist ϕ.

If the beam cross section has only a single axis of symmetry, it is not sufficient to express the twisting moment $M_{x^* x^*}$ in terms of the gross moment M_{yy} acting on the entire cross section, as in Eq. (2.77a). Instead, the $M_{yy} v'$ term in Eq. (2.77a) must be replaced by an expression for the moment of the force $\sigma_x \, dA$ relative to the x^* axis. (Compare Sec. 2.3. For the moment M_{yy}, the stress $\sigma_x = M_{yy} z / I_{yy}$.) Cross sections with only a single axis of symmetry are treated in, for example, Ref. 2.3.

Two examples of beams with doubly symmetric cross sections will be considered, a simply supported beam subjected to bending moments M_0 at its ends and a cantilever beam subjected to a concentrated load P at its free end.

The first example is illustrated in Fig. 2.23. From the figure the moments M_{yy} and M_{xx} acting at section mn on the portion of the beam to the left of the section are, approximately,

$$M_{yy} = -M_0 \qquad M_{xx} = 0$$

Introduction of these values into Eqs. (2.75) and (2.77) gives

$$EC_W \phi''' - GJ \phi' - M_0 v' = 0$$
$$EI_{zz} v'' - M_0 \phi = 0 \tag{2.78}$$

Differentiation of the first equation and elimination of v'' by use of the second one yields the expression

$$EC_W \phi^{iv} - GJ \phi'' - \frac{M_0^2}{EI_{zz}} \phi = 0 \tag{2.79}$$

Equation (2.79) is seen to be a homogeneous equation in the variable ϕ. It has nontrivial solutions only for discrete values of the parameter M_0. The smallest such value is M_{cr}.

The general solution of the constant-coefficient differential equation is readily obtained. The associated characteristic equation is seen to be

$$\lambda^4 - \frac{GJ}{EC_W}\lambda^2 - \frac{M_0{}^2}{E^2 C_W I_{zz}} = 0$$

The roots of the characteristic equation are

$$\lambda = \alpha,\ -\alpha,\ i\beta,\ -i\beta$$

where α, β are real positive numbers given by

$$\alpha = \left\{ \frac{GJ}{2EC_W} + \left[\left(\frac{GJ}{2EC_W}\right)^2 + \frac{M_0{}^2}{E^2 C_W I_{zz}} \right]^{1/2} \right\}^{1/2} \qquad (2.80a)$$

$$\beta = \left\{ -\frac{GJ}{2EC_W} + \left[\left(\frac{GJ}{2EC_W}\right)^2 + \frac{M_0{}^2}{E^2 C_W I_{zz}} \right]^{1/2} \right\}^{1/2} \qquad (2.80b)$$

In this notation the general solution to Eq. (2.79) may be written

$$\phi = A_1 e^{-\alpha x} + A_2 e^{\alpha x} + A_3 \cos \beta x + A_4 \sin \beta x \qquad (2.81)$$

where A_1, A_2, A_3, A_4 are constants to be evaluated from the boundary conditions.

If the simply supported beam is free to warp at its ends, $\sigma_x = 0$ at $x = 0, L$. Then, from Eq. (2.63), $\phi'' = 0$ at $x = 0, L$. If, in addition, the ends are prevented from rotating about the longitudinal axis, $\phi = 0$ at the ends. Thus the boundary-condition equations are

$$\phi = \phi'' = 0 \qquad \text{at } x = 0, L \qquad (2.82)$$

From the conditions at $x = 0$, we find that $A_3 = 0$ and $A_1 = -A_2$. Then Eq. (2.81) may be written in the more convenient form

$$\phi = A \sinh \alpha x + B \sin \beta x \qquad (2.83)$$

where A and B are new constants. Introduction into the boundary conditions at $x = L$ gives

$$A \sinh \alpha L + B \sin \beta L = 0 \qquad (2.84a)$$

$$\alpha^2 A \sinh \alpha L - \beta^2 B \sin \beta L = 0 \qquad (2.84b)$$

For a nontrivial solution to this homogeneous equation system, the determinant of the coefficients of A, B must equal zero. Accordingly,

$$(\sin \beta L)(\sinh \alpha L)(\alpha^2 + \beta^2) = 0$$

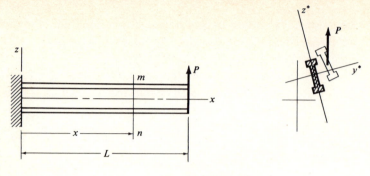

FIGURE 2.24
Beam subjected to transverse load P.

But α, β, and $\sinh \alpha L$ are nonzero quantities. Therefore $\sin \beta L = 0$. Then $\beta L = m\pi$, where m is a positive integer (cf. the column analysis in Sec. 1.2). Furthermore, from Eq. (2.84a), $A = 0$. Consequently,

$$\phi = B \sin \frac{m\pi x}{L} \qquad (2.85)$$

Equation (2.80b) and the relation $\beta = m\pi/L$ now give

$$\left(\frac{m\pi}{L}\right)^2 = -\frac{GJ}{2EC_W} + \left[\left(\frac{GJ}{2EC_W}\right)^2 + \frac{M_0{}^2}{E^2 C_W I_{zz}}\right]^{1/2}$$

The smallest value of M_0 is seen to correspond to $m = 1$. Accordingly, the critical value of M_0 is given by the expression

$$M_{cr} = \left(\frac{\pi^4 EI_{zz} EC_W}{L^4} + \frac{\pi^2 EI_{zz} GJ}{L^2}\right)^{1/2} \qquad (2.86)$$

Finally, for $m = 1$, Eq. (2.85) gives, for the mode shape, the relation

$$\phi = B \sin \frac{\pi x}{L} \qquad (2.87)$$

Equation (2.87) indicates that the maximum angle of twist occurs at midspan.

A sketch for the second example is shown in Fig. 2.24. The load is applied to the cantilever beam at the centroid of the end cross section and remains parallel to the z axis during the deformation. From the figure the moments M_{yy} and M_{xx} acting at section mn on the portion of the beam to the left of the section are seen to be

$$M_{yy} = -P(L - x) \qquad M_{xx} = P[v(L) - v(x)] \qquad (2.88)$$

Introduction of these values into Eqs. (2.75) and (2.77) gives

$$EC_W \phi''' - GJ\phi' - P(L-x)v' + P(v_L - v) = 0$$
$$EI_{zz} v'' - P(L-x)\phi = 0 \qquad (2.89)$$

where $v_L \equiv v$ at $x = L$. As in the first example, the variable v may be eliminated by differentiation of the first equation and substitution of the second. The resulting equation is

$$EC_W \phi^{iv} - GJ\phi'' - \frac{P^2}{EI_{zz}}(L-x)^2\phi = 0 \qquad (2.90)$$

This expression may be simplified somewhat by introduction of the variable

$$\xi \equiv L - x$$

to give

$$EC_W \frac{d^4\phi}{d\xi^4} - GJ \frac{d^2\phi}{d\xi^2} - \frac{P^2}{EI_{zz}} \xi^2\phi = 0 \qquad (2.91)$$

The homogeneous equation has nontrivial solutions only for discrete values of the parameter P. The smallest such value is P_{cr}.

Unlike the other differential equations in Chap. 2, Eq. (2.91) is a variable-coefficient equation. A solution for a particular set of boundary conditions may be obtained by use of numerical methods such as those discussed in Chap. 8. Numerical results based on a series solution are given in Ref. 2.5, p. 259.

If the cantilever beam has a narrow rectangular cross section, the variable $\bar{r} \equiv 0$ in the expression for the warping functions in Eq. (2.61). Then, from Eq. (2.64), the warping constant $C_W = 0$, and Eq. (2.91) simplifies to the expression

$$GJ \frac{d^2\phi}{d\xi^2} + \frac{P^2}{EI_{zz}} \xi^2\phi = 0 \qquad (2.92)$$

Now let $k_1^2 \equiv P^2/(EI_{zz} GJ)$. Then

$$\frac{d^2\phi}{d\xi^2} + k_1^2\xi^2\phi = 0 \qquad (2.93)$$

Thus, for a narrow rectangular cross section, or any other cross section for which the warping rigidity is negligibly small, Eq. (2.91) is replaced by the second-order equation (2.93).

Equation (2.93) also has a variable coefficient, and a solution for a particular set of boundary conditions again may be obtained by use of numerical methods. From Ref. 2.5, eq. (6.22), however, the general solution of this equation is

$$\phi = \xi^{1/2}\left[A_1 J_{1/4}\left(\frac{k_1}{2}\xi^2\right) + A_2 J_{-1/4}\left(\frac{k_1}{2}\xi^2\right)\right] \qquad (2.94)$$

where A_1, A_2 are constants and $J_{1/4}$, $J_{-1/4}$ are Bessel functions of the first kind of order 1/4 and $-1/4$, respectively. From Ref. 2.9, eqs. (6.32) and (6.33), we obtain the useful relationships

$$\frac{d}{d\xi}\left[\left(\frac{k_1}{2}\xi^2\right)^{1/4} J_{1/4}\left(\frac{k_1}{2}\xi^2\right)\right] = \left[\left(\frac{k_1}{2}\xi^2\right)^{1/4} J_{-3/4}\left(\frac{k_1}{2}\xi^2\right)\right]k_1\xi$$

$$\frac{d}{d\xi}\left[\left(\frac{k_1}{2}\xi^2\right)^{1/4} J_{-1/4}\left(\frac{k_1}{2}\xi^2\right)\right] = \left[-\left(\frac{k_1}{2}\xi^2\right)^{1/4} J_{3/4}\left(\frac{k_1}{2}\xi^2\right)\right]k_1\xi$$

$$(2.95)$$

and

$$\frac{d}{d\xi}\left[\left(\frac{k_1}{2}\xi^2\right)^{3/4} J_{3/4}\left(\frac{k_1}{2}\xi^2\right)\right] = \left[\left(\frac{k_1}{2}\xi^2\right)^{3/4} J_{-1/4}\left(\frac{k_1}{2}\xi^2\right)\right]k_1\xi$$

$$\frac{d}{d\xi}\left[\left(\frac{k_1}{2}\xi^2\right)^{3/4} J_{-3/4}\left(\frac{k_1}{2}\xi^2\right)\right] = \left[-\left(\frac{k_1}{2}\xi^2\right)^{3/4} J_{1/4}\left(\frac{k_1}{2}\xi^2\right)\right]k_1\xi$$

$$(2.96)$$

For the cantilever beam the boundary conditions are [see Eq. (2.58)]

$$\phi = 0 \text{ at } \xi = L \qquad \text{and} \qquad \frac{d\phi}{d\xi} = 0 \text{ at } \xi = 0 \qquad (2.97)$$

From Eqs. (2.94) and (2.95),

$$\frac{d\phi}{d\xi} = 2\left(\frac{k_1}{2}\right)^{1/4}\left[A_1\left(\frac{k_1}{2}\xi^2\right)^{3/4} J_{-3/4}\left(\frac{k_1}{2}\xi^2\right) - A_2\left(\frac{k_1}{2}\xi^2\right)^{3/4} J_{3/4}\left(\frac{k_1}{2}\xi^2\right)\right]$$

For $\xi = 0$ the coefficients of A_2 and A_1 inside the square brackets are found to be zero and approximately 0.47, respectively [Ref. 2.10, p. 164, and Ref. 2.9, eq. (6.5)]. Consequently, the boundary condition at $\xi = 0$ requires that $A_1 = 0$. Then, from Eq. (2.94),

$$\phi = \xi^{1/2} A_2 J_{-1/4}\left(\frac{k_1}{2}\xi^2\right) \qquad (2.98)$$

Introduction of Eq. (2.98) into the boundary condition at $\xi = L$ now gives

$$J_{-1/4}\left(\frac{k_1}{2}L^2\right) = 0$$

From Ref. 2.10, p. 167, the smallest value of the parameter for which the Bessel function of the first kind of order $-1/4$ is equal to zero is $k_1 L^2/2 = 2.0063$. Then $k_1 = 4.013/L^2$. But $k_1{}^2 \equiv P^2/EI_{zz} GJ$. Consequently,

$$P_{cr} = 4.01 \frac{(EI_{zz} GJ)^{1/2}}{L^2} \qquad (2.99)$$

Equation (2.99) is the final expression for the critical load for the cantilever beam with narrow, rectangular cross section. The corresponding critical mode is given by Eq. (2.98) for $k_1 = 4.01/L^2$.

As a numerical example, let us consider a 24-ft long beam that has an I-shaped cross section and is subjected to a uniform bending moment $M_{yy} = -M_0$, as shown in Fig. 2.23. Let $E = 29 \times 10^6$ psi, $G = 11.2 \times 10^6$ psi, the beam depth $d = 23.7$ in., the cross-sectional area $A = 20.0$ in.^{2}, $I_{yy} = 1,820$ in.^{4}, $I_{zz} = 70.0$ in.^{4}, $C_W = 9,350$ in.^{6}, and $J = 1.86$ in.^{4} Equation (2.86) may be re-arranged and written in the form

$$M_{cr} = \left[\frac{\pi^2 EI_{zz}}{L^2} \left(\frac{\pi^2 EC_W}{L^2} + GJ \right) \right]^{1/2}$$

Then introduction of the numerical values gives

$$M_{cr} = 10^6 \left\{ \frac{\pi^2 (29)(70.0)}{(288)^2} \left[\frac{\pi^2 (29)(9,350)}{(288)^2} + (11.2)(1.86) \right] \right\}^{1/2}$$

$$= 10^6 \left[0.242(32.3 + 20.8) \right]^{1/2}$$

$$= 3.58 \times 10^6 \text{ in.-lb}$$

The corresponding maximum stress is given by the equation

$$\sigma_{cr} = \frac{M_{cr}}{S}$$

where
$$S = \frac{I_{yy}}{d/2} = \frac{1,820}{23.7/2} = 153 \text{ in.}^3$$

Then $\sigma_{cr} = (3.58 \times 10^6)/153 = 23,300$ psi.

Solutions for the lateral instability of doubly symmetrical beams for many other kinds of loading and support are given in, for example, Ref. 2.5. Equations and extensive numerical results for cross sections with one or two axes of symmetry are given in Ref. 2.11.

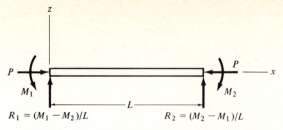

FIGURE 2.25
Beam column.

2.5 BEAM COLUMNS

A slender member that is subjected to both bending and longitudinal compression is called a beam column. As an example of beam-column analysis let us consider a member that has a doubly symmetrical cross section and is subjected to compressive loads P and end moments M_1, M_2, as shown in Fig. 2.25. The loads P are applied at the centroid of the cross section, and $|M_1| < |M_2|$. Let $I_{yy} \gg I_{zz}$. Then the member can bend in the xz plane or bend and twist out of that plane.

The elastic relations for bending and twisting are given by Eqs. (2.75). Differentiation of the first equation with respect to x gives

$$EC_W \phi^{iv} - GJ\phi'' = -M'_{x*x*} \qquad (2.100a)$$

$$EI_{yy} w'' = -M_{y*y*} \qquad (2.100b)$$

$$EI_{zz} v'' = M_{z*z*} \qquad (2.100c)$$

where M_{x*x*}, M_{y*y*}, M_{z*z*} are moments in the deformed configuration. The bending and twisting moments due to the load P are $M_{y*y*} = Pw$, $M_{z*z*} = -Pv$, and from Eq. (2.56), $M'_{x*x*} = P(I_o/A)\phi''$. (For a cross section with double symmetry, $y_o = z_o = 0$.) From Eqs. (2.77), the corresponding quantities due to the end moments M_1, M_2 are

$$M_{x*x*} = -R_2 v + (M_1 - R_1 x)v'$$

$$M_{y*y*} = M_1 - R_1 x$$

$$M_{z*z*} = -(M_1 - R_1 x)\phi$$

Introduction into Eqs. (2.100) and rearrangement gives, for the beam column, the final equations

$$EC_W \phi^{iv} - GJ\phi'' + P\frac{I_o}{A}\phi'' + M_1 v'' + \frac{M_2 - M_1}{L} xv'' = 0 \qquad (2.101a)$$

$$EI_{yy} w'' + Pw = -\frac{M_2 - M_1}{L} x - M_1 \qquad (2.101b)$$

$$EI_{zz} v'' + Pv + \left(M_1 + \frac{M_2 - M_1}{L} x\right)\phi = 0 \qquad (2.101c)$$

Equation (2.101b) is uncoupled from Eqs. (2.101a) and (2.101c), and it is an inhomogeneous equation. It governs bending in the xz plane, and because of the presence of the term Pw, the bending is nonlinear. In this respect Eq. (2.101b) is similar to Eq. (1.26) for a slightly crooked column. Equations (2.101a) and (2.101c), on the other hand, are coupled homogeneous equations in the lateral displacement v and angle of twist ϕ. They govern loss of stability in a lateral-torsional mode.

Let us first examine the solution to the equation for nonlinear bending in the xz plane. The general solution of Eq. (2.101b) is

$$w = C_1 \sin kx + C_2 \cos kx - \frac{M_2 - M_1}{L} x - \frac{M_1}{P}$$

where $k^2 \equiv P/EI_{yy}$ and C_1, C_2 are constants. For simply supported ends, $w = 0$ at $x = 0, L$. Therefore the solution for the displacement w is

$$w = \frac{M_2 - M_1 \cos kL}{P \sin kL} \sin kx + \frac{M_1}{P} \cos kx - \frac{M_2 - M_1}{PL} x - \frac{M_1}{P} = 0 \qquad (2.102)$$

As the load P approaches the value $P_y = \pi^2 EI_{yy}/L^2$ for loss of stability by bending in the xz plane, the factor $\sin kL$ in the first term in Eq. (2.102) approaches zero and the magnitude of the displacement w approaches infinity. In this respect Eq. (2.102) is similar to Eq. (1.27) for a slightly crooked column, and the equilibrium path for nonlinear bending of the beam column has the general form of the path for a slightly crooked column in Fig. 1.10.

For values of P smaller than P_y, the allowable values of P, M_1, and M_2 are limited by the strength of the beam-column material. In general, the maximum stress σ_x is given by the relationship

$$\sigma_{max} = \frac{P}{A} + \frac{M_{max}}{S} \qquad (2.103)$$

where $S = 2I_{yy}/d$, as noted. Since $|M_1| \le |M_2|$, the maximum moment occurs either at the right end of the beam (i.e., at $x = L$) or at a point between the ends of the beam (i.e., in the interval $0 < x < L$), depending on the relative magnitude of the nonlinear bending moment due to the Pw term in Eq. (2.101b). For the latter case the magnitude of the maximum moment may be determined from Eq. (2.102), as follows. Introduction into Eq. (2.100b) gives

$$M_{y^*y^*} = \frac{M_2 - M_1 \cos kL}{\sin kL} \sin kx + M_1 \cos kx$$

For the maximum moment, $dM_{y^*y^*}/dx = 0$. Thus the maximum occurs at $x = x_1$, where

$$\tan kx_1 = \frac{M_2 - M_1 \cos kL}{M_1 \sin kL}$$

Introduction of this expression into the preceding equation gives

$$M_{max} = M_1 \tan kx_1 \sin kx_1 + M_1 \cos kx_1$$

By use of the trigonometric identities $\sin^2 kx_1 + \cos^2 kx_1 = 1$ and $\sec kx_1 = (\tan^2 kx_1 + 1)^{1/2}$, this equation can be written in the form

$$M_{max} = M_1(\tan^2 kx_1 + 1)^{1/2}$$

Introduction of the expression for $\tan kx_1$ and rearrangement gives the final equation for M_{max}:

$$M_{max} = \frac{M_2[(M_1/M_2)^2 - 2(M_1/M_2) \cos kL + 1]^{1/2}}{\sin kL} \qquad (2.104)$$

For a beam column subjected to a uniform moment $M_1 = M_2 = M_0$, Eq. (2.104) simplifies to the expression

$$M_{max} = \frac{M_0[2(1 - \cos kL)]^{1/2}}{\sin kL} \qquad (2.105)$$

For this case the maximum moment occurs at midspan, and the coefficient $[2(1 - \cos kL)]^{1/2}/\sin kL$ represents a so-called *amplification factor* by which the applied moment M_0 must be multiplied to account for the additional moment due to the applied load P in nonlinear bending. As P approaches the value $P_y = \pi^2 EI_{yy}/L^2$, $\sin kL$ approaches 0, and the amplification factor approaches ∞. For small values of P, on the other hand, $\sin kL \approx kL$ and $(1 - \cos kL) \approx (kL)^2/2$. Consequently, as P approaches 0 the amplification factor approaches 1.

Equation (2.104) often is written in the form of Eq. (2.105) as follows:

$$M_{max} = \frac{C_m M_2[2(1 - \cos kL)]^{1/2}}{\sin kL} \qquad (2.106)$$

In this form C_m is a reduction factor introduced to account for the fact that $|M_1| \leq |M_2|$. From Eqs. (2.104) and (2.105),

$$C_m = \left[\frac{(M_1/M_2)^2 - 2(M_1/M_2)\cos kL + 1}{2(1 - \cos kL)} \right]^{1/2} \quad (2.107)$$

For $M_1 = M_2 = M_0$, $C_m = 1$. For $|M_1| < |M_2|$, $0 < C_m < 1$.

Introduction of Eq. (2.106) into Eq. (2.103) gives, for the maximum stress in the interval $0 < x < L$, the expression

$$\sigma_{max} = \frac{P}{A} + \frac{[2(1 - \cos kL)]^{1/2}}{\sin kL} \frac{C_m M_2}{S} \quad (2.108)$$

The amplification factor $[2(1 - \cos kL)]^{1/2}/\sin kL$ often is replaced by the simpler approximate expression $1/(1 - P/P_y)$, where $P_y = \pi^2 EI_{yy}/L^2$ is the critical load for loss of stability by bending in the xz plane. Then Eq. (2.106) appears in the form

$$M_{max} = \frac{C_m M_2}{1 - P/P_y} \quad (2.109)$$

and Eq. (2.108) becomes

$$\sigma_{max} = \frac{P}{A} + \frac{C_m}{1 - P/P_y} \frac{M_2}{S} \quad (2.110)$$

If the maximum moment occurs at $x = L$ rather than between supports, Eq. (2.106) is replaced by

$$M_{max} = M_2 \quad (2.111)$$

and Eq. (2.108) by

$$\sigma_{max} = \frac{P}{A} + \frac{M_2}{S} \quad (2.112)$$

Now let us turn to the solution of Eqs. (2.101a) and (2.101c) for lateral-torsional instability of the beam column. For the general case $M_1 \neq M_2$, the equations have variable coefficients and numerical methods must be used to obtain a solution. For a beam column subjected to a compressive load P and a uniform bending moment $M_1 = M_2 = M_0$, however, the equations simplify to the form

$$EC_W \phi^{iv} - GJ\phi'' + P\frac{I_o}{A}\phi'' + M_0 v'' = 0$$
$$\quad (2.113)$$
$$EI_{zz} v'' + Pv + M_0 \phi = 0$$

Equations (2.113) are readily solved for the boundary conditions for simply supported ends given by Eqs. (2.68). Introduction of the displacement functions in Eqs. (2.69) into Eqs. (2.113) and rearrangement gives

$$(P_\phi - P)\frac{I_o}{A} A_1 - M_0 A_2 = 0$$

$$- M_0 A_1 + (P_z - P)A_2 = 0$$

where P_ϕ and P_z are as defined by Eqs. (2.71). For a nontrivial solution, the determinant of the coefficients of A_1, A_2 must equal zero. Consequently, the equation for lateral-torsional instability of the beam column is

$$(P_\phi - P)(P_z - P)\frac{I_o}{A} - M_0{}^2 = 0 \qquad (2.114)$$

For given values of M_0 or for a beam column subjected to eccentric compressive loads P that cause end moments $M_0 = Pe$, the load P_{cr} is the smaller of the two roots of Eq. (2.114). Conversely, for given values of P, the critical value of M_0 is

$$M_{cr} = \left[(P_\phi - P)(P_z - P)\frac{I_o}{A}\right]^{1/2}$$

As a numerical example let us examine again the doubly symmetrical beam considered in the preceding section. Let us suppose that the end moments are caused by compressive loads P applied at the point $y = 0$, $z = 8$ in. in the cross section. Then

$$M_0 = 8P$$

From Eqs. (2.71) the parameters P_ϕ and P_y are

$$P_\phi = \frac{20.0}{1,820 + 70}\left[\frac{\pi^2(29)(9,350)}{(288)^2} + (11.2)(1.86)\right](10^6)$$
$$= 562,000 \text{ lb}$$

$$P_z = \left[\frac{\pi^2(29)(70.0)}{(288)^2}\right](10^6)$$
$$= 242,000 \text{ lb}$$

Therefore Eq. (2.114) gives

$$(562,000 - P)(242,000 - P)\left(\frac{1,820 + 70}{20.0}\right) - (8P)^2 = 0$$

The smaller root is found to be $P_{cr} = 182,300$ lb. Since a load applied at the centroid would cause flexural instability in the xy plane at a load $P_z = 242,000$ lb, the 8-in. eccentricity reduces the critical load by 25 percent.

The maximum stress is given approximately by Eq. (2.110), where $M_2 = 8P$ and $C_m = 1$. The critical load for bending in the xz plane is $P_y = \pi^2(29 \times 10^6)(1,820)/(288)^2 = 6,280,000$ lb. Consequently,

$$1 - \frac{P}{P_y} = 1 - \frac{182,300}{6,280,000} = 0.97$$

Thus nonlinear bending increases the calculated bending stress by only 3 percent. Introduction into Eq. (2.110) gives

$$\sigma_{cr} = \frac{182,300}{20.0} + \frac{(8)(182,300)}{(0.97)(153)}$$

$$= 9,110 + 9,830$$

$$= 18,940 \text{ psi}$$

By comparison the critical stress for flexural instability of the column under a concentric compressive load is $\sigma_{cr} = P_z/A = 242,000/20 = 12,100$ psi. For the eccentric load the allowable elastic buckling load P is lower, but, due to bending, the corresponding maximum elastic stress is higher.

2.6   FAILURE OF SLENDER BARS

Equilibrium paths for the flexural instability of an initially straight column are shown in Fig. 1.8 (for moderately large rotations), and the corresponding curve for a slightly crooked column is shown in Fig. 1.10. The illustrations show that the paths for the two columns converge for $P = \pi^2 EI/L^2$. Since a real column cannot be perfectly straight, let us refer to the crooked and straight models as the real and ideal cases, respectively. Very large lateral displacements of the real column occur as the load approaches the bifurcation-point load of the ideal column. Therefore well-designed tests of very slender laboratory models give experimental buckling loads that are in close agreement with the theoretical bifurcation-point values of Eq. (2.26).

The reason for the close agreement between theoretical and experimental results is that the load P in the differential equation for equilibrium of an initially straight column in a slightly deformed configuration [Eq. (2.1)] is an independent load parameter rather than a function of the displacement variable w. Therefore the secondary path in the P-w plot is a horizontal straight line (for the intermediate class of deformations), and when the equilibrium paths of the real and the ideal columns converge, they do so at the bifurcation-point load, $P = P_{cr}$. An example of a secondary path that is not horizontal is provided by the path for a cylindrical panel in Fig. 1.11, as noted.

In spite of the agreement in critical loads, the two models are markedly different in behavior. The ideal column shortens as the load P is applied, but remains perfectly straight until P reaches P_{cr}. At the critical load it deflects to one side if a lateral disturbance is present. In the bent configuration the load P is the same for all values of lateral displacement amplitude, and the equilibrium no longer is stable. The real column, on the other hand, begins to bend as soon as P is applied. Both shortening and bending increase as P increases, but the bending moments remain small until P approaches P_{cr}. There is no bifurcation point, and since the equilibrium path never becomes horizontal (barring onset of inelastic action), there is no loss of stability. Nevertheless, the maximum load is limited to the value $P = \pi^2 EI/L^2$.

The preceding remarks pertain to the *elastic* behavior of columns. For slightly crooked columns the initiation of inelastic action at any load level leads to a progressive decrease in the bending stiffness EI. There is an accompanying progressive decrease in the slope of the equilibrium path, and sooner or later, the path reaches a maximum point that signifies loss of stability. If the column is more than slightly crooked, it must be treated as a beam column and the limit-point load must be determined through nonlinear bending analysis. For nominally straight columns, however, equations for the load P_{cr} may be obtained by modification of the preceding elastic stability equations to take into account the influence of inelastic action, as follows.

Prior to buckling, the bending stresses in a nominally straight column are much smaller than the average stress $\sigma = P/A$. An expression for average stress when $P = P_{cr}$ may be obtained by dividing the equation for the critical load in Eq. (2.26) by the column cross-sectional area A. The resulting equation may be written in the form

$$\frac{P_{cr}}{A} = \frac{\pi^2 E}{(KL/r)^2} \qquad (2.115)$$

where $r \equiv (I/A)^{1/2}$. The variable r is called the *radius of gyration*. The ratio L/r is termed the *slenderness ratio*, and KL/r is the *effective slenderness ratio*. For a given material, and therefore a given value of Young's modulus, the relationship between average stress and effective slenderness ratio represented by Eq. (2.115) is shown in Fig. 2.26a. For P_{cr}/A values smaller than the propor-

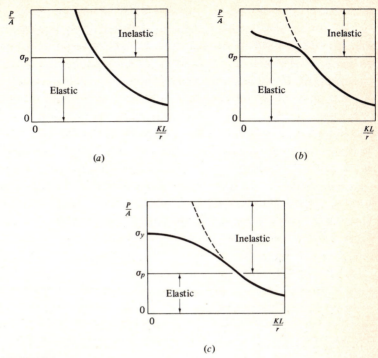

FIGURE 2.26
Column curves representing Eqs. (2.115), (2.116), and (2.117), respectively.

tional limit stress (denoted by σ_p in the figure), the column material is elastic, as assumed in the derivation of Eq. (2.115). For higher stresses it is not. Consequently, Eq. (2.115) is valid only in the region labeled "elastic" in Fig. 2.26a.

Many attempts have been made to obtain a suitable modification of Eq. (2.115) for inelastic buckling. An excellent history of the efforts is given by Hoff in Ref. 2.13. It is now generally agreed that the most rational modification is represented by the so-called *tangent modulus formula* (Ref. 2.14):

$$\frac{P_{cr}}{A} = \frac{\pi^2 E_t}{(KL/r)^2} \qquad (2.116)$$

where E_t, called the *tangent modulus*, is the local slope of the stress-strain curve in the inelastic range, i.e., the slope of the stress-strain curve at the stress $\sigma = P_{cr}/A$. Unlike Young's modulus, the tangent modulus is a function of the stress level. A typical plot of average stress versus slenderness ratio based on Eq. (2.116) is shown in Fig. 2.26b.

When there is inelastic action, several factors can cause actual column strength to be different from theoretical bifurcation-point values. Because of the presence of bending stresses, inelastic action begins at a lower load for a relatively crooked column than for a relatively straight one (compare Prob. 1.2). Consequently, initial imperfections and unintentional load eccentricity can affect the magnitude of the buckling load. Other factors include residual stresses due to rolling or welding, and variations in yield stress over the column cross section. The influence of such factors is discussed authoritatively and extensively in the Column Research Council's "Guide to Design Criteria for Metal Compression Members" (Ref. 2.12).

Design recommendations that take all these factors into account are offered by the Column Research Council in Ref. 2.12, chap. 2. For aluminum alloys Eq. (2.116) is recommended (with an appropriate factor of safety) for P_{cr}/A stresses between the proportional limit and the 0.2 percent offset yield strength, and the yield strength itself is recommended for shorter columns. For structural steels it is suggested that the effective proportional limit σ_p for columns arbitrarily be set at $\sigma_p = \sigma_y/2$, where σ_y is the yield stress or 0.2 percent offset yield strength. This conservative reduction in proportional limit is recommended in order to account for the influence of residual stresses and variations in yield stress over the cross section. Then, for P_{cr}/A stresses greater than such proportional limit values, the following approximate equation based on the tangent modulus formula is offered [Ref 2.12, eq. (2.10)]:

$$\frac{P_{cr}}{A} = \sigma_y - \frac{\sigma_p^2}{\pi^2 E/(KL/r)^2} \qquad (2.117)$$

A plot of stress versus slenderness ratio based on Eq. (2.117) is shown in Fig. 2.26c. Figure 2.26b and c illustrates a principal advantage of the effective-length concept. By use of this concept a single curve representing the influence of inelastic action for a column with simply supported ends can be used for all possible end conditions.

Since the stress level is not known in advance in column design, selection of the appropriate working-load formula entails trial and error. The effort to obtain a minimum-weight design usually leads to a column whose critical stress is higher than the proportional limit stress.

The characteristics of flexural instability failure discussed in the preceding paragraphs also apply to the flexural-torsional instability of columns and the lateral-torsional instability of beams and beam columns. Slightly imperfect columns, beams, and beam columns bend and twist from the outset when load is applied. The equilibrium path is curved from the outset, and it eventually converges with the secondary path of the corresponding initially straight model. In all the differential equations in this chapter for equilibrium of an initially straight model in a slightly deformed configuration, the loads P and M are independent load parameters rather than functions of the displacement variables. Consequently, the secondary equilibrium paths are horizontal straight lines, and the buckling loads of sufficiently slender, slightly imperfect models are in close agreement with the theoretical bifurcation-point values for perfect models. For less slender models the elastic stability equations may be modified for inelastic buckling by use of the tangent modulus concept. A detailed discussion and approximate analyses of the influence of inelastic action on the buckling behavior of columns, beams, and beam columns are presented in Ref. 2.3. A comprehensive discussion of the plastic buckling of columns, plates, and shells is given by Hutchinson in Ref. 2.15.

Design recommendations that are based on the tangent modulus and that take into account residual stresses, variation in yield stress, and other factors are included in Ref. 2.12 for beams and beam columns, as well as for columns.

REFERENCES

2.1 NEWMARK, N. M.: A Simple Approximate Formula for Effective End-Fixity of Columns, *J. Aeronaut. Sci.*, vol. 16, p. 116, February 1949.

2.2 ZIEGLER, H.: "Principles of Structural Stability," Blaisdell, Waltham, Mass., 1968.

2.3 GALAMBOS, T. V.: "Structural Members and Frames," Prentice-Hall, Englewood Cliffs, N.J., 1968.

2.4 TIMOSHENKO, S. P.: "Strength of Materials," Part II, 3d ed., Van Nostrand, New York, 1956.

2.5 TIMOSHENKO, S. P., and J. M. GERE: "Theory of Elastic Stability," 2d ed., McGraw-Hill, New York, 1961.

2.6 CRANDALL, S. H., N. C. DAHL, and T. J. LARDNER: "An Introduction to the Mechanics of Solids," 2d ed., McGraw-Hill, New York, 1972.

2.7 BLEICH, F.: "Buckling Strength of Metal Structures," McGraw-Hill, New York, 1952.

2.8 CHAJES, A., and G. WINTER: Torsional-flexural Buckling of Thin-walled Members, *J. Struct. Div.*, *ASCE*, vol. 91, no. ST4, pp. 103–124, 1965.

2.9 BOWMAN, F.: "Introduction to Bessel Functions," Dover, New York, 1958.

2.10 JANKE, E., and F. EMDE: Tables of Functions, 4th ed., Dover, New York, 1945.

2.11 CLARK, J. W., and H. N. HILL: Lateral Buckling of Beams, *J. Struct. Div.*, *ASCE*, vol. 86, no. ST7, pp. 175–196, 1960.

2.12 B. G. JOHNSTON (ED.): "Guide to Design Criteria for Metal Compression Members," 2d ed., Column Research Council, Wiley, New York, 1966.

2.13 HOFF, N. J.: Buckling and Stability, *J. Roy. Aeronaut. Soc., Aeronaut. Reprint* 123, London, January 1954.

2.14 SHANLEY, F. R.: Inelastic Column Theory, *J. Aeronaut. Sci.*, vol. 14, no. 5, pp. 261–268, 1947.

2.15 HUTCHINSON, J. W., Plastic Buckling, to appear in "Advances in Applied Mechanics," vol. 14, edited by C. S. Yih, Academic Press, New York, 1974.

PROBLEMS

2.1 A steel pipe has inner and outer diameters of 3.55 and 4.00 in., respectively. (*a*) If a 15-ft section is used as a column with simply supported ends, determine the critical load for elastic buckling. Let $E = 29 \times 10^6$ psi. (*b*) Calculate the corresponding average stress $\sigma = P_{cr}/A$. If the proportional limit of the column material is 18,000 psi, is the Euler column formula applicable?

2.2 A 42-in.-long aluminum bar has a rectangular cross section 1.8 in. wide and 0.40 in. thick. The ends are simply supported in relation to bending in the "strong" direction, but clamped in relation to bending in the "weak" direction. Determine the critical load for elastic buckling if $E = 10 \times 10^6$ psi.   *Ans. $P_{cr} = 2,150$ lb*

2.3 The general solution of the differential equation governing loss of stability of a centrally loaded column is given in Eq. (2.2). Using that expression, determine P_{cr} for a column clamped at one end and free to displace laterally without rotation at the other, i.e., for the boundary conditions $w = w' = 0$ at $x = 0$ and $w' = Q = 0$ at $x = L$.
          Ans. $P_{cr} = \pi^2 EI/L^2$

2.4 (*a*) By treating n as a continuous variable, show that Eq. (2.42) for the column on an elastic foundation gives, for the critical load, the approximate expression $P_{cr} = 2(EIk_f)^{1/2}$. Note that the equation is independent of column length. (*b*) Using this expression, calculate P_{cr} for $EI = 1,200$ lb-in.^{2} and $k_f = 1,400$ psi. For $L = 28$ in., compare the result with the value for P_{cr} calculated from Eq. (2.42) for integral values of n.

2.5 The differential equation for a column on an elastic foundation is given in Eq. (2.41). (*a*) Derive the general solution in terms of four arbitrary constants. Note that there are two cases: $[P/(2EI)]^2 = k_f/(EI)$ and $[P/(2EI)]^2 > k_f/(EI)$. (Compare Prob. 2.4.) (*b*) By substitution into the four boundary condition equations, show that for simply supported ends, $w = C_1 \sin n\pi x/L$.

2.6 In Sec. 2.2*b* the effective-length factor for the frame in Fig. 2.2 was found to be $K = 0.84$. Show that if the horizontal member is clamped at its right end rather than

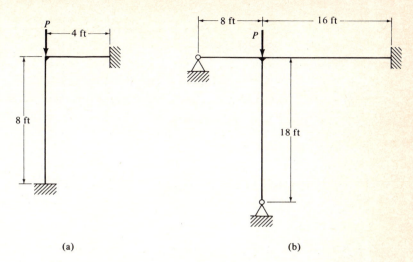

(a)   (b)

FIGURE 2.27

simply supported, the load parameter kL is increased from 3.72 to 3.83 and the effective-length factor is reduced to $K = 0.82$.

2.7 A steel column 8 ft long is clamped at the bottom and rigidly attached to a 4-ft-long horizontal steel beam at the top, as shown in Fig. 2.27a. Determine the critical load for elastic buckling in the plane of the frame. The bending stiffness EI of the column is 8×10^6 lb-in.^{2}, and that for the beam is 2×10^6 lb-in.^{2}.

Ans. $P_{cr} = 22,100$ lb

2.8 A steel column 18 ft long is simply supported at the bottom and rigidly attached to a horizontal beam at the top, as shown in Fig. 2.27b. The beam is simply supported at its left end and clamped at its right. Determine the critical load for elastic buckling in the plane of the frame. The moments of inertia for the column and beam are 84.0 and 66.0 in.^{4}, respectively. Use $E = 29 \times 10^6$ psi.

2.9 A simply supported aluminum column 34 in. long has a rectangular cross section 1.2 in. wide and 0.36 in. thick. An elastic spring of stiffness $k_s = 1,200$ lb/in. is attached to the column at a distance of 13.5 in. from one end. The spring resists bending in the weak direction only. Determine the critical load for elastic buckling. Use $E = 10 \times 10^6$ psi. *Ans.* $P_{cr} = 1,454$ lb

2.10 A 30-ft-long steel column with simply supported ends is supported at an intermediate point by a 40-ft-long steel beam that is clamped at both ends, as shown in Fig. 2.28. The beam acts as an elastic lateral support for the column. Determine the critical load for elastic buckling out of the plane of the frame. The moments of inertia for the column and beam are 140 and 40 in.^{4}, respectively. Use $E = 29 \times 10^6$ psi.

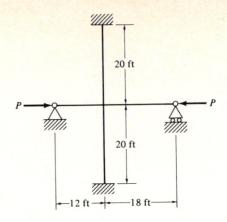

FIGURE 2.28

2.11   A simply supported aluminum column 84 in. long is supported laterally at mid-span by an elastic spring. (*a*) Determine the minimum spring stiffness, in pounds per inch, necessary to force a node at midspan. The bending stiffness EI of the column is 1.8×10^6 lb-in.^{2}. (*b*) From Fig. 2.7, estimate the minimum stiffness necessary to force a node at a spring located 21 in. from an end.

Ans. (*a*) $k_s = 481$ lb/in., (*b*) $k_s = 30,000$ lb/in.

2.12   The cross section of a 15-ft-long steel column has a minimum radius of gyration $r = 1.22$ in. (*a*) Determine the slenderness ratio L/r for the column simply supported at both ends. (*b*) For the column clamped at one end and supported at the other by a rotational spring for which the nondimensional stiffness parameter $\lambda = 0.35$, determine the effective slenderness ratio KL/r of the equivalent simply supported column.

Ans (*a*) $L/r = 147$, (*b*) $KL/r = 89$

2.13   For torsional-flexural instability of a column with simply supported ends, the critical load P is given by Eq. (2.72). Derive the corresponding expression for a column with clamped ends, i.e., for the boundary conditions $v = w = \phi = v' = w' = \phi' = 0$ at $x = 0, L$. Use solution functions of the form

$$\phi = A_1\left(1 - \cos\frac{2\pi x}{L}\right) \quad v = A_2\left(1 - \cos\frac{2\pi x}{L}\right) \quad w = A_3\left(1 - \cos\frac{2\pi x}{L}\right)$$

2.14   Show that for an I-shaped cross section of depth d, flange width b, and flange thickness h, Eq. (2.64) gives for the warping constant the expression $C_W = b^3 h(d - h)^2/24$.

2.15   A 15-ft-long simply supported steel column has the thin-walled open cross section shown in Fig. 2.29*a*. The area $A = 3.75$ in.^{2}, the moments of inertia are $I_{yy} = 8.83$ in.^{4} and $I_{zz} = 2.29$ in.^{4}, and $y_o = 1.67$ in. Show that the column buckles in pure bending rather than in combined bending and torsion, and determine the elastic-buckling load. Let $E = 29 \times 10^6$ psi and $G = 11.2 \times 10^6$ psi. Note that the shear center lies at the intersection of the legs, so that the warping constant $C_W = 0$.   *Ans.* $P_{cr} = 20,200$ lb

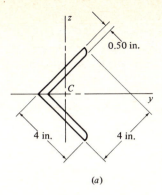

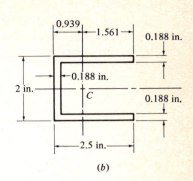

(a)   (b)

FIGURE 2.29

2.16 Show that if the 4×4-in. angle section in Prob. 2.12 had a wall thickness $h = 0.25$ in. and a length $L = 5$ ft, it would buckle in combined bending and torsion. Determine the elastic-buckling load. Use $A = 1.94$ in.^{2}, $I_{yy} = 4.85$ in.^{4}, $I_{zz} = 1.226$ in.^{4}, and $y_o = 1.54$ in.

2.17 A 60-in.-long simply supported aluminum column has the channel-shaped cross section shown in Fig. 2.29b. Determine the elastic-buckling load for combined torsional-flexural instability. Use $y_o = 1.833$ in., $E = 10 \times 10^6$ psi and $G = 3.8 \times 10^6$ psi. Determine C_W from Eq. (2.64).

2.18 A simply supported steel beam is subjected to bending moments M_0 applied at its ends. The beam cross section is shown in Fig. 2.30. For a span length of 20 ft, determine the critical value of M_0 for elastic lateral instability. Let $E = 29 \times 10^6$ psi and $G = 11.2 \times 10^6$ psi.

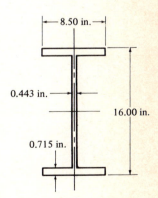

FIGURE 2.30

2.19 In structural applications an approximate expression for the stress σ_{cr} for lateral instability of a beam sometimes is obtained from Eq. (2.86) by neglecting in each particular example the smaller of the two terms on the right-hand side of the equation. (*a*) Show that if the first term is the smaller and is neglected the critical stress is approximately

$$\sigma_{cr} = \frac{\pi}{SL} (EI_{zz} GJ)^{1/2}$$

where $S = 2I_{yy}/d$. (*b*) Show that if the second term is the smaller and is neglected the critical stress is approximately

$$\sigma_{cr} = \left(\frac{\pi}{L}\right)^2 \frac{E}{S} (I_{zz} C_w)^{1/2}$$

(*c*) Show that in any given example the larger of these two approximate values under-estimates the critical stress by not more than 29 percent [relative to the value based on the complete Eq. (2.86)]. (*d*) Using this approximate procedure calculate the critical stress for the beam in the preceding problem and determine the percent error due to the approximation.

2.20 A cantilever beam is subjected to a concentrated load P at its free end. The beam has a rectangular cross section 0.45 in. wide and 6 in. deep. It is 10 ft long, and the load is applied at the centroid of the cross section as in Fig. 2.24. Determine the critical value of the applied load for elastic lateral instability. Let $E = 29 \times 10^6$ psi and $G = 11.2 \times 10^6$ psi.

2.21 If the beam in Prob. 2.18 were to be subjected to bending moments $M_0 = 1,200,000$ in.-lb, determine the magnitude of the additional concentric compressive loads P at which elastic instability in a lateral-torsional mode would occur, and calculate the corresponding maximum stress σ_x.

2.22 Rederive the differential equation of equilibrium for the column on an elastic foundation [Eq. (2.41)] by application of the principle of stationary potential energy, i.e., by introduction of the integrand in the potential energy expression into the appropriate Euler equation of the calculus of variations (Sec. A.2). (See also Prob. 1.7.) For the elastic foundation the strain energy expression may be written $U_f = \frac{1}{2}k_f \int w^2 \, dx$.

RECTANGULAR FLAT PLATES

3.1 INTRODUCTION

Equilibrium and stability equations for flat plates, circular rings, and cylindrical shells are independently derived in Chaps. 3, 4, and 5, respectively. The equations alternatively may be obtained by specialization of the equations in Chap. 6 for shells of general shape. Readers who are familiar with general shell theory may prefer to pass over the repeated derivations of the equations for plates, rings, and cylinders and to treat the applications sections of Chaps. 3 to 5 as applications of the general shell equations.

The earliest solution of a flat-plate stability problem apparently was given by Bryan in 1891 (Ref. 3.1). A great many publications on the subject have appeared since that time. Treatments of flat-plate stability analysis that are much more extensive than that given here may be found in, for example, Refs. 3.2 and 3.3.

Because of the relative simplicity of the governing equations, the rectangular flat plate is an excellent model for illustration of the systematic unity among the balance-of-forces and stationary potential energy criteria for equilibrium, and the adjacent-equilibrium and minimum potential energy criteria for loss of stability. As background for this chapter as well as later ones, the kinematic and constitutive equations of two-dimensional thin-plate theory are derived from the corresponding equations of three-dimensional solid mechanics in Sec. 3.2. In Sec. 3.3 the nonlinear differential equations governing equilibrium of the plate for configurations in the intermediate class of deformations are derived first by summation of forces and moments and then by the stationary potential energy criterion. The corresponding linear equations for loss of stability are then developed in the following section by application of the adjacent-equilibrium and the minimum potential energy criteria. In the interest of simplicity, the stability equations for the plate are derived only for in-plane edge loading; the problem of nonlinear prebuckling deformation is deferred to later chapters.

Applications to specific examples are treated in Sec. 3.5. Only rectangular flat plates are considered in this chapter. Circular flat plates are examined in Chap. 6 as a special case of the shell of revolution.

3.2 ELEMENTS OF PLATE BENDING THEORY

This sketch of the essential elements of thin-plate bending theory is presented in an abbreviated but complete form. Detailed treatments are given in books on the theory of plates (for example, Ref. 3.4).

We initially consider a rectangular thin flat plate of length a, width b, and thickness h, subjected to edge loading and to a surface load component p, in pounds per square inch, normal to the plane of the plate. The plate is referred

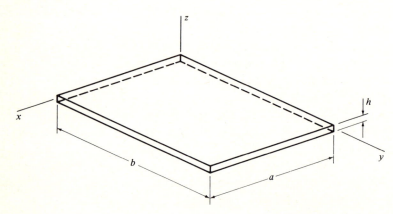

FIGURE 3.1
Rectangular flat plate.

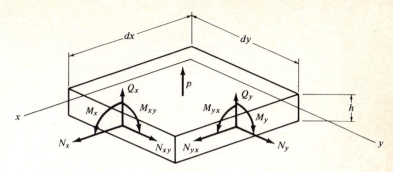

FIGURE 3.2
Plate element $dx\, dy$ in undeformed configuration.

to rectangular cartesian coordinates x, y, z, where x and y lie in the middle plane of the plate and z is measured from the middle plane, as shown in Fig. 3.1.

The object of thin-plate theory is to reduce a three-dimensional problem to an approximate two-dimensional one. Internal forces and moments acting on the edges of a plate element $dx\, dy$, as shown in Fig. 3.2, are expressed in terms of forces and moments per unit distance along the element edge. The force and moment intensities are related to the internal stresses by the equations

$$N_x = \int_{-h/2}^{h/2} \bar{\sigma}_x \, dz \qquad N_y = \int_{-h/2}^{h/2} \bar{\sigma}_y \, dz$$

$$N_{xy} = \int_{-h/2}^{h/2} \bar{\tau}_{xy} \, dz \qquad N_{yx} = \int_{-h/2}^{h/2} \bar{\tau}_{yx} \, dz$$

$$Q_x = \int_{-h/2}^{h/2} \bar{\tau}_{xz} \, dz \qquad Q_y = \int_{-h/2}^{h/2} \bar{\tau}_{yz} \, dz \qquad (3.1)$$

$$M_x = \int_{-h/2}^{h/2} \bar{\sigma}_x z \, dz \qquad M_y = \int_{-h/2}^{h/2} \bar{\sigma}_y z \, dz$$

$$M_{xy} = \int_{-h/2}^{h/2} \bar{\tau}_{xy} z \, dz \qquad M_{yx} = \int_{-h/2}^{h/2} \bar{\tau}_{yx} z \, dz$$

where

N_x, N_y, N_{xy}, N_{yx} = in-plane normal and shearing force intensities, lb/in.

Q_x, Q_y = transverse shearing force intensities, lb/in.

M_x, M_y = bending moment intensities, in.-lb/in.

M_{xy}, M_{yx} = twisting moment intensities, in.-lb/in.

The symbols $\bar{\sigma}_x$, $\bar{\tau}_{xy}$, etc., denote stress components at any point through the plate thickness, as distinguished from σ_x, τ_{xy}, etc., which refer to corresponding

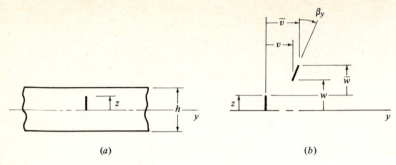

FIGURE 3.3
Normal to plate middle surface before and after deformation.

quantities on the middle plane ($z = 0$) only. Since $\bar{\tau}_{xy} = \bar{\tau}_{yx}$, it follows from Eqs. (3.1) that $N_{xy} = N_{yx}$ and $M_{xy} = M_{yx}$. In general, the force and moment intensities are functions of the coordinates x, y.

Thin-plate theory may be derived in terms of the following simplifying approximations:

1 Normals to the undeformed middle plane are assumed to remain straight, normal, and inextensional during the deformation, so that transverse normal and shearing strains may be neglected in deriving the plate kinematic relations.

2 Transverse normal stresses are assumed to be small compared with the other normal stress components, so that they may be neglected in the stress-strain relations.

These approximations are known as the *Kirchhoff assumptions* (Ref. 3.5, p. 194). As a consequence of the first approximation, the displacement components at any point in the plate, $\bar{u}$, $\bar{v}$, $\bar{w}$, may be expressed in terms of the corresponding middle-plane quantities, u, v, w by the relations (see Fig. 3.3)

$$
\begin{aligned}
\bar{u} &= u + z\beta_x \\
\bar{v} &= v + z\beta_y \\
\bar{w} &= w
\end{aligned}
\qquad (3.2)
$$

where β_x and β_y are rotations relative to the y and x coordinate directions, respectively.

The intermediate class of deformations is defined by the limitations that the strains be small compared with unity, the rotations relative to the x and y

directions moderately small, and rotations relative to the z direction negligibly small. For such deformations the $\bar{\varepsilon}_x$, $\bar{\varepsilon}_y$, and $\bar{\gamma}_{xy}$ components of the strain-displacement relations for a three-dimensional medium are (Ref. 3.5, p. 52)

$$\bar{\varepsilon}_x = \bar{u}_{,x} + \tfrac{1}{2}\bar{w}_{,x}{}^2$$
$$\bar{\varepsilon}_y = \bar{v}_{,y} + \tfrac{1}{2}\bar{w}_{,y}{}^2 \tag{3.3}$$
$$\bar{\gamma}_{xy} = \bar{u}_{,y} + \bar{v}_{,x} + \bar{w}_{,x}\bar{w}_{,y}$$

where subscripts x and y preceded by commas denote differentiation with respect to x or y. These expressions are the counterpart for flat plates of Eq. (1.7) for straight bars. Furthermore, for this class of deformations, $\beta_x = -w_{,x}$ and $\beta_y = -w_{,y}$ (Ref. 3.5, p. 48). Introduction of Eqs. (3.2) and rearrangement gives

$$\bar{\varepsilon}_x = \varepsilon_x + z\kappa_x$$
$$\bar{\varepsilon}_y = \varepsilon_y + z\kappa_y \tag{3.4}$$
$$\bar{\gamma}_{xy} = \gamma_{xy} + 2z\kappa_{xy}$$

where $\bar{\varepsilon}_x$, $\bar{\varepsilon}_y$, $\bar{\gamma}_{xy}$ are extensional and shearing strain components at any point through the plate thickness and ε_x, ε_y, γ_{xy} denote the corresponding quantities at points on the plate middle plane only, and where

$$\varepsilon_x = u_{,x} + \tfrac{1}{2}\beta_x{}^2 \qquad \beta_x = -w_{,x} \qquad \kappa_x = \beta_{x,x}$$
$$\varepsilon_y = v_{,y} + \tfrac{1}{2}\beta_y{}^2 \qquad \beta_y = -w_{,y} \qquad \kappa_y = \beta_{y,y} \tag{3.5}$$
$$\gamma_{xy} = (u_{,y} + v_{,x}) + \beta_x\beta_y \qquad\qquad\qquad \kappa_{xy} = \tfrac{1}{2}(\beta_{x,y} + \beta_{y,x})$$

Equations (3.5) are the kinematic relations for the plate. They are the same as the expressions given for the von Kármán plate theory in Ref. 3.5, eqs. (VI.13). They are also the expressions obtained by specialization for plates of the Sanders kinematic relations in Eqs. (6.6) and (6.7) of this book. All the variables in Eqs. (3.5) are middle-plane quantities and are functions of x and y alone.

The generalized Hooke's law for the strain components $\bar{\varepsilon}_x$, $\bar{\varepsilon}_y$, and $\bar{\gamma}_{xy}$ in a three-dimensional isotropic medium has the form

$$\bar{\varepsilon}_x = \frac{1}{E}\left[\bar{\sigma}_x - v(\bar{\sigma}_y + \bar{\sigma}_z)\right]$$

$$\bar{\varepsilon}_y = \frac{1}{E}\left[\bar{\sigma}_y - v(\bar{\sigma}_z + \bar{\sigma}_x)\right] \tag{3.6}$$

$$\bar{\gamma}_{xy} = \frac{2(1+v)}{E}\bar{\tau}_{xy}$$

where v is Poisson's ratio. As a consequence of the second approximation of thin-plate theory, $\bar{\sigma}_z$ is negligibly small. Omission from Eqs. (3.6) and rearrangement gives the relations

$$\bar{\sigma}_x = \frac{E}{1 - v^2}(\bar{\varepsilon}_x + v\bar{\varepsilon}_y)$$

$$\bar{\sigma}_y = \frac{E}{1 - v^2}(\bar{\varepsilon}_y + v\bar{\varepsilon}_x) \qquad (3.7)$$

$$\bar{\tau}_{xy} = \frac{E}{2(1 + v)}\bar{\gamma}_{xy}$$

Introduction of Eqs. (3.7) and (3.4) into Eqs. (3.1) and integration yields the equations

$$N_x = C(\varepsilon_x + v\varepsilon_y) \qquad M_x = D(\kappa_x + v\kappa_y)$$
$$N_y = C(\varepsilon_y + v\varepsilon_x) \qquad M_y = D(\kappa_y + v\kappa_x) \qquad (3.8)$$
$$N_{xy} = C\frac{1 - v}{2}\gamma_{xy} \qquad M_{xy} = D(1 - v)\kappa_{xy}$$

where
$$C \equiv \frac{Eh}{1 - v^2} \quad \text{and} \quad D \equiv \frac{Eh^3}{12(1 - v^2)}$$

Equations (3.8) are the constitutive relations for the plate. The coefficients C and D are termed extensional and bending stiffness parameters, respectively.

In plate equilibrium analysis, the eight kinematic relations and six constitutive relations are added to five equilibrium equations to give a system of 19 equations in 19 unknowns, as will be seen in the following section. All 19 variables are functions of x and y alone.

3.3 NONLINEAR EQUILIBRIUM EQUATIONS

3.3a Summation of Forces and Moments

To take into account the nonlinear interaction between forces and rotations, the equations representing equilibrium of forces and moments must be derived for the plate element in a slightly deformed configuration, as shown in Fig. 3.4. To simplify the diagrams, force and moment intensities are shown in separate sketches, and the plate element is drawn as an element of surface without thick-

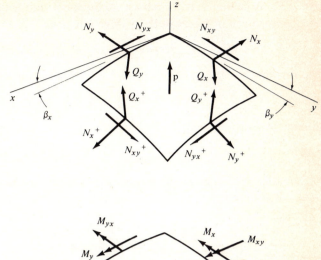

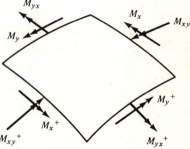

FIGURE 3.4
Plate element in deformed configuration.

ness. The double-headed vector notation is used to denote directions of the moment couples (right-hand rule). The force and moment intensities are shown in Figs. 3.2 and 3.4 in their positive directions; the sign conventions are chosen so that all the terms on the right sides of Eqs. (3.1) are positive. The rotations β_x and β_y in Fig. 3.4, each represent the angle between a coordinate direction and the corresponding tangent to the middle surface of the plate element at its upper corner, as shown. The force and moment intensities and rotations vary across the element, and the notation N_x^+ is used to denote $(N_x + N_{x,x}\, dx)$, etc.

The angles of rotation, β_x and β_y, are small, and sines and cosines of the angles may be replaced by the angles themselves and by unity, respectively. Quadratic terms representing nonlinear interaction between the small transverse shearing forces and the rotations are assumed to be negligibly small, as in

Sec. 1.2. Recalling that N_x represents force per unit length along the edge dy, etc., summation of forces in the x direction gives

$$-N_x\,dy + (N_x + N_{x,x}\,dx)\,dy - N_{yx}\,dx + (N_{yx} + N_{yx,y}\,dy)\,dx = 0 \qquad (3.9)$$

On dividing out the quantity $dx\,dy$, Eq. (3.9) becomes

$$N_{x,x} + N_{yx,y} = 0 \qquad (3.10)$$

Similarly, from summation of forces in the y direction, we obtain

$$N_{xy,x} + N_{y,y} = 0 \qquad (3.11)$$

Summation of forces in the z direction is somewhat more complicated. From Fig. 3.4 we obtain

$$
\begin{aligned}
&N_y\,dx\beta_y - (N_y + N_{y,y}\,dy)\,dx(\beta_y + \beta_{y,y}\,dy) \\
&\quad + N_x\,dy\beta_x - (N_x + N_{x,x}\,dx)\,dy(\beta_x + \beta_{x,x}\,dx) \\
&\quad - Q_y\,dx + (Q_y + Q_{y,y}\,dy)\,dx - Q_x\,dy + (Q_x + Q_{x,x}\,dx)\,dy \\
&\quad + N_{xy}\,dy\beta_y - (N_{xy} + N_{xy,x}\,dx)\,dy(\beta_y + \beta_{y,x}\,dx) \\
&\quad + N_{yx}\,dx\,\beta_x - (N_{yx} + N_{yx,y}\,dy)\,dx(\beta_x + \beta_{x,y}\,dy) + p\,dx\,dy = 0 \qquad (3.12)
\end{aligned}
$$

Regrouping and omission of higher-order terms in Eq. (3.12) gives

$$
\begin{aligned}
&-(N_{x,x} + N_{xy,y})\beta_x - (N_{yx,x} + N_{y,y})\beta_y - N_x\beta_{x,x} \\
&\quad - N_{xy}\beta_{y,x} - N_{yx}\beta_{x,y} - N_y\beta_{y,y} + Q_{x,x} + Q_{y,y} = -p \qquad (3.13)
\end{aligned}
$$

The terms containing derivatives of N_x, N_{xy}, N_{yx}, and N_y in Eq. (3.13) may be seen to add to zero as a consequence of Eqs. (3.10) and (3.11), and they are omitted in subsequent equations.

Summation of moments relative to the x and y coordinate directions, respectively, gives the equations

$$
\begin{aligned}
-M_{xy,x} - M_{y,y} + Q_y &= 0 \\
M_{yx,y} + M_{x,x} - Q_x &= 0
\end{aligned} \qquad (3.14)
$$

The sixth equilibrium equation, $\sum M_z = 0$, leads to an identity and yields no new information, as may be seen by derivation of the equation and comparison of the result with the expressions for N_{xy} and N_{yx} in Eqs. (3.1).

As noted, $N_{yx} = N_{xy}$ and $M_{yx} = M_{xy}$. Collection of the five equilibrium equations and rearrangement gives

$$N_{x,x} + N_{xy,y} = 0 \qquad (3.15a)$$

$$N_{xy,x} + N_{y,y} = 0 \qquad (3.15b)$$

$$Q_{x,x} + Q_{y,y} - N_x\beta_{x,x} - N_{xy}(\beta_{y,x} + \beta_{x,y}) - N_y\beta_{y,y} = -p \qquad (3.15c)$$

$$Q_y = M_{y,y} + M_{xy,x} \qquad (3.15d)$$

$$Q_x = M_{x,x} + M_{xy,y} \qquad (3.15e)$$

These five equations, together with the constitutive and kinematic relations in Eqs. (3.5) and (3.8), constitute the set of 19 equations referred to in Sec. 3.2.

Several of the variables are readily eliminated. Introduction of Eqs. (3.15d) and (3.15e) into Eq. (3.15c) gives, for the latter equation, the expression

$$M_{x,xx} + 2M_{xy,xy} + M_{y,yy} - N_x\beta_{x,x} - N_{xy}(\beta_{y,x} + \beta_{x,y}) - N_y\beta_{y,y} = -p \qquad (3.16)$$

Introduction of the appropriate constitutive and kinematic relations for the moment intensities and rotations reduces Eq. (3.16) to the form

$$D\nabla^4 w - (N_x w_{,xx} + 2N_{xy} w_{,xy} + N_y w_{,yy}) = p \qquad (3.17)$$

where
$$\nabla^4 w = w_{,xxxx} + 2w_{,xxyy} + w_{,yyyy}$$

With these simplifications the equations of equilibrium may be written in the relatively compact form

$$N_{x,x} + N_{xy,y} = 0 \qquad (3.18a)$$

$$N_{xy,x} + N_{y,y} = 0 \qquad (3.18b)$$

$$D\nabla^4 w - (N_x w_{,xx} + 2N_{xy} w_{,xy} + N_y w_{,yy}) = p \qquad (3.18c)$$

These equations are the counterpart for the plate of Eq. (1.12) for the column. They are a form of the widely used von Kármán plate equations, and they are the nonlinear equilibrium equations for all flat or bent configurations of the plate within the scope of the intermediate class of deformations.

3.3b Stationary Potential Energy

In this section the preceding nonlinear equilibrium equations are rederived on the basis of the stationary potential energy criterion. A brief development of energy criteria for equilibrium and stability is given in the Appendix. A loaded plate is in equilibrium if its total potential energy V is stationary, and V is stationary if the integrand in the expression for V satisfies the Euler equations of the calculus of variations (Sec. A.4).

The total potential energy of a plate subjected to lateral pressure $p(x,y)$ and edge loading is the sum of the strain energy U and the potential energy of the applied loads, Ω:

$$V = U + \Omega \qquad (3.19)$$

The strain energy for a three-dimensional isotropic medium referred to arbitrary orthogonal coordinates may be written [Ref. 3.6, eq. (3.49)]

$$U = \frac{1}{2} \iiint (\bar\sigma_x \bar\varepsilon_x + \bar\sigma_y \bar\varepsilon_y + \bar\sigma_z \bar\varepsilon_z + \bar\tau_{xy} \bar\gamma_{xy} + \bar\tau_{yz} \bar\gamma_{yz} + \bar\tau_{zx} \bar\gamma_{zx})\, dx\, dy\, dz$$

Omission of $\bar\gamma_{yz}$, $\bar\gamma_{zx}$, and $\bar\sigma_z$ in accordance with the basic approximations of thin-plate theory (Sec. 3.2), introduction of Eqs. (3.7), and rearrangement gives

$$U = \frac{E}{2(1 - v^2)} \iiint \left(\bar\varepsilon_x{}^2 + \bar\varepsilon_y{}^2 + 2v\bar\varepsilon_x \bar\varepsilon_y + \frac{1 - v}{2} \bar\gamma_{xy}{}^2 \right) dx\, dy\, dz$$

Introduction of Eqs. (3.4) and integration with respect to z leads to the relations

$$U = U_m + U_b \qquad (3.20)$$

where

$$U_m = \frac{C}{2} \iint \left(\varepsilon_x{}^2 + \varepsilon_y{}^2 + 2v\varepsilon_x \varepsilon_y + \frac{1 - v}{2} \gamma_{xy}{}^2 \right) dx\, dy \qquad (3.21)$$

and

$$U_b = \frac{D}{2} \iint [\kappa_x{}^2 + \kappa_y{}^2 + 2v\kappa_x \kappa_y + 2(1 - v)\kappa_{xy}{}^2]\, dx\, dy \qquad (3.22)$$

The expressions in Eqs. (3.21) and (3.22), respectively, are referred to as the *membrane strain energy* and the *bending strain energy* of the plate.

The potential energy of the applied loads for a conservative system is the negative of the work done by the loads as the structure is deformed. Consequently, for the lateral pressure p,

$$\Omega = -\iint pw\, dx\, dy$$

The form of the expression for the potential energy of the edge loads depends on the nature of the loads. As an example, we consider an in-plane compressive edge load P_x, in pounds, uniformly distributed along the edges $x = 0$, a, as shown in Fig. 3.5. For such a load the potential energy may be written

$$\Omega = P_x[u(a) - u(0)]$$

or

$$\Omega = P_x \frac{1}{b} \int_0^b \int_0^a u_{,x}\, dx\, dy$$

Thus, for the lateral and edge loads together, the expression for Ω is

$$\Omega = \iint \left(\frac{1}{b} P_x u_{,x} - pw \right) dx\, dy \qquad (3.23)$$

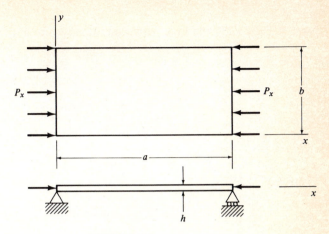

FIGURE 3.5
Plate subjected to in-plane compressive loading.

Collection of the various components of the total potential energy now gives

$$V = \iint F \, dx \, dy \qquad (3.24)$$

where

$$F = \frac{C}{2} \left(\varepsilon_x^{\,2} + \varepsilon_y^{\,2} + 2v\varepsilon_x \varepsilon_y + \frac{1-v}{2} \gamma_{xy}^{\,2} \right)$$

$$+ \frac{D}{2} [\kappa_x^{\,2} + \kappa_y^{\,2} + 2v\kappa_x \kappa_y + 2(1-v)\kappa_{xy}^{\,2}]$$

$$+ \left(\frac{1}{b} P_x u_{,x} - pw \right) \qquad (3.25)$$

and where, from Eqs. (3.5),

$$\varepsilon_x = u_{,x} + \tfrac{1}{2} w_{,x}^{\,2} \qquad\qquad \kappa_x = -w_{,xx}$$
$$\varepsilon_y = v_{,y} + \tfrac{1}{2} w_{,y}^{\,2} \qquad\qquad \kappa_y = -w_{,yy}$$
$$\gamma_{xy} = (u_{,y} + v_{,x}) + w_{,x} w_{,y} \qquad \kappa_{xy} = -w_{,xy}$$

For equilibrium the potential energy V must be stationary; i.e., its first variation δV must equal zero (Sec. A.2). Accordingly, the integrand F must satisfy the Euler equations of the calculus of variations. For an integrand of the form of Eq. (3.25), the Euler equations are given by Eqs. (A.35):

$$\frac{\partial F}{\partial u} - \frac{\partial}{\partial x}\frac{\partial F}{\partial u_{,x}} - \frac{\partial}{\partial y}\frac{\partial F}{\partial u_{,y}} = 0$$

$$\frac{\partial F}{\partial v} - \frac{\partial}{\partial x}\frac{\partial F}{\partial v_{,x}} - \frac{\partial}{\partial y}\frac{\partial F}{\partial v_{,y}} = 0$$

$$\frac{\partial F}{\partial w} - \frac{\partial}{\partial x}\frac{\partial F}{\partial w_{,x}} - \frac{\partial}{\partial y}\frac{\partial F}{\partial w_{,y}} + \frac{\partial^2}{\partial x^2}\frac{\partial F}{\partial w_{,xx}} + \frac{\partial^2}{\partial x\,\partial y}\frac{\partial F}{\partial w_{,xy}} + \frac{\partial^2}{\partial y^2}\frac{\partial F}{\partial w_{,yy}} = 0$$

From Eq. (3.25)

$$\frac{\partial F}{\partial u} = 0 \qquad \frac{\partial F}{\partial u_{,x}} = \frac{C}{2}(2\varepsilon_x + 2\nu\varepsilon_y) + \frac{P_x}{b}$$

$$\frac{\partial F}{\partial u_{,y}} = \frac{C}{2}[(1-\nu)\gamma_{xy}] \qquad \frac{\partial F}{\partial v} = 0$$

$$\frac{\partial F}{\partial v_{,x}} = \frac{C}{2}[(1-\nu)\gamma_{xy}] \qquad \frac{\partial F}{\partial v_{,y}} = \frac{C}{2}(2\varepsilon_y + 2\nu\varepsilon_x) \qquad \frac{\partial F}{\partial w} = -p$$

$$\frac{\partial F}{\partial w_{,x}} = \frac{C}{2}[(2\varepsilon_x + 2\nu\varepsilon_y)w_{,x} + (1-\nu)\gamma_{xy}w_{,y}]$$

$$\frac{\partial F}{\partial w_{,y}} = \frac{C}{2}[(2\varepsilon_y + 2\nu\varepsilon_x)w_{,y} + (1-\nu)\gamma_{xy}w_{,x}]$$

$$\frac{\partial F}{\partial w_{,xx}} = -\frac{D}{2}(2\kappa_x + 2\nu\kappa_y) \qquad \frac{\partial F}{\partial w_{,xy}} = -\frac{D}{2}[4(1-\nu)\kappa_{xy}]$$

$$\frac{\partial F}{\partial w_{,yy}} = -\frac{D}{2}(2\kappa_y + 2\nu\kappa_x)$$

Introduction into the Euler equations and simplification gives the equations

$$C\left[(\varepsilon_x + \nu\varepsilon_y)_{,x} + \frac{1-\nu}{2}(\gamma_{xy})_{,y}\right] = 0$$

$$C\left[\frac{1-\nu}{2}(\gamma_{xy})_{,x} + (\varepsilon_y + \nu\varepsilon_x)_{,y}\right] = 0$$

$$-D[(\kappa_x + \nu\kappa_y)_{,xx} + 2(1-\nu)(\kappa_{xy})_{,xy} + (\kappa_y + \nu\kappa_x)_{,yy}]$$

$$-C\left[(\varepsilon_x + \nu\varepsilon_y)w_{,x} + \frac{1-\nu}{2}(\gamma_{xy})w_{,y}\right]_{,x}$$

$$-C\left[(\varepsilon_y + \nu\varepsilon_x)w_{,y} + \frac{1-\nu}{2}(\gamma_{xy})w_{,x}\right]_{,y} = p$$

Introduction of the constitutive relations from Eqs. (3.8) now gives

$$N_{x,x} + N_{xy,y} = 0$$
$$N_{xy,x} + N_{y,y} = 0$$
$$D\nabla^4 w - (N_x w_{,xx} + 2N_{xy} w_{,xy} + N_y w_{,yy}) = p$$

The term containing the edge load P_x is seen to drop out of the equilibrium equations; as an edge load, it enters the analysis only in the boundary condition $N_x = -P_x/b$ at $x = 0, L$. These equations for equilibrium of the plate element are seen to be the same as Eqs. (3.18).

The nonlinear equations of equilibrium appear as a coupled set of three nonlinear differential equations in the four variables N_x, N_{xy}, N_y, and w. Three equations in three unknowns u, v, w may be obtained by introduction of the constitutive and kinematic relations. The resulting equations are

$$[(u_{,x} + \tfrac{1}{2}w_{,x}^2) + v(v_{,y} + \tfrac{1}{2}w_{,y}^2)]_{,x} + \frac{1-v}{2}(u_{,y} + v_{,x} + w_{,x}w_{,y})_{,y} = 0$$

$$\frac{1-v}{2}(u_{,y} + v_{,x} + w_{,x}w_{,y})_{,x} + [(v_{,y} + \tfrac{1}{2}w_{,y}^2) + v(u_{,x} + \tfrac{1}{2}w_{,x}^2)]_{,y} = 0 \qquad (3.26)$$

$$D\nabla^4 w - C[(u_{,x} + \tfrac{1}{2}w_{,x}^2) + v(v_{,y} + \tfrac{1}{2}w_{,y}^2)]w_{,xx} - (1-v)C[u_{,y} + v_{,x}$$
$$+ w_{,x}w_{,y}]w_{,xy} - C[(v_{,y} + \tfrac{1}{2}w_{,y}^2) + v(u_{,x} + \tfrac{1}{2}w_{,x}^2)]w_{,yy} = p$$

A much simpler set of two equations in two unknowns may be obtained by introduction of a stress function f defined by the relations (see, for example, Ref. 3.4, p. 417)

$$N_x = f_{,yy} \qquad N_y = f_{,xx} \qquad N_{xy} = -f_{,xy} \qquad (3.27)$$

where $f = f(x,y)$. These expressions are seen to satisfy Eqs. (3.18a) and (3.18b) identically. Introduction into Eq. (3.18c) gives

$$D\nabla^4 w - (f_{,yy} w_{,xx} - 2f_{,xy} w_{,xy} + f_{,xx} w_{,yy}) = p \qquad (3.28)$$

For geometric compatibility, however, from Eqs. (3.5),

$$\varepsilon_{x,yy} + \varepsilon_{y,xx} - \gamma_{xy,xy} = w_{,xy}^2 - w_{,xx} w_{,yy}$$

and from Eqs. (3.8),

$$\varepsilon_x = \frac{1}{Eh}(f_{,yy} - vf_{,xx}) \qquad \varepsilon_y = \frac{1}{Eh}(f_{,xx} - vf_{,yy}) \qquad \gamma_{xy} = -\frac{2(1+v)}{Eh}f_{,xy}$$

Therefore
$$\nabla^4 f - Eh(w_{,xy}^2 - w_{,xx} w_{,yy}) = 0 \qquad (3.29)$$

Equations (3.28) and (3.29) form two equations in the two variables w and f. They are called the *equilibrium* and *compatibility equations*, respectively, for the plate. These equations are known as the von Kármán equations for large deflections of flat plates.

To obtain the equilibrium equations of *linear* plate theory it is necessary only to omit quadratic and cubic terms in the displacement components. The linear equations corresponding to Eqs. (3.18) are found to be

$$N_{x,x} + N_{xy,y} = 0 \qquad (3.30a)$$

$$N_{xy,x} + N_{y,y} = 0 \qquad (3.30b)$$

$$D\nabla^4 w = p \qquad (3.30c)$$

where now

$$N_x = C(\varepsilon_x + v\varepsilon_y) \qquad \varepsilon_x = u_{,x}$$

$$N_y = C(\varepsilon_y + v\varepsilon_x) \qquad \varepsilon_y = v_{,y} \qquad (3.31)$$

$$N_{xy} = C\frac{1-v}{2}\gamma_{xy} \qquad \gamma_{xy} = u_{,y} + v_{,x}$$

As may be seen, Eq. (3.30c) is uncoupled from the rest. This equation refers to transverse load and displacement; the others, to in-plane quantities. Much of the relative simplicity of classical thin-plate theory is a consequence of this uncoupling.

For the in-plane edge loading treated in the following sections on plate stability, $p = w = 0$ for the prebuckling equilibrium analysis. Consequently, the prebuckling deformation is governed by Eqs. (3.30a) and (3.30b).

3.3c Nonlinear Equilibrium Paths

Equations (3.26) determine all linear and nonlinear equilibrium configurations of the plate within the scope of the intermediate class of deformations. The differential equations contain linear, quadratic, and cubic terms in the dependent variables u, v, w, and hence are nonlinear. Approximate numerical solutions of the nonlinear equations have been obtained for particular cases (for example, Ref. 3.7). Load-displacement plots based on such solutions for a plate subjected to the edge load P_x, shown in Fig. 3.5, are illustrated in Fig. 3.6a and b (Refs. 3.7 and 3.8). In Fig. 3.6a, load is plotted versus the longitudinal displacement u of one end of the plate relative to the other (often called the *shortening*). In

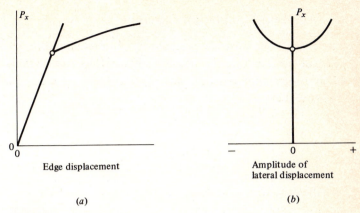

FIGURE 3.6
Equilibrium paths for a plate subjected to in-plane compressive loading.

Fig. 3.6*b* the same results are represented in terms of load versus amplitude of lateral displacement *w*. Points on the primary and secondary equilibrium paths denote flat and bent configurations of the plate, respectively. The symmetry of Fig. 3.6*b* signifies only that the plate may deflect in either direction. The linear equilibrium equations (3.30) govern the primary equilibrium path. The nonlinear equations (3.18) govern both primary and secondary paths.

The equilibrium paths determined by solution of the *equilibrium* equations show immediately the existence of a bifurcation point and the magnitude of the corresponding critical load. Consequently, a separate solution of *stability* equations is not necessary for the determination of the critical load. However, the equilibrium paths in Fig. 3.6 are based on numerical solution of nonlinear differential equations. The purpose of stability analyses such as those in the following section is to permit determination of the bifurcation-point load by solution of *linear* differential equations.

3.4   LINEAR STABILITY EQUATIONS

The linear differential equations for determination of the bifurcation-point load of a rectangular flat plate subjected to in-plane edge loading (for lateral pressure $p = 0$) are derived in Sec. 3.4*a* by application of the adjacent-equilibrium criterion. The equations are rederived in Sec. 3.4*b* by use of the minimum potential energy criterion.

3.4*a* Adjacent-Equilibrium Criterion

To investigate the possible existence of adjacent equilibrium *configurations*, we give small increments to the *displacement* variables and examine the two adjacent configurations represented by the displacements before and after the increment, as follows. Let

$$u \to u_0 + u_1$$
$$v \to v_0 + v_1 \qquad (3.32)$$
$$w \to w_0 + w_1$$

where the incremental displacement (u_1, v_1, w_1) is arbitrarily small and, tentatively, (u_0, v_0, w_0) and (u, v, w) are any two adjacent equilibrium configurations. Introduction into Eqs. (3.26) is seen to give terms that are linear, quadratic, and cubic in the u_0, v_0, w_0 and u_1, v_1, w_1 displacement components. In the new equations, the terms in u_0, v_0, w_0 alone add to zero because (u_0, v_0, w_0) is an equilibrium configuration, and terms that are quadratic and cubic in u_1, v_1, w_1 may be omitted because of the smallness of the incremental displacement. Thus the resulting equation is homogeneous and linear in u_1, v_1, w_1, with variable coefficients in u_0, v_0, w_0.

The coefficients u_0, v_0, w_0, however, are governed by the original nonlinear equations. For this reason it is desirable to limit the range of applicability of the linearized equations by requiring that u_0, v_0, w_0 be confined to configurations that are governed by the linear equilibrium equations (3.30). For the flat plate subjected to in-plane edge loading, this limitation has the added advantage that w_0 and its derivatives equal zero.

The actual derivation of the linearized equations is substantially shortened by use of the nonlinear equations in the form of Eqs. (3.18) rather than Eqs. (3.26), as follows. The increment in u, v, w causes a corresponding change in the internal force intensities. Accordingly, in Eqs. (3.18), we let

$$N_x \to N_{x0} + \Delta N_x$$
$$N_y \to N_{y0} + \Delta N_y \qquad (3.33)$$
$$N_{xy} \to N_{xy0} + \Delta N_{xy}$$

where terms with 0 subscripts correspond to the u_0, v_0, w_0 displacement, and the ΔN_x, ΔN_y, ΔN_{xy} are increments corresponding to u_1, v_1, w_1. We also let N_{x1}, N_{y1}, N_{xy1} represent the parts of $\Delta N_x, \Delta N_y, \Delta N_{xy}$, respectively, that are linear in u_1, v_1, w_1. For example, from Eqs. (3.5) and (3.8),

$$N_x = C[(u_{,x} + \tfrac{1}{2}w_{,x}^2) + v(v_{,y} + \tfrac{1}{2}w_{,y}^2)]$$

Then, since w_0 and its derivatives are equal to zero for in-plane edge loading,

$$N_x + \Delta N_x = C[(u_{0,x} + u_{1,x} + \tfrac{1}{2}w_{1,x}{}^2) + v(v_{0,y} + v_{1,y} + \tfrac{1}{2}w_{1,y}{}^2)]$$

Therefore

$$N_{x0} = C(u_{0,x} + vv_{0,y})$$
$$\Delta N_x = C[(u_{1,x} + \tfrac{1}{2}w_{1,x}{}^2) + v(v_{1,y} + \tfrac{1}{2}w_{1,y}{}^2)]$$
$$N_{x1} = C(u_{1,x} + vv_{1,y})$$

Then introduction into Eqs. (3.18) gives

$$N_{x1,x} + N_{xy1,y} = 0 \qquad (3.34a)$$
$$N_{xy1,x} + N_{y1,y} = 0 \qquad (3.34b)$$
$$D\nabla^4 w_1 - (N_{x0}\,w_{1,xx} + 2N_{xy0}\,w_{1,xy} + N_{y0}\,w_{1,yy}) = 0 \qquad (3.34c)$$

where

$$N_{x0} = C(u_{0,x} + vv_{0,y}) \qquad\qquad N_{x1} = C(u_{1,x} + vv_{1,y})$$
$$N_{y0} = C(v_{0,y} + vu_{0,x}) \qquad\qquad N_{y1} = C(v_{1,y} + vu_{1,x}) \qquad (3.35)$$
$$N_{xy0} = C\,\frac{1-v}{2}\,(u_{0,y} + v_{0,x}) \qquad N_{xy1} = C\,\frac{1-v}{2}\,(u_{1,y} + v_{1,x})$$

Equations (3.34) are the stability equations for the plate subjected to in-plane edge loading. They are the counterpart for the plate of Eq. (1.23) for the straight column.

As in the linear equilibrium equations, Eq. (3.34c) is uncoupled from Eqs. (3.34a) and (3.34b). This uncoupling greatly simplifies the analysis of particular cases in Sec. 3.5. The corresponding equations for shells are coupled.

Equation (3.34c) is a homogeneous linear equation in w_1 with variable coefficients in N_{x0}, N_{xy0}, N_{y0}. The coefficients are determined by the linear equations in Eqs. (3.30a) and (3.30b). As a homogeneous equation, Eq. (3.34c) has solutions only for discrete values of the applied load. For each of these values two adjacent equilibrium configurations exist, a flat one on the primary equilibrium path and a slightly bent one on a secondary equilibrium path. The linearized analysis gives no information about the initial slope or shape of the secondary equilibrium paths, but the bifurcation-point loads given by Eq. (3.34c) nevertheless represent loss of stability. The smallest such load is defined as the critical load, P_{cr}.

It should be emphasized that this procedure for formulation of the eigenvalue problem with the applied load fixed in magnitude leads to correct eigenvalues and eigenmodes, not only when the secondary path is horizontal at the bifurcation point, but also when it is not.

3.4b Minimum Potential Energy Criterion

In this section the plate stability equations (3.34) are rederived by application of the minimum potential energy criterion. The flat form of the plate is an equilibrium configuration for all values of the applied load. For sufficiently small loads the equilibrium is stable. The equilibrium changes from stable to neutral when the expression for the total potential energy V ceases to be a relative minimum. The criterion for the loss of stability is that the integrand in the expression for the second variation of V satisfy the Euler equations of the calculus of variations (Sec. A.5).

An expression for the total potential energy of the edge-loaded plate is given by Eqs. (3.24) and (3.25). To obtain the second variation, let

$$\begin{aligned} u &\to u_0 + u_1 \\ v &\to v_0 + v_1 \\ w &\to w_0 + w_1 \end{aligned} \tag{3.36}$$

where (u_0, v_0, w_0) is a configuration on the primary equilibrium path and (u_1, v_1, w_1) is a virtual increment. The second variation of the total potential energy is the sum of all terms in the expression for the potential energy increment that are quadratic in u_1, v_1, w_1. The derivation may be carried out term by term. For example,

$$\varepsilon_x = u_{,x} + \tfrac{1}{2} w_{,x}{}^2$$

and therefore

$$\varepsilon_x{}^2 = u_{,x}{}^2 + u_{,x} w_{,x}{}^2 + \tfrac{1}{4} w_{,x}{}^4$$

Introduction of Eqs. (3.36) and collection of second-order terms gives, for $w_0 \equiv 0$,

$$\tfrac{1}{2} \delta^2(\varepsilon_x{}^2) = u_{1,x}{}^2 + u_{0,x} w_{1,x}{}^2$$

Introduction of similar expressions for ε_y and γ_{xy} gives, for the second variation of the membrane strain energy, the expression

$$\begin{aligned} \tfrac{1}{2} \delta^2 U_m = \frac{C}{2} \iint \Big\{ &\Big[u_{1,x}{}^2 + v_{1,y}{}^2 + 2v u_{1,x} v_{1,y} + \frac{1-v}{2}(u_{1,y} + v_{1,x})^2 \Big] \\ &+ [(u_{0,x} + v v_{0,y}) w_{1,x}{}^2 + (v_{0,y} + v u_{0,x}) w_{1,y}{}^2 \\ &+ (1-v)(u_{0,y} + v_{0,x}) w_{1,x} w_{1,y}] \Big\} \, dx \, dy \end{aligned} \tag{3.37}$$

By use of the constitutive relations in Eqs. (3.35), Eqs. (3.37) can be written in the simpler form

$$\tfrac{1}{2}\delta^2 U_m = \frac{C}{2}\iint \left[u_{1,x}^2 + v_{1,y}^2 + 2vu_{1,x}v_{1,y} + \frac{1-v}{2}(u_{1,y}+v_{1,x})^2 \right] dx\,dy$$

$$+ \frac{1}{2}\iint (N_{x0}\,w_{1,x}^2 + 2N_{xy0}\,w_{1,x}w_{1,y} + N_{y0}\,w_{1,y}^2)\,dx\,dy \qquad (3.38)$$

Similarly, the second variation of the bending strain energy is found to be

$$\tfrac{1}{2}\delta^2 U_b = \frac{D}{2}\iint [w_{1,xx}^2 + w_{1,yy}^2 + 2vw_{1,xx}w_{1,yy} + 2(1-v)w_{1,xy}^2]\,dx\,dy \qquad (3.39)$$

Equation (3.23) for the potential energy of the applied loads is seen to have no quadratic or higher-order terms in the displacement components, so that $\delta^2\Omega = 0$. Consequently,

$$\delta^2 V = \delta^2 U_m + \delta^2 U_b \qquad (3.40)$$

Thus the final expression for the second variation may be written

$$\delta^2 V = C\iint F\,dx\,dy \qquad (3.41)$$

where

$$F = \left[u_{1,x}^2 + v_{1,y}^2 + 2vu_{1,x}v_{1,y} + \frac{1-v}{2}(u_{1,y}+v_{1,x})^2 \right]$$

$$+ \frac{1-v^2}{Eh}(N_{x0}\,w_{1,x}^2 + 2N_{xy0}\,w_{1,x}w_{1,y} + N_{y0}\,w_{1,y}^2)$$

$$+ \frac{h^2}{12}[w_{1,xx}^2 + w_{1,yy}^2 + 2vw_{1,xx}w_{1,yy} + 2(1-v)w_{1,xy}^2] \qquad (3.42)$$

According to the Trefftz criterion (Sec. A.5), the equations governing loss of stability are the Euler equations for the integrand in the second-variation expression. For an integrand of the form of Eq. (3.42), the Euler equations are [compare Eqs. (A.35)]

$$\frac{\partial F}{\partial u_1} - \frac{\partial}{\partial x}\frac{\partial F}{\partial u_{1,x}} - \frac{\partial}{\partial y}\frac{\partial F}{\partial u_{1,y}} = 0$$

$$\frac{\partial F}{\partial v_1} - \frac{\partial}{\partial x}\frac{\partial F}{\partial v_{1,x}} - \frac{\partial}{\partial y}\frac{\partial F}{\partial v_{1,y}} = 0 \qquad (3.43)$$

$$\frac{\partial F}{\partial w_1} - \frac{\partial}{\partial x}\frac{\partial F}{\partial w_{1,x}} - \frac{\partial}{\partial y}\frac{\partial F}{\partial w_{1,y}} + \frac{\partial^2}{\partial x^2}\frac{\partial F}{\partial w_{1,xx}} + \frac{\partial^2}{\partial x\,\partial y}\frac{\partial F}{\partial w_{1,xy}} + \frac{\partial^2}{\partial y^2}\frac{\partial F}{\partial w_{1,yy}} = 0$$

Introduction of Eqs. (3.42) and simplification gives

$$(u_{1,x} + vv_{1,y})_{,x} + \frac{1-v}{2}(u_{1,y} + v_{1,x})_{,y} = 0$$

$$(v_{1,y} + vu_{1,x})_{,y} + \frac{1-v}{2}(u_{1,y} + v_{1,x})_{,x} = 0 \qquad (3.44)$$

$$D\nabla^4 w_1 - (N_{x0}\,w_{1,xx} + 2N_{xy0}\,w_{1,xy} + N_{y0}\,w_{1,yy}) = 0$$

Equations (3.44) may be written in the more compact form

$$N_{x1,x} + N_{xy1,y} = 0$$
$$N_{xy1,x} + N_{y1,y} = 0$$
$$D\nabla^4 w_1 - (N_{x0}\,w_{1,xx} + 2N_{xy0}\,w_{1,xy} + N_{y0}\,w_{1,yy}) = 0$$

where N_{x1}, N_{xy1}, N_{y1} are defined by Eqs. (3.35). These equations are the same as Eqs. (3.34).

3.5   APPLICATIONS OF THE STABILITY EQUATION

Equation (3.34c) applies for all possible in-plane edge loads. In the most general case the coefficients N_{x0}, N_{xy0}, and N_{y0} are functions of the coordinate variables x and y. The applications in this chapter, however, are limited to cases in which these coefficients are constants. Examples of the method of analysis for variable-coefficient stability equations are included in Chap. 8.

To simplify the notation, the 1 subscripts are omitted from the incremental quantities (w_1, M_{x1}, etc.) in the remainder of Sec. 3.5.

3.5a   Plate Simply Supported on Four Edges

As a first example of the application of the linear stability equation, we consider a flat plate simply supported on four edges and subjected to an in-plane compressive load P_x uniformly distributed along the edges $x = 0$, a, as shown in Fig. 3.5. From an equilibrium analysis of the plate, using Eqs. (3.30a) and (3.30b),

$$N_{x0} = -\frac{P_x}{b} \qquad N_{xy0} = N_{y0} = 0$$

Introduction into Eq. (3.34c) simplifies that expression to the form

$$D\nabla^4 w + \frac{P_x}{b}\,w_{,xx} = 0 \qquad (3.45)$$

The simple-support boundary conditions are $w = M_x = 0$ on $x = 0, a$ and $w = M_y = 0$ on $y = 0, b$, where, from Eqs. (3.5) and (3.8), $M_x = -EI(w_{,xx} + vw_{,yy})$ and $M_y = -EI(w_{,yy} + vw_{,xx})$. Therefore the boundary-condition equations may be written

$$
\begin{aligned}
w = w_{,xx} = 0 &\quad \text{on} \quad x = 0, a \\
w = w_{,yy} = 0 &\quad \text{on} \quad y = 0, b
\end{aligned}
\tag{3.46}
$$

Equation (3.45) is a constant-coefficient equation. A solution of the form

$$
w = C_1 \sin \frac{m\pi x}{a} \sin \frac{n\pi y}{b} \qquad m, n = 1, 2, 3, \dots \tag{3.47}
$$

where C_1 is a constant, is seen to satisfy both the differential equation and the boundary conditions. Introduction into Eq. (3.45) gives

$$
D\left[\left(\frac{m\pi}{a}\right)^4 + 2\left(\frac{m\pi}{a}\right)^2\left(\frac{n\pi}{b}\right)^2 + \left(\frac{n\pi}{b}\right)^4\right] - \frac{P_x}{b}\left(\frac{m\pi}{a}\right)^2 = 0
$$

The discrete values of P_x for which Eq. (3.45) has nontrivial solutions therefore are

$$
\frac{P_x}{b} = \left(\frac{\pi a}{m}\right)^2 D\left[\left(\frac{m}{a}\right)^2 + \left(\frac{n}{b}\right)^2\right]^2 \tag{3.48}
$$

The critical load corresponds to the smallest eigenvalue. For all values of a, b, the smallest eigenvalue is given by $n = 1$. Accordingly,

$$
\frac{P_x}{b} = \left(\frac{\pi a}{m}\right)^2 D\left[\left(\frac{m}{a}\right)^2 + \left(\frac{1}{b}\right)^2\right]^2 \qquad m = 1, 2, 3, \dots \tag{3.49}
$$

Equation (3.49) may be expressed in the form

$$
P_x = k_c \frac{\pi^2 D}{b} \tag{3.50}
$$

where

$$
k_c = \left(\frac{mb}{a} + \frac{a}{mb}\right)^2 \tag{3.51}
$$

The coefficient k_c is seen to be a function of the aspect ratio a/b and the wavelength parameter m. For a given a/b, the value of m may be chosen by trial to yield the smallest eigenvalue.

As an example, let $a = 20$ in., $b = 10$ in., $h = 0.1$ in., $E = 10 \times 10^6$ psi, and $v = 0.3$. Then $D = 916$ lb-in., and for $m = 1, 2,$ and 3, respectively, Eq. (3.51) gives the values $k_c = 6.25, 4.00,$ and 4.70. For other values of m, the coefficient k is larger. The smallest eigenvalue corresponds to $m = 2$. For that value, Eq. (3.50) gives, for the critical load, $P_{cr} = 3,620$ lb.

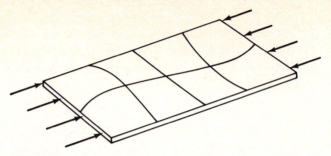

FIGURE 3.7
Buckled form of plate subjected to in-plane compressive loading.

A sketch of the plate in the bent equilibrium configuration is shown in Fig. 3.7. The configuration is said to have one half sine wave in the transverse direction and two in the longitudinal direction. The amplitude of the waves is infinitesimally small because, in the linearized equation, w_1 is infinitesimally small. The load $P_{cr} = 3,620$ lb is the lowest load at which such a plate can lose its stability.

A plot of k_c versus a/b from Eq. (3.51) for other values of m is shown in Fig. 3.8. The solid curves represent lowest eigenvalues, and the dotted ones higher eigenvalues, for given aspect ratios.

3.5b   Other Boundary Conditions

The simple form of the solution in Eq. (3.47) is not suitable for other boundary conditions. As a more general example, let us consider the load P_x applied to a plate that is simply supported on the edges $x = 0$, a but that has other, as yet unspecified, boundary conditions on $y = 0$, b. The smallest critical load for such boundary conditions evidently corresponds to free edges on $y = 0$, b. Such a plate is simply a column whose bending stiffness EI is replaced by Db. Then, from Eq. (1.17),

$$P_x = m^2 \, \frac{\pi^2 Db}{a^2} \qquad (3.52)$$

Such a plate is called a *wide column*. The smallest eigenvalues are seen to correspond to $m = 1$.

In the more general case, the differential equation and the boundary conditions on $x = 0$, a are satisfied by solutions of the form

$$w = f(y) \sin \frac{m\pi x}{a} \qquad m = 1, 2, 3, \ldots \qquad (3.53)$$

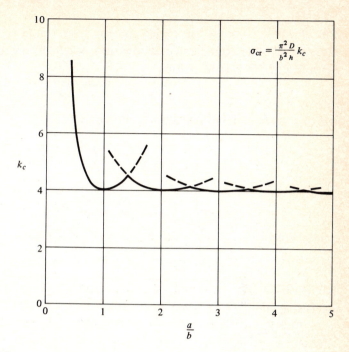

$$\sigma_{cr} = \frac{\pi^2 D}{b^2 h} k_c$$

FIGURE 3.8
Critical values of axial stress for simply supported plates subjected to in-plane compressive loading.

Introduction of Eq. (3.53) into Eq. (3.45) reduces the latter to the ordinary differential equation:

$$\frac{d^4 f}{dy^4} - 2\left(\frac{m\pi}{a}\right)^2 \frac{d^2 f}{dy^2} + \left[\left(\frac{m\pi}{a}\right)^4 - \frac{P_x}{Db}\left(\frac{m\pi}{a}\right)^2\right] f = 0 \qquad (3.54)$$

Equation (3.54) is a constant-coefficient equation. Consequently, its solution is readily expressed for arbitrary conditions on $y = 0, b$. The characteristic equation associated with Eq. (3.54) is seen to be

$$\lambda^4 - 2\left(\frac{m\pi}{a}\right)^2 \lambda^2 + \left[\left(\frac{m\pi}{a}\right)^4 - \frac{P_x}{Db}\left(\frac{m\pi}{a}\right)^2\right] = 0 \qquad (3.55)$$

The roots of Eq. (3.55) are

$$\lambda = \pm\left[\frac{m\pi}{a}\left(\frac{m\pi}{a} \pm \sqrt{\frac{P_x}{Db}}\right)\right]^{1/2} \qquad (3.56)$$

But $(P_x/Db)^{1/2} = m\pi/a$ for a wide column, from Eq. (3.52), and $(P_x/Db)^{1/2} >$ $m\pi/a$ for all other boundary conditions on $y = 0, b$. Consequently, for all other cases, Eq. (3.56) may be written

$$\lambda = \alpha, \ -\alpha, \ i\beta, \ -i\beta$$

where α, β are real and positive and are given by

$$\alpha = \left[\left(\frac{m\pi}{a}\right)^2 + \frac{m\pi}{a}\sqrt{\frac{P_x}{Db}}\right]^{1/2}$$

$$\beta = \left[-\left(\frac{m\pi}{a}\right)^2 + \frac{m\pi}{a}\sqrt{\frac{P_x}{Db}}\right]^{1/2}$$

(3.57)

In this notation, the solution to Eq. (3.54) may be written

$$f = C_1 e^{-\alpha y} + C_2 e^{\alpha y} + C_3 \cos \beta y + C_4 \sin \beta y \qquad (3.58)$$

[compare Eq. (2.81)], where C_1, C_2, C_3, C_4 are constants to be determined from the four boundary conditions on $y = 0, b$.

As a particular example, consider a plate that is simply supported on the edge $y = 0$ and free on $y = b$. The equations for the condition on $y = 0$ are

$$w = w_{,yy} = 0 \qquad (3.59)$$

as before. On $y = b$ the equations may be written (Ref. 3.4, p. 214)

$$w_{,yy} + vw_{,xx} = w_{,yyy} + (2 - v)w_{,xxy} = 0 \qquad (3.60)$$

where the first equation represents zero moment M_y, and the second represents zero transverse shear Q_y and twist M_{yx}. Equations (3.58) and (3.59) give

$$C_3 = 0 \qquad C_1 = -C_2$$

Then Eq. (3.58) can be rewritten in the more convenient form

$$f = A \sinh \alpha y + B \sin \beta y$$

where A and B are new constants. Introduction into Eq. (3.60) gives

$$\left[\left(\alpha^2 - v\frac{m^2\pi^2}{a^2}\right)\sinh \alpha b\right]A - \left[\left(\beta^2 + v\frac{m^2\pi^2}{a^2}\right)\sin \beta b\right]B = 0$$

$$\left\{\alpha\left[\alpha^2 - (2-v)\frac{m^2\pi^2}{a^2}\right]\cosh \alpha b\right\}A - \left\{\beta\left[\beta^2 + (2-v)\frac{m^2\pi^2}{a^2}\right]\cos \beta b\right\}B = 0$$

For a nontrivial solution the determinant of the coefficients of A and B in the homogeneous-equation system must equal zero. Setting the determinant equal to zero and rearrangement gives

$$\beta b\left[(\alpha b)^2 - v\left(\frac{m\pi b}{a}\right)^2\right]^2 \tanh \alpha b = \alpha b\left[(\alpha b)^2 + v\left(\frac{m\pi b}{a}\right)^2\right]^2 \tan \beta b \qquad (3.61)$$

where, from Eqs. (3.57),

$$\alpha b = \left[\left(\frac{m\pi b}{a}\right)^2 + \frac{m\pi b}{a}\sqrt{\frac{P_x b}{D}}\right]^{1/2}$$

$$\beta b = \left[-\left(\frac{m\pi b}{a}\right)^2 + \frac{m\pi b}{a}\sqrt{\frac{P_x b}{D}}\right]^{1/2}$$

These equations form an implicit expression for the nondimensional load parameter $P_x b/D$ in terms of the wavelength parameter m, Poisson's ratio v, and aspect ratio a/b.

Calculations show that for all values of a/b the minimum load occurs for $m = 1$. Results can be expressed in the form

$$P_x = k_c \frac{\pi^2 D}{b} \qquad (3.62)$$

where k_c is a nondimensional buckling coefficient for compressive loading. Numerical results of such calculations are given in, for example, Refs. 3.9 and 3.10. Results from Ref. 3.10, fig. 14, are shown in Fig. 3.9.

Also shown in Fig. 3.9 are corresponding values for other sets of boundary conditions on both the x and y edges of the plate. Extensive numerical results for additional sets of boundary conditions, including elastic edge restraint, are given in Ref. 3.10 in graphical form. The results are based on analyses which utilize numerical methods. Also to be found in Ref. 3.10 is a comprehensive list of references on the stability of plates.

3.5c Shear Loading

For a plate subjected to uniformly distributed shear loading as illustrated in Fig. 3.10, Eq. (3.34c) specializes to the form

$$D\nabla^4 w - 2N_{xy0}\, w_{,xy} = 0 \qquad (3.63)$$

This expression, like that for uniform compression loading in Eq. (3.45), is a constant-coefficient equation. Its simple appearance is deceptive, however. One

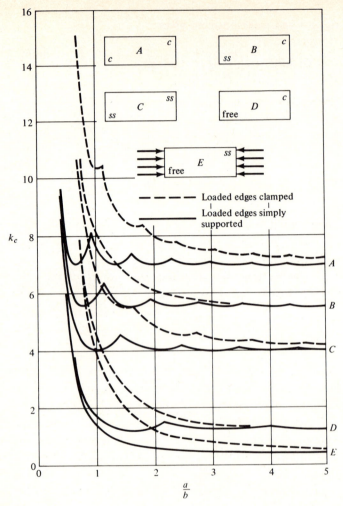

FIGURE 3.9
Influence of boundary conditions on the buckling coefficients of plates subjected to in-plane compressive loading.

term of the equation contains even-ordered derivatives with respect to each of the coordinate variables, and the other term, odd-ordered derivatives. Consequently, a deflection function of the form of that in Eq. (3.47) is not a solution.

Exact solutions of Eq. (3.63) are available only for the case of an infinitely long strip. Such a solution is given by Southwell and Skan (Ref. 3.11). The analysis is similar to that in Sec. 3.5b. A solution is assumed of the form

$$w = f(y)e^{ikx/b}$$

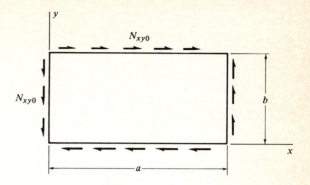

FIGURE 3.10
Plate subjected to in-plane shear loading.

where k is a longitudinal wavelength parameter and b is plate width. Introduction into Eq. (3.63) and rearrangement gives the ordinary differential equation

$$\left[\frac{d^2 f}{dy^2} - \left(\frac{k}{b}\right)^2 f\right]^2 - \frac{2N_{xy0}}{D} \frac{ik}{b} \frac{df}{dy} = 0$$

If the function $f(y)$ is then assumed to be of the form

$$f(y) = C_\lambda e^{i\lambda y/b}$$

where C_λ is a constant, the equation is replaced by the fourth-degree polynomial equation

$$\lambda^4 + 2k^2\lambda^2 + \frac{2N_{xy0}}{D} k\lambda + k^4 = 0$$

For each specified value of k, this equation has four roots, which may be designated λ_1, λ_2, λ_3, and λ_4. Then the solution of the ordinary differential equation may be written

$$f = C_1 e^{i\lambda_1 y/b} + C_2 e^{i\lambda_2 y/b} + C_3 e^{i\lambda_3 y/b} + C_4 e^{i\lambda_4 y/b}$$

and the solution for w is of the form

$$w = (C_1 e^{i\lambda_1 y/b} + C_2 e^{i\lambda_2 y/b} + C_3 e^{i\lambda_3 y/b} + C_4 e^{i\lambda_4 y/b})e^{ikx/b}$$

where C_1, C_2, C_3, C_4 are arbitrary constants. Introduction of this expression for w into four homogeneous boundary-condition equations leads to a determinantal equation from which the critical values of N_{xy0} may be determined, as in Sec. 3.5b.

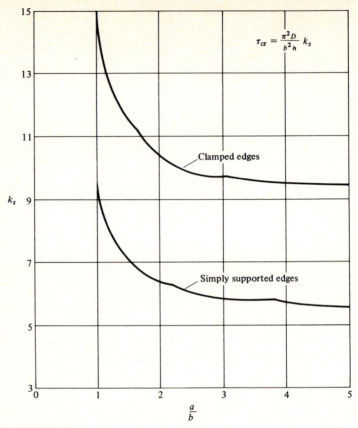

FIGURE 3.11
Critical values of shear stress for plates subjected to in-plane shear loading.

Southwell and Skan (Ref. 3.11) carry out such an analysis for simply supported edges ($w = w_{,yy} = 0$ on $y = 0, b$) and for clamped edges ($w = w_{,y} = 0$ on $y = 0, b$). Their results may be expressed in the form

$$N_{xy0} = k_s \frac{\pi^2 D}{b^2} \qquad (3.64)$$

where k_s is a nondimensional shear buckling coefficient. In this notation their results for simply supported and clamped edges, respectively, are $k_s = 5.35$ and $k_s = 8.98$.

Corresponding solutions for infinitely long strips with elastically restrained edges also are available (Ref. 3.10).

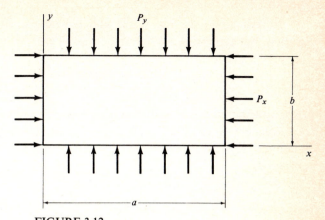

FIGURE 3.12
Plate subjected to in-plane compression in two directions.

For plates of finite dimensions, available solutions of Eq. (3.63) utilize numerical methods such as those discussed in Chap. 8. Results of such analyses for plates with simply supported edges and with clamped edges are shown in Fig. 3.11 (Ref. 3.10, fig. 22).

3.5d Combined Loading

Consider a plate subjected to in-plane compressive loading in two directions, as shown in Fig. 3.12. From an equilibrium analysis of the plate in the flat form,

$$N_{x0} = -\frac{P_x}{b} \qquad N_{xy0} = 0 \qquad N_{y0} = -\frac{P_y}{a}$$

Introduction into Eq. (3.34c) gives

$$D\nabla^4 w + \frac{P_x}{b} w_{,xx} + \frac{P_y}{a} w_{,yy} = 0 \qquad (3.65)$$

Equation (3.65) is a homogeneous equation, but there are two independent load parameters. The equation can be changed to a single-parameter equation simply by letting

$$\frac{P_y}{a} \equiv R \frac{P_x}{b} \qquad (3.66)$$

where R is a nondimensional constant. The resulting equation then may be solved for a series of selected values of R.

Introduction into Eq. (3.65) and rearrangement gives for the stability equation, the expression

$$DV^4w + \frac{P_x}{b}(w_{,xx} + Rw_{,yy}) = 0 \qquad (3.67)$$

For simplicity, we again treat only simply supported edges; then the boundary conditions are expressed by Eqs. (3.46), and solutions that satisfy the boundary conditions are of the form of Eq. (3.47). Introduction of Eq. (3.47) and rearrangement gives

$$P_x = k_{cc}\frac{\pi^2 D}{b} \qquad (3.68)$$

where

$$k_{cc} = \frac{[(mb/a)^2 + n^2]^2}{(mb/a)^2 + Rn^2} \qquad (3.69)$$

For given values of the load ratio R and plate aspect ratio a/b, the values of m and n may be chosen by trial to give the smallest eigenvalue P_{cr}.

For a square plate, for example, $a/b = 1$. Then, for $R = 1$, 0, and -1, respectively, Eq. (3.69) gives, for the m, n, and k_{cc} for minimum k_{cc}, the values

R	m	n	k_{cc}
1	1	1	2
0	1	1	4
-1	2	1	8.33

Negative values of R of course signify tensile loading in the y direction. As would be expected on intuitive grounds, the addition of a tensile load in the transverse direction is seen to have a stabilizing influence.

Results of stability analyses for combined loading frequently are presented in terms of so-called *interaction curves*. Such curves for the present analysis are illustrated in Fig. 3.13 for the case $a/b = 1$. The coordinates in the graph are $\sigma_x/\sigma_{x_{cr}}$ and $\sigma_y/\sigma_{y_{cr}}$, where

$$\sigma_x \equiv \frac{P_x}{bh} \qquad \sigma_y \equiv \frac{P_y}{ah} \qquad (3.70)$$

and $\sigma_{x_{cr}}$ and $\sigma_{y_{cr}}$ are the critical values of σ_x and σ_y, respectively, when each is acting alone. From Eqs. (3.66) and (3.70), $R = \sigma_y/\sigma_x$. The solid lines in the

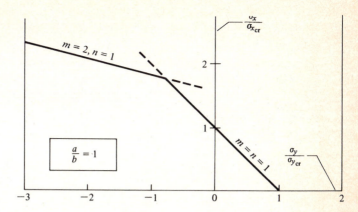

FIGURE 3.13
Interaction curves for plate subjected to in-plane compression in two directions.

interaction plot represent minimum eigenvalues, and the dotted ones represent higher eigenvalues. Negative values of the parameter $\sigma_y/\sigma_{y_{cr}}$ represent tensile loading in the y direction.

Interaction curves for an extensive variety of combinations of compression, bending, and shear loading are given in Ref. 3.10 in graphical form.

3.6   ORTHOTROPIC PLATES

In the development of the plate equations the plate material has been assumed to be isotropic. The equations are readily generalized to apply to orthotropic construction as well. Examples of plate wall construction that may be treated as orthotropic include corrugated sheets, fiber-reinforced plastic plates, and plates with closely spaced stiffeners. For orthotropic construction, only the constitutive equations in the foregoing analysis of isotropic plates need be replaced; the kinematic and equilibrium relationships in Eqs. (3.5) and (3.15) are independent of the constitution of the plate material.

Let us consider a plate reinforced by closely spaced stiffeners that are parallel to the edges of the plate. For simplicity, the stiffeners are assumed to be symmetrical relative to the plate middle plane, as shown in Fig. 3.14; i.e., the influence of stiffener eccentricity is not considered. (A more general analysis of orthotropic construction that takes into account the influence of stiffener eccentricity is included in Chap. 5.)

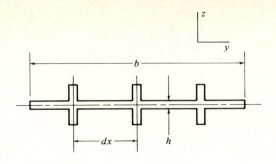

FIGURE 3.14
Integrally stiffened plate.

For symmetrical stiffeners the isotropic constitutive relations in Eqs. (3.8) may be replaced by the more general relations

$$N_x = C_{11}\varepsilon_x + C_{12}\varepsilon_y \qquad M_x = C_{44}\kappa_x + C_{45}\kappa_y$$
$$N_y = C_{12}\varepsilon_x + C_{22}\varepsilon_y \qquad M_y = C_{45}\kappa_x + C_{55}\kappa_y \qquad (3.71)$$
$$N_{xy} = C_{33}\gamma_{xy} \qquad M_{xy} = C_{66}\kappa_{xy}$$

where the C_{ij}'s are constants called *stiffness parameters*. The magnitudes of the constants in Eqs. (3.71) may be determined experimentally, or they may be calculated in terms of the elastic constants E and v of the plate and stiffener materials. The latter procedure is followed in this example. From Fig. 3.14 the extensional stiffness parameters C_{11} and C_{22} for uniformly spaced stiffeners made of the same material as the plate may be seen to be

$$C_{11} = C + \frac{EA_x}{d_x} \qquad C_{22} = C + \frac{EA_y}{d_y}$$

where

C = extensional stiffness of the plate itself, lb/in.

A_x, A_y = cross-sectional areas of stiffeners parallel to the x and y directions, respectively

d_x, d_y = corresponding stiffener spacings.

Because the stiffeners are discrete, there is no Poisson ratio effect between them, and $C_{12} = vC$ as in the isotropic case. The corresponding expressions for the bending and twisting stiffness parameters are seen to be

$$C_{44} = D + \frac{EI_x}{d_x} \qquad C_{55} = D + \frac{EI_y}{d_y}$$

$$C_{45} = vD \qquad C_{66} = (1 - v)D + \frac{1}{2}\left(\frac{GJ_x}{d_x} + \frac{GJ_y}{d_y}\right)$$

$$(3.72)$$

where

I_x = stiffener moment of inertia relative to the plate middle plane
G = shear modulus
J_x = torsional constant
etc.

[An approximate expression for the torsional constant is given in Eq. (2.59).] For a plate stiffened in only, say, the x direction, $A_y = I_y = J_y = 0$.

The constitutive relations in Eqs. (3.71) lead to a set of stability equations analogous to Eqs. (3.34), in which w_1 again is uncoupled from u_1, v_1. Therefore, in the present analysis, we need utilize only the last three of the five equilibrium equations in Eqs. (3.15). As noted in Sec. 3.3a, those three equations combine to give the single equilibrium expression in Eq. (3.16). Then Eqs. (3.16) and (3.71) yield the relationship

$$C_{44}\,\kappa_{x,xx} + C_{45}(\kappa_{x,yy} + \kappa_{y,xx}) + 2C_{66}\,\kappa_{xy,xy} + C_{55}\,\kappa_{y,yy}$$
$$- N_x\,\beta_{x,x} - N_{xy}(\beta_{y,x} + \beta_{x,y}) - N_y\,\beta_{y,y} = -p$$

With the kinematic relations in Eqs. (3.5), this equation may be written in the form

$$C_{44}\,w_{,xxxx} + 2(C_{45} + C_{66})w_{,xxyy} + C_{55}\,w_{,yyyy}$$
$$- (N_x\,w_{,xx} + 2N_{xy}\,w_{,xy} + N_y\,w_{,yy}) = p$$

This expression corresponds to Eq. (3.18c). Linearization in accordance with the adjacent-equilibrium criterion now gives

$$C_{44}\,w_{1,xxxx} + 2(C_{45} + C_{66})w_{1,xxyy} + C_{55}\,w_{1,yyyy}$$
$$- (N_{x0}\,w_{1,xx} + 2N_{xy0}\,w_{1,xy} + N_{y0}\,w_{1,yy}) = 0 \qquad (3.73)$$

Equation (3.73) is the counterpart for the orthotropic plate of Eq. (3.34c) for the isotropic one. For the isotropic case, $A_x = A_y = I_x = I_y = J_x = J_y = 0$, and Eq. (3.73) reduces to Eq. (3.34c).

As an example let us consider an orthotropic plate that is simply supported on all four edges and subjected to a uniformly distributed compressive load $P_x = -N_{x0}b$. Then $N_{xy0} = N_{y0} = 0$, and Eq. (3.73) simplifies to the expression

$$C_{44}\,w_{1,xxxx} + 2(C_{45} + C_{66})w_{1,xxyy} + C_{55}\,w_{1,yyyy} + \frac{P_x}{b}\,w_{1,xx} = 0 \qquad (3.74)$$

Equation (3.74) is a constant-coefficient equation. A solution of the form of Eq. (3.47) again is seen to satisfy the differential equation and boundary conditions. Introduction of that expression and rearrangement yields the equation

$$\frac{P_x}{b} = \left(\frac{\pi a}{m}\right)^2 \left[\left(\frac{m}{a}\right)^4 C_{44} + 2\left(\frac{m}{a}\frac{n}{b}\right)^2 (C_{45} + C_{66}) + \left(\frac{n}{b}\right)^4 C_{55}\right]$$

where m and n are positive integers. Again, the smallest value of P_x is seen to correspond to $n = 1$. Accordingly,

$$\frac{P_x}{b} = \left(\frac{\pi a}{m}\right)^2 \left[\left(\frac{m}{a}\right)^4 C_{44} + 2\left(\frac{m}{ab}\right)^2 (C_{45} + C_{66}) + \left(\frac{1}{b}\right)^4 C_{55}\right] \qquad (3.75)$$

As with isotropic plates, the value of the wavelength parameter m is chosen by trial to give the smallest value of P_x. For the isotropic case, Eq. (3.75) reduces to Eq. (3.49).

As a numerical example consider a short wide aluminum plate stiffened by closely spaced longitudinal stiffeners, and let $E = 10.0 \times 10^6$ psi, $v = 0.3$, the plate length $a = 38.0$ in., the width $b = 60.0$ in., the thickness $h = 0.0283$ in., the stiffener spacing $d_x = 1.00$ in., and the stiffener height and width be 0.300 and 0.097 in., respectively. Then $I_x = (\frac{1}{12})(0.097)(0.300)^3 = 0.000218$ in.^{4}, $J_x = (\frac{1}{3})(0.300)(0.097)^3 = 0.0000913$ in.^{4}, and $D = 20.7$ lb-in. Introduction of these values into Eqs. (3.72) gives $C_{44} = 2,200$, $C_{45} = 6.23$, $C_{55} = 20.7$, and $C_{66} = 190.1$ lb-in. Then, from Eq. (3.75),

$$\frac{P_x}{60.0} = \left(\frac{38.0\pi}{m}\right)^2 \left[\left(\frac{m}{38.0}\right)^4 (2,200) + 2\left(\frac{m}{38.0 \times 60.0}\right)^2 (196.3) + \left(\frac{1}{60.0}\right)^4 (20.7)\right]$$

In this example the smallest value of P_x corresponds to $m = 1$. For $m = 1$ the critical load turns out to be $P_{cr} = 971$ lb.

The "average" plate thickness in this example is $h_{avg} = 0.0283 + (0.300) \times (0.097)/(1.00) = 0.0574$ in. Introduction of this value into the equation for isotropic plates in Eq. (3.49) leads to a critical load of only 139 lb. Thus this stiffened plate is seen to represent a much more efficient use of material.

Extensive numerical results for orthotropic plates subjected to in-plane compression and shear loading are given in graphical form in Ref. 3.12. Also included are results for plates stiffened by one, two, or three discrete stiffeners, and a comprehensive list of references.

3.7   SANDWICH PLATES

Sandwich plates consist of two thin load-bearing sheets separated by a lightweight low-stiffness core, as illustrated in Fig. 3.15. The face sheets usually are made of metal or of fiber-reinforced plastic. Common core materials include foamed plastic, lightweight metallic honeycomb, and lightweight corrugated sheet. The core serves to increase the overall bending resistance of the composite cross section by keeping the load-bearing areas at a relatively large distance from

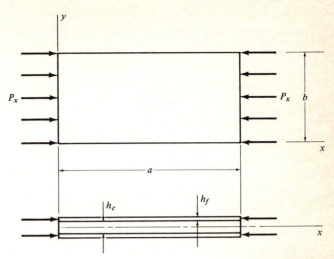

FIGURE 3.15
Sandwich plate.

the middle plane of the sandwich plate. Thus sandwich construction is the analog for plates of wide-flange sections for beams.

The present analysis of sandwich plates follows in some respects the analysis of orthotropic sandwich plates in Ref. 3.13. Only isotropic construction is treated here, for simplicity.

One of the fundamental approximations of the thin-plate theory in Sec. 3.2 is that lines normal to the undeformed middle surface of the plate remain straight and normal during deformation. Because of the low stiffness of the lightweight core material, that approximation may not be retained in sandwich-plate analysis; transverse shearing strains are too large to be neglected. The necessary modification may be introduced into the analysis as follows. In sandwich construction, rotations and curvatures are caused not only by the moment intensities M_x, M_y, M_{xy}, but also by the transverse shearing force intensities Q_x, Q_y. For isotropic sandwiches the rotations β_{xQ}, β_{yQ} due to Q_x and Q_y, respectively, are (see Fig. 3.4)

$$\beta_{xQ} = -\frac{Q_x}{C_Q} \qquad \beta_{yQ} = -\frac{Q_y}{C_Q}$$

where C_Q, in pounds per inch, is a transverse shearing stiffness parameter analogous to the extensional stiffness parameter C. The magnitude of C_Q may be determined experimentally, or it may be calculated in terms of the elastic constants E and v of the face and core materials (as in Ref. 3.14, for example). The associated curvatures κ_{xQ}, κ_{yQ} and average twist κ_{xyQ} corresponding to β_{xQ}, β_{yQ} are

$$\kappa_{xQ} = -\frac{Q_{x,x}}{C_Q} \qquad \kappa_{yQ} = -\frac{Q_{y,y}}{C_Q} \qquad \kappa_{xyQ} = -\frac{Q_{x,y} + Q_{y,x}}{2C_Q}$$

From Eqs. (3.5) the total curvatures and twist are given by the relations

$$\kappa_x = -w_{,xx} \qquad \kappa_y = -w_{,yy} \qquad \kappa_{xy} = -w_{,xy}$$

Then the corresponding curvatures and twist $\kappa_{xM}, \kappa_{yM}, \kappa_{xyM}$ due to the moments alone are obtained by subtracting from the total curvatures and twist the portions due to shear, as follows:

$$\kappa_{xM} = -\left(w_{,xx} - \frac{Q_{x,x}}{C_Q}\right) \qquad \kappa_{yM} = -\left(w_{,yy} - \frac{Q_{y,y}}{C_Q}\right)$$

$$\kappa_{xy} = -\left(w_{,xy} - \frac{Q_{x,y} + Q_{y,x}}{2C_Q}\right)$$

Consequently, for sandwich construction, the constitutive relations for moments in Eqs. (3.8) may be replaced by the following expressions:

$$M_x = -D\left[\left(w_{,xx} - \frac{Q_{x,x}}{C_Q}\right) + v\left(w_{,yy} - \frac{Q_{y,y}}{C_Q}\right)\right]$$

$$M_y = -D\left[\left(w_{,yy} - \frac{Q_{y,y}}{C_Q}\right) + v\left(w_{,xx} - \frac{Q_{x,x}}{C_Q}\right)\right] \qquad (3.76)$$

$$M_{xy} = -(1 - v)D\left[w_{,xy} - \frac{1}{2C_Q}(Q_{x,y} + Q_{y,x})\right]$$

For $C_Q \to \infty$ these equations reduce to the earlier expressions for homogeneous plates. Corresponding constitutive relations for orthotropic sandwich plates are given in Ref. 3.13, eqs. (12). For the isotropic case, the present equations and those in Ref. 3.13 are the same.

The kinematic and equilibrium relations in Eqs. (3.5) and (3.15) remain applicable for sandwich construction. Those equations plus the constitutive relations in Eqs. (3.76) lead to a stability equation analogous to Eqs. (3.34c), in which the displacement component w_1 is uncoupled from the components u_1, v_1. Accordingly, it is necessary to utilize only the last three of the equilibrium equations in Eqs. (3.15) for the present analysis. Linearization of those equations and introduction of the kinematic relations $\beta_{x1} = -w_{1,x}$ and $\beta_{y1} = -w_{1,y}$ gives, for $w_0 \equiv \beta_{x0} \equiv \beta_{y0} \equiv 0$,

$$Q_{x1,x} + Q_{y1,y} + N_{x0}w_{1,xx} + 2N_{xy0}w_{1,xy} + N_{y0}w_{1,yy} = 0 \qquad (3.77a)$$

$$Q_{y1} = M_{y1,y} + M_{xy1,x} \qquad (3.77b)$$

$$Q_{x1} = M_{x1,x} + M_{xy1,y} \qquad (3.77c)$$

Equations (3.76) and (3.77) form a set of six linear equations in the six variables M_{x1}, M_{xy1}, M_{y1}, Q_{x1}, Q_{y1}, w_1. Five of the variables may be eliminated as follows. Introduction of Eqs. (3.77b) and (3.77c) into Eq. (3.77a) gives

$$M_{x1,xx} + 2M_{xy1,xy} + M_{y1,yy} + N_{x0}w_{1,xx} + 2N_{xy0}w_{1,xy} + N_{y0}w_{1,yy} = 0$$

Subsequent introduction of Eqs. (3.76) and rearrangement gives

$$D\nabla^4 w_1 - \frac{D}{C_Q}\nabla^2(Q_{x1,x} + Q_{y1,y}) - (N_{x0}w_{1,xx} + 2N_{xy0}w_{1,xy} + N_{y0}w_{1,yy}) = 0$$

Then, with Eq. (3.77a), this equation may be written in the final form

$$D\nabla^4 w_1 - \left(1 - \frac{D}{C_Q}\nabla^2\right)(N_{x0}w_{1,xx} + 2N_{xy0}w_{1,xy} + N_{y0}w_{1,yy}) = 0 \qquad (3.78)$$

Equation (3.78) is seen to be a homogeneous equation in the single variable w_1. For $C_Q \to \infty$ the equation reduces to the corresponding expression for homogeneous plates in Eq. (3.34c).

As an example we consider a sandwich plate whose faces are subjected to a uniformly distributed load $P_x = -N_{x0}b$, as shown in Fig. 3.15. Then Eq. (3.78) specializes to the form

$$D\nabla^4 w_1 + \frac{P_x}{b}\left(1 - \frac{D}{C_Q}\nabla^2\right)w_{1,xx} = 0 \qquad (3.79)$$

Equation (3.79) is a constant-coefficient equation. Let us suppose that the plate is simply supported on all four edges. Then the boundary-condition equations are given by Eqs. (3.46), and the solution is of the form of Eq. (3.47). Introduction of that expression for w_1 and rearrangement gives

$$\frac{P_x}{b} = \frac{\pi a}{m}\frac{D[(m/a)^2 + (n/b)^2]^2}{1 + \pi^2(D/C_Q)[(m/a)^2 + (n/b)^2]}$$

where the m, n are positive integers. Again, the minimum P_x is seen to correspond to $n = 1$. Introduction of that value and rearrangement gives

$$P_x = \frac{\pi^2 D}{b}\frac{(mb/a + a/mb)^2}{1 + r[1 + (mb/a)^2]} \qquad (3.80)$$

where r is a nondimensional shear stiffness parameter defined by the relation

$$r \equiv \frac{\pi^2 D}{b^2 C_Q} \qquad (3.81)$$

Equation (3.80) may be written in the alternative form

$$P_x = k \frac{\pi^2 D}{b} \qquad (3.82)$$

where the nondimensional buckling coefficient k is defined by the expression

$$k = \frac{(mb/a + a/mb)^2}{1 + r[1 + (mb/a)^2]} \qquad (3.83)$$

For given values of the plate aspect ratio a/b and the shear stiffness parameter r, the value of the wavelength parameter m may be chosen by trial to give the smallest value of P_x. Equation (3.83) is the same as eq. (1) in Ref. 3.15. For $r \to \infty$, this equation reduces to the corresponding expression for homogeneous plates in Eq. (3.51).

In sandwich construction it is commonly assumed that the bending stiffness of the core and the bending stiffnesses of the individual faces relative to their own centroids may be neglected. Then, for sandwich-plate cross sections, the moment of inertia per unit width is seen to be $2h_f[(h_c + h_f)/2]^2$, where h_f and h_c denote face and core thicknesses, respectively (Fig. 3.15). Then the bending stiffness parameter D for the sandwich plate becomes

$$D = \frac{E_f h_f (h_c + h_f)^2}{2}$$

where E_f is Young's modulus for the face material; the influence of anticlastic bending is neglected. For a homogeneous core, furthermore, $C_Q = G_c h_c$, where G_c is the shearing modulus of the core material.

As a numerical example, let $a = 20$ in., $b = 10$ in., $h_f = 0.020$ in., $h_c = 0.200$ in., $E_f = 10 \times 10^6$ psi, and $G_c = 12,000$ psi. Then $D = 4,840$ lb-in., $C_Q = 2,400$ lb/in., and $r = 0.199$. Introduction into Eq. (3.83) reveals that the smallest value of the coefficient k corresponds to $m = 3$. For this value, Eq. (3.82) gives $P_{cr} = 13,620$ lb.

If the core were omitted and the two face sheets bonded together to form a thin homogeneous plate 0.040 in. thick, the corresponding critical load would be only 231 lb.

Numerical results for other values of the parameters are given in Ref. 3.15 in graphical form.

3.8   FAILURE OF PLATES

The equilibrium paths in Fig. 3.6 for an initially perfect plate subjected to in-plane compression are shown again in Fig. 3.16. Also included in Fig. 3.16 are corresponding curves for a slightly imperfect plate (Ref. 3.8). Two important results are immediately apparent from the diagrams: (1) Buckling of real (and therefore imperfect) plates is so gradual that it is difficult to decide at precisely what load the buckling may be said to occur. Therefore comparisons of theoretical and experimental values for plate critical loads entail an element of arbitrariness. (2) In any case, the plate continues to accept additional load after buckling. Therefore the load P_{cr} for the plate, unlike that for the column, does not represent its ultimate strength. These conclusions, based on Fig. 3.16 for in-plane compression, apply for other kinds of in-plane loading as well.

The fact that plates with supported edges can accept additional load after buckling was discovered in the late 1920s through experimental studies made in connection with the structural design of airplanes. In 1929, Wagner (Ref. 3.16) established a criterion for the postbuckling strength of a shear web, i.e., a thin plate supported on all four edges and subjected to in-plane shear loading (compare Sec. 3.5c). The prebuckling state of stress in a shear web consists of compressive stresses on diagonal lines oriented at 45° to the plate edges, and tensile stresses at right angles to the compressive stresses. The compressive stresses cause buckles to form in the web along diagonal lines. Before buckling the tensile and compressive stresses are equal in magnitude, but after buckling the

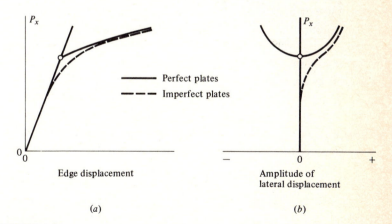

(a)   (b)

FIGURE 3.16
Equilibrium paths for initially perfect and imperfect plates subjected to in-plane compression.

tensile stresses are larger. The imbalance in forces is carried by the edge supports. Wagner assumed in his approximate analysis that for a thin web the compressive stresses could be neglected entirely in the buckled configuration. Such models are sometimes called *diagonal-tension beams*. A more refined method of analysis for shear webs and extensive test results were reported by Kuhn (Ref. 3.17).

For efficient design the postbuckling strength of plates must be taken into account. As an example of the procedure, let us consider a case that is simpler than the shear web, namely, a plate subjected to a uniformly distributed in-plane compressive load P_x, as shown in Fig. 3.17a. The applied load P_x is related to the stress σ_x by the equation

$$P_x = h \int_0^b \sigma_x \, dy$$

where h and b are plate thickness and width, respectively. For $P_x \leq P_{cr}$ the stress is uniform across the plate width, as illustrated by the lines marked 1-1 and 2-2 in Fig. 3.17b. Then $P_x = hb\sigma_x$. In particular, for $P_x = P_{cr}$,

$$P_{cr} = hb\sigma_{cr} \qquad (3.84)$$

For $P_x > P_{cr}$, on the other hand, the stress near the plate edges $y = 0, b$ is larger than that near the center, because of the stabilizing influence of the edge supports. In such cases the stress distribution is nonuniform, as illustrated by the curves 3-3 and 4-4 in Fig. 3.17b.

For configurations within the scope of the intermediate class of deformations, the postbuckling distribution of stress can be determined from the nonlinear equilibrium equations in Eqs. (3.18). Such an analysis is included in Chap. 9. For design purposes it is convenient to express the results of the analysis in terms of an *effective width* over which the stress is considered to be uniform, as illustrated in Fig. 3.17c (Ref. 3.8). From the figure,

$$P_x = hb_{eff} \, \sigma_{max} \qquad (3.85)$$

where σ_{max} is the maximum stress at the plate edges $y = 0, b$. A widely used approximate expression for b_{eff} is [Ref. 3.8, eq. (7)]

$$b_{eff} = b\left(\frac{\sigma_{cr}}{\sigma_{max}}\right)^{1/2} \qquad (3.86)$$

where σ_{cr} is the classical critical stress for the given boundary conditions. Equation (3.86) is called the *von Kármán effective-width formula*. Somewhat more accurate expressions are available (Ref. 3.8) but are less widely used in design.

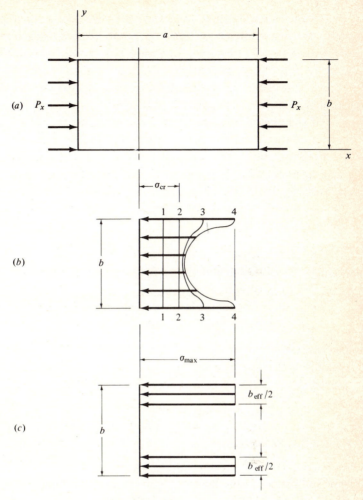

FIGURE 3.17
Stress distribution in plate before and after buckling.

For $\sigma_{max} = \sigma_{cr}$, $2\sigma_{cr}$, and $4\sigma_{cr}$, respectively, Eqs. (3.84) to (3.86) give $P_x = P_{cr}$, $1.4P_{cr}$, and $2.0P_{cr}$.

The maximum allowable stress σ_{max} at the plate edges is governed by the yield strength of the plate material or by the buckling strength of the members supporting the plate at its edges. The value of σ_{cr} in Eq. (3.86) of course is given by Eq. (3.84), where P_{cr} is determined from Eq. (3.50) or (3.62), depending on the plate boundary conditions.

Similar design procedures have been developed for plates subjected to in-plane bending or shear loading. A comprehensive discussion of the ultimate strength of plates in bending, in shear, and in combined bending and shear is given in Ref. 3.18, chap. 5.

Occasionally, the initial buckling load P_{cr} itself is used as a conservative estimate of the ultimate strength of the plate.

If yielding of the plate material occurs before the plate buckles, the critical load of course is smaller than the value given by an elastic stability analysis. Because the state of stress is biaxial, inelastic stability analysis is more complicated for plates than for columns. A discussion of the influence of plasticity on the buckling and postbuckling behavior of plates is given in Ref. 3.19.

Results of plastic stability analyses of plates often are expressed in the form

$$\bar{\sigma}_{cr} = \eta \sigma_{cr}$$

where $\bar{\sigma}_{cr}$ = plastic buckling stress for a particular system of loads and boundary conditions,

σ_{cr} = corresponding elastic critical stress

η = a *plasticity reduction factor*.

As a conservative approximation, the value $\eta = (E_t/E)^{1/2}$, where E_t is the tangent modulus, sometimes is used. In structural applications plastic buckling is less common for plates and shells than it is for columns.

REFERENCES

3.1 BRYAN, G. H.: On the Stability of a Plane Plate under Thrusts in Its Own Plane, with Applications to the Buckling of the Sides of a Ship, *Proc. London Math. Soc.*, vol. 22, pp. 54–67, 1891.

3.2 BLEICH, F.: "Buckling Strength of Metal Structures," McGraw-Hill, New York, 1952.

3.3 BULSON, P. S.: "The Stability of Flat Plates," American Elsevier, New York, 1969.

3.4 TIMOSHENKO, S. P., and S. WOINOWSKY-KRIEGER: "Theory of Plates and Shells," 2d ed., McGraw-Hill, New York, 1959.

3.5 NOVOZHILOV, V. V.: "Foundations of the Nonlinear Theory of Elasticity," Greylock Press, Rochester, N.Y., 1953.

3.6 WANG, C.-T.: "Applied Elasticity," McGraw-Hill, New York, 1953.

3.7 STEIN, M.: Loads and Deformations of Buckled Rectangular Plates, *NASA Tech. Rep.* R-40, 1959.

3.8 van der neut, a.: Postbuckling Behavior of Structures, *NATO AGARD Rep.* 60, 1956.

3.9 timoshenko, s. p., and j. m. gere: "Theory of Elastic Stability," 2d ed., McGraw-Hill, New York, 1961.

3.10 gerard, g., and h. becker: Handbook of Structural Stability, Part I, Buckling of Flat Plates, *NACA TN* 3781, 1957.

3.11 southwell, r. v., and s. w. skan: On the Stability under Shearing Forces of a Flat Elastic Strip, *Proc. Roy. Soc. London*, ser. A, vol. 105, pp. 582–607, 1924.

3.12 becker, h.: Handbook of Structural Stability, Part II, Buckling of Composite Elements, *NACA TN* 3782, 1957.

3.13 libove, c., and s. b. batdorf: A General Small-Deflection Theory for Flat Sandwich Plates, *NACA Rep.* 899, 1948.

3.14 libove, c., and r. e. hubka: Elastic Constants for Corrugated Core Sandwich Plates, *NACA TN* 2289, 1951.

3.15 seide, p., and e. z. stowell: Elastic and Plastic Buckling of Simply Supported Solid-Core Sandwich Plates in Compression, *NACA Rep.* 967, 1950.

3.16 wagner, h.: Ebene Blechwandträger mit sehr dünnem Stegblech, *Z. Flugtech. Motorluftschiffahrt*, vol. 20, pp. 200, 227, 256, 279, and 306, 1929.

3.17 kuhn, p.: Investigations on the Incompletely Developed Plane Diagonal Tension Field, *NACA Rep.* 697, 1940.

3.18 b. g. johnston (ed.): "Guide to Design Criteria for Metal Compression Members," 2d ed., Column Research Council, Wiley, New York, 1966.

3.19 hutchinson, j. w.: Plastic Buckling, to appear in "Advances in Applied Mechanics," vol. 14, edited by C. S. Yih, Academic Press, New York, 1974.

PROBLEMS

3.1 Show that the nonlinear equilibrium equations in Eqs. (3.18) can be obtained from those in Eqs. (3.15) by appropriate substitutions and rearrangement.

3.2 By introduction of the appropriate constitutive and kinematic relations, derive the nonlinear equilibrium equations in terms of u, v, w in Eqs. (3.26) from those in terms of N_x, N_{xy}, N_y, and w in Eqs. (3.18).

3.3 Show that if the rotation terms $w_{,x}$ and $w_{,y}$ are omitted from the expressions for ε_x, ε_y, and γ_{xy} in the kinematic relations, introduction of the integrand in Eq. (3.25) into the Euler equations of the calculus of variations leads to the *linear* equilibrium equations in Eqs. (3.30).

3.4 Show that application of the adjacent-equilibrium criterion to the nonlinear equilibrium equations in Eqs. (3.18) yields the stability equations in Eqs. (3.34).

3.5 Derive the second-variation expressions in Eqs. (3.38) and (3.39) from the potential energy expression in Eqs. (3.19) to (3.22).

3.6 Show that introduction of the integrand of the second-variation expression in Eq. (3.42) into the Euler equations (3.43) yields the stability equations in Eqs. (3.44). Then show by introduction of the kinematic and constitutive relations that the latter equations are the same as those in Eqs. (3.34).

3.7 A general analysis of a plate subjected to an in-plane compressive load P_x and simply supported on the loaded edges $x = 0$, a is given in Sec. 3.5b. Show by substitution of the expression in Eq. (3.58) into the four boundary condition equations on $y = 0$, b that the solution $w(x,y)$ for a plate that is simply supported on the unloaded edges as well is of the form of Eq. (3.47).

3.8 A 0.020-in.-thick aluminum plate of length $a = 24$ in. and width $b = 18$ in. is simply supported on all four edges and subjected to a uniform compressive load P_x. Determine the critical value of P_x for elastic buckling. Use $E = 10 \times 10^6$ psi and $v = 0.3$. Compare results based on Eq. (3.49) with those based on Fig. 3.8.

3.9 A thin flat plate is subjected to a uniform compressive load P_x in the longitudinal direction. The plate is simply supported on the loaded edges and clamped on the unloaded edges. Show that the characteristic equation for the critical load is

$$2\alpha\beta + (\alpha^2 - \beta^2)\sinh \alpha b \sin \beta b - 2\alpha\beta(\cosh \alpha b \cos \beta b) = 0$$

where α, β are as defined in Eqs. (3.57).

3.10 Derive the characteristic equation for the plate in Prob. 3.9 if only one unloaded edge is clamped and the other one is simply supported.

3.11 Derive the characteristic equation for a plate subjected to a compressive load P_x if the plate is clamped on the loaded edges $x = 0$, a and simply supported on the unloaded edges $y = 0$, b.

3.12 An aluminum plate is subjected to a uniform compressive load P_x. The length and thickness of the plate are $a = 10.0$ in. and $h = 0.040$ in., respectively. Using Fig. 3.9, determine the critical load, in pounds, and stress, in pounds per square inch, for the following plate widths and boundary conditions: (*a*) width $b = 5.00$ in., simply supported on four edges; (*b*) width $b = 5.00$ in., simply supported on the loaded edges and one unloaded edge, free on the remaining edge; and (*c*) width $b = 1.00$ in., simply supported on the loaded edges and one unloaded edge, free on the remaining edge. Let $E = 10 \times 10^6$ psi and $v = 0.3$.

Ans. (*a*) $P_{cr} = 463$ lb, (*b*) $P_{cr} = 81.0$ lb, (*c*) $P_{cr} = 261$ lb

3.13 For a plate subjected to a compressive load P_x, Eq. (3.62) gives for the average stress $\sigma_x = P_x/bh$ the relationship

$$\sigma_x = k_c \frac{\pi^2 E}{12(1 - v^2)(b/h)^2}$$

For an infinitely long plate that is simply supported on one unloaded edge and free on the other, $k_c = 0.425$ for $v = 0.3$ (Ref. 3.2, p. 330). Then for $E = 29 \times 10^3$ ksi,

$$\sigma_{cr} = 11{,}100\left(\frac{h}{b}\right)^2$$

Using this expression determine the elastic critical stress for independent buckling of one of the 2.5-in.-wide legs of the channel section in Fig. 2.29(b) if the web is assumed to furnish only simple support to the legs, and if general instability of the entire cross section in a torsional-flexural mode is prevented.   *Ans.* $\sigma_{cr} = 63.$ ksi.

3.14 (*a*) Show that if an infinitely long steel plate that is simply supported on one unloaded edge and free on the other (cf. Prob. 3.13) is to reach its full proportional limit stress σ_p (ksi) before buckling occurs, the b/h ratio must satisfy the criterion

$$\frac{b}{h} \leq \frac{105}{\sqrt{\sigma_p}}$$

(*b*) Show that if the cross section in Fig. 2.30 is made of steel with $\sigma_p = 30$ ksi and the web is assumed to furnish simple support to the compression flange, the b/h ratio for the flange satisfies this criterion.

3.15 A 0.050-in.-thick aluminum plate of length $a = 32$ in. and width $b = 24$ in. is simply supported on all four edges. It is subjected to uniform compressive loads in both directions. If $P_y = 0.6P_x$, determine the critical value of P_x for elastic buckling. Let $E = 10 \times 10^6$ psi and $\nu = 0.3$.

3.16 A stiffened aluminum plate of length $a = 24$ in. and width $b = 18$ in. is simply supported on all four edges and subjected to a uniform compressive load P_x. The plate thickness is 0.050 in., and there are closely spaced longitudinal stiffeners at intervals of 3 in. across the plate width. The stiffeners are 0.080 in. wide and 0.40 in. high ($A_s = 0.032$ in.^{2}) and are symmetrically located relative to the plate middle plane. The values of Young's modulus and Poisson's ratio for both plate and stiffeners are $E = 10 \times 10^6$ psi and $\nu = 0.3$. Determine the critical load for elastic buckling.

3.17 If the stiffened plate in Prob. 3.16 is simultaneously subjected to the compressive load P_x and a load $P_y = 0.2P_x$, determine the critical load for elastic buckling. Use Eq. (3.47).

3.18 A short wide sandwich plate of length $a = 38.0$ in. and width $b = 60.0$ in. is simply supported on all four edges and subjected to a uniform compressive load P_x. The face sheets are made of fiber-reinforced plastic for which $E = 2.0 \times 10^6$ psi and $\nu = 0.3$. The core is a homogeneous material for which the transverse shear stiffness parameter $C_Q = 1,800$ lb/in. The face and core thicknesses are $h_f = 0.030$ in. and $h_c = 0.400$ in., respectively. Determine the critical load for elastic buckling.

3.19 Rederive Eq. (3.49) for the critical load of a simply supported plate by substituting the displacement function in Eq. (3.47) directly into the second-variation expression in Eqs. (3.41) and (3.42). (This procedure is known as a *direct* energy method because it bypasses the differential equation. See the discussion of the Rayleigh-Ritz procedure in Chap. 8.) Note that u_1 and v_1 are not coupled with w_1 in the Euler equations for this second-variation expression [Eqs. (3.44)]. Consequently, Eq. (3.49) for the critical load is the solution for all functions u_1, v_1 that satisfy the first two of Eqs. (3.44), including $u_1 \equiv v_1 \equiv 0$. Accordingly, use $u_1 \equiv v_1 \equiv 0$ in this exercise.

4

CIRCULAR RINGS

4.1 INTRODUCTION

This chapter on circular rings is placed after the one on flat plates because the contents are intended to serve primarily as an introduction to the analysis of shell stability. Many variations of the equilibrium and stability equations for circular rings are given in the literature. The particular forms derived here are those to which the cylindrical-shell equations in the following chapter reduce on appropriate specialization.

The relative simplicity of the ring equations makes them especially well suited for an introductory examination of the widely used Donnell approximations. Those approximations are utilized in the analysis of circular cylindrical shells in Chap. 5 and, in a generalized form, for the shell of arbitrary shape in Chap. 6. Also introduced in anticipation of the analyses in the chapters on shells is the distinction between fluid-pressure loading, which remains normal to the structure's surface during deformation, and so-called dead loads, which remain constant in magnitude and direction as the structure deforms.

For the straight bars and flat plates considered in earlier chapters, the lateral displacement $w \equiv 0$ for the unbuckled form. The ring serves as our first example in which the ideal structure undergoes a lateral displacement prior to loss of stability.

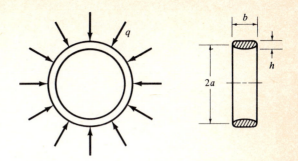

FIGURE 4.1
Thin circular ring subjected to uniform external pressure.

4.2 THIN-RING DEFORMATION GEOMETRY

Before beginning an examination of the equilibrium and stability equations for circular rings, let us determine kinematic relations for a thin ring such as shown in Fig. 4.1. For simplicity, the ring cross section is assumed to be symmetrical, and only in-plane bending is considered. The constant a represents the radius of the undeformed centroidal surface, and the maximum thickness h is taken to be much smaller than a. Points in the ring are referred to polar coordinates r and θ, as shown in a sketch of a portion of the undeformed ring in Fig. 4.2. For convenience, an additional coordinate variable is defined by the relation $z \equiv r - a$. Thus z is measured positive outward from the centroidal surface.

Consider a circumferential line element of length dS referred to rectangular cartesian coordinates x, y, as shown in Fig. 4.3. After deformation the length of the line element is dS^*, and the element is referred to new coordinates x^*, y^*. Let $\bar{v}$, $\bar{w}$ denote components of the displacement vector in the θ and z directions, respectively. Then, from Fig. 4.3,

$$
\begin{aligned}
x &= r \cos \theta \\
y &= r \sin \theta \\
x^* &= r \cos \theta - \bar{v} \sin \theta + \bar{w} \cos \theta \\
y^* &= r \sin \theta + \bar{v} \cos \theta + \bar{w} \sin \theta
\end{aligned}
\tag{4.1}
$$

and

$$
\frac{dx^*}{d\theta} = -r \sin \theta - \bar{v}' \sin \theta - \bar{v} \cos \theta + \bar{w}' \cos \theta - \bar{w} \sin \theta
$$

$$
\frac{dy^*}{d\theta} = r \cos \theta + \bar{v}' \cos \theta - \bar{v} \sin \theta + \bar{w}' \sin \theta + \bar{w} \cos \theta
$$

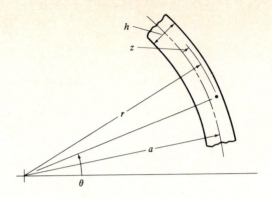

FIGURE 4.2
Coordinate system.

where $\bar{v}' \equiv d\bar{v}/d\theta$, etc. In terms of polar coordinates,

$$(dS)^2 = (r\,d\theta)^2$$

and

$$(dS^*)^2 = (dx^*)^2 + (dy^*)^2$$
$$= [r^2 + 2r(\bar{v}' + \bar{w}) + (\bar{v}' + \bar{w})^2 + (\bar{v} - \bar{w}')^2]\,(d\theta)^2$$

Therefore

$$\left(\frac{dS^*}{dS}\right)^2 - 1 = 2\left[\frac{\bar{v}' + \bar{w}}{r} + \frac{1}{2}\left(\frac{\bar{v}' + \bar{w}}{r}\right)^2 + \frac{1}{2}\left(\frac{\bar{v} - \bar{w}'}{r}\right)^2\right] \qquad (4.2)$$

Equation (4.2) is exact.

Now let the extensional strain of the circumferential line element be denoted by $\bar{\varepsilon}$ and defined by the relation

$$\bar{\varepsilon} = \frac{dS^* - dS}{dS}$$

Rearrangement and squaring gives

$$\bar{\varepsilon} + \tfrac{1}{2}\bar{\varepsilon}^2 = \frac{1}{2}\left[\left(\frac{dS^*}{dS}\right)^2 - 1\right]$$

For $\bar{\varepsilon}$ small compared with unity, the equation may be replaced by the approximate relation

$$\bar{\varepsilon} = \frac{1}{2}\left[\left(\frac{dS^*}{dS}\right)^2 - 1\right]$$

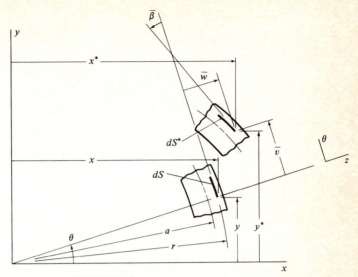

FIGURE 4.3
Circumferential line element before and after deformation.

Introduction of Eq. (4.2) gives, for the extensional strain, the approximate expression

$$\bar{\varepsilon} = \frac{\bar{v}' + \bar{w}}{r} + \frac{1}{2}\left(\frac{\bar{v}' + \bar{w}}{r}\right)^2 + \frac{1}{2}\left(\frac{\bar{v} - \bar{w}'}{r}\right)^2 \qquad (4.3)$$

Equation (4.3) is valid for arbitrarily large rotations $\bar{\beta}$.

For the intermediate class of deformations, the angles of rotation $\bar{\beta}$ are small. From Fig. 4.3, small rotations of circumferential line elements are seen to be composed of two parts, a clockwise component due to the rotation $d\bar{w}/dS = d\bar{w}/(r\,d\theta)$ and a counterclockwise one due to the circumferential displacement of the element and given approximately by $\bar{v}/r$. Combination of the two parts gives

$$\bar{\beta} = \frac{\bar{v} - \bar{w}'}{r} \qquad (4.4)$$

The rotation $\bar{\beta}$ is positive in the counterclockwise sense.

For both $\bar{\varepsilon}$ and $\bar{\beta}$ small, the square of the $(\bar{v}' + \bar{w})/r$ term may be neglected in Eq. (4.3). (We exclude from consideration cases in which $\bar{v}'$ or $\bar{w}$ is of the order of magnitude of the radius r or larger.) Then that equation simplifies to the form

$$\bar{\varepsilon} = \frac{\bar{v}' + \bar{w}}{r} + \frac{1}{2}\left(\frac{\bar{v} - \bar{w}'}{r}\right)^2 \qquad (4.5)$$

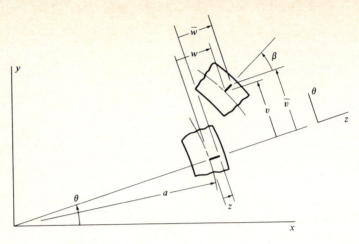

FIGURE 4.4
Normal to centroidal surface before and after deformation.

Equation (4.5) is the strain-displacement relation for the intermediate class of deformations in terms of polar coordinates. It corresponds to Eq. (1.7) for rectangular coordinates.

Ring-bending theory is based on the simplifying approximation that normals to the undeformed centroidal surface remain straight, normal, and in-extensional during the deformation. Therefore, from Fig. 4.4, displacement components $\bar{v}$, $\bar{w}$ may be expressed in terms of the corresponding displacements v, w of a point on the centroidal surface by the relations

$$\bar{v} = v + z\beta$$
$$\bar{w} = w \qquad (4.6)$$

where β represents rotation at a point on the centroidal surface and is a small quantity. From Eq. (4.4),

$$\beta = \frac{v - w'}{a} \qquad (4.7)$$

Introduction into Eq. (4.5), rearrangement, and observation that $r \approx a$ for a thin ring gives

$$\bar{\varepsilon} = \frac{v' + w}{a} + \tfrac{1}{2}\beta^2 + z\frac{\beta'}{a}$$

[It may be seen from Eqs. (4.4), (4.6), and (4.7) that $\beta = \bar{\beta}$.] The extensional strain ε of a circumferential line element on the centroidal surface is obtained by setting $z = 0$ in the expression for $\bar{\varepsilon}$. The curvature change κ of a centroidal line element may be defined as the rate of change of the rotation β in the circumferential direction. Therefore,

$$\bar{\varepsilon} = \varepsilon + z\kappa \qquad (4.8)$$

where

$$\varepsilon = \frac{v' + w}{a} + \tfrac{1}{2}\beta^2 \qquad (4.9)$$

and

$$\kappa = \frac{\beta'}{a} \qquad (4.10)$$

Equations (4.7), (4.9), and (4.10) are centroidal-surface kinematic relations for the thin circular ring. They are the expressions given by specialization for polar coordinates of the Sanders kinematic relations in Eqs. (6.6) and (6.7).

4.3 NONLINEAR EQUILIBRIUM EQUATIONS FOR FLUID-PRESSURE LOADING

Let us consider a ring subjected to external fluid pressure. Following the usual notation, we let p, in pounds per square inch, denote load per unit area and q, in pounds per inch, denote load per unit circumferential length. Then, for a ring of width b (Fig. 4.1), $|q| = |bp|$. Equations for equilibrium of a ring element in a slightly deformed configuration may be derived from the principle of stationary potential energy, as follows. The total potential energy is the sum of the strain energy U of the ring itself and the potential energy Ω of the applied pressure. The strain energy of the ring may be written

$$U = \frac{E}{2} \iint \bar{\varepsilon}^2 \, dA \, (r \, d\theta)$$

where dA is an element of cross-sectional area. Introduction of Eq. (4.8) and observation that $r \approx a$ for a thin ring gives

$$U = \frac{Ea}{2} \iint (\varepsilon + z\kappa)^2 \, dA \, d\theta$$

The variables ε and κ are functions of θ alone, and the coordinate z is measured from the centroidal axis. Therefore integration gives, for the strain energy, the expression

$$U = \frac{EAa}{2} \int \varepsilon^2 \, d\theta + \frac{EIa}{2} \int \kappa^2 \, d\theta \qquad (4.11)$$

The two terms are the membrane and bending strain energy, respectively, for the ring.

A ring subjected to uniform external pressure is a conservative system (Ref. 4.1). For a conservative system the change in potential energy of the applied loads as the structure deforms is the negative of the work of the loads during the deformation. For fluid-pressure loading the pressure at each point on the ring surface remains normal to the surface as the ring deforms. Thus the potential energy of the applied pressure may be expressed in terms of the product of the pressure q times the change in the area enclosed by the outer surface of the ring. Since the ring thickness h does not change appreciably during the deformation, the change in the area enclosed by the outer surface is approximately equal to the change in the area enclosed by the centroidal surface. Consequently, the expression for the potential energy Ω is, approximately,

$$\Omega = -q(\pi a^2 - A^*) \qquad (4.12)$$

where A^* is the area enclosed by the centroidal surface of the ring after deformation, and q is positive inward.

From Eqs. (4.1) the coordinates of a point on the centroidal surface after deformation are

$$x^* = (a + w) \cos \theta - v \sin \theta$$
$$y^* = (a + w) \sin \theta + v \cos \theta \qquad (4.13)$$

The area enclosed by the centroidal surface is conveniently expressed in terms of the line integral around the periphery, as follows (see, for example, Ref. 4.2, p. 242):

$$A^* = \frac{1}{2} \oint_C (-y^* \, dx^* + x^* \, dy^*)$$

Therefore

$$A^* = \frac{1}{2} \int_0^{2\pi} \left(-y^* \frac{dx^*}{d\theta} + x^* \frac{dy^*}{d\theta}\right) d\theta$$

Introduction of Eqs. (4.13) and rearrangement gives

$$A^* = \frac{1}{2} \int_0^{2\pi} (a^2 + av' + 2aw + v^2 - vw' + v'w + w^2) \, d\theta$$

But

$$\frac{1}{2} \int_0^{2\pi} a^2 \, d\theta = \pi a^2$$

and because of periodicity,

$$\int_0^{2\pi} v' \, d\theta = 0$$

Therefore

$$A^* = \pi a^2 + \frac{1}{2} \int_0^{2\pi} (2aw + v^2 - vw' + v'w + w^2) \, d\theta \qquad (4.14)$$

Introduction into Eq. (4.12) and rearrangement gives

$$\Omega = qa \int_0^{2\pi} \left[w + \frac{1}{2a} (v^2 - vw' + v'w + w^2) \right] d\theta \qquad (4.15)$$

Equations (4.11) and (4.15) now give, for the total potential energy, the expression

$$V = a \int_0^{2\pi} F \, d\theta \qquad (4.16)$$

where

$$F = \frac{EA}{2} \varepsilon^2 + \frac{EI}{2} \kappa^2 + q \left[w + \frac{1}{2a} (v^2 - vw' + v'w + w^2) \right]$$

and

$$\varepsilon = \frac{v' + w}{a} + \frac{1}{2} \left(\frac{v - w'}{a} \right)^2 \qquad \kappa = \frac{v' - w''}{a^2}$$

For equilibrium, V must be stationary. Consequently, the integrand in Eq. (4.16) must satisfy the Euler equations of the calculus of variations (Sec. A.4). The Euler equations for an integrand of the form of that in Eq. (4.16) are [compare Eqs. (A.34)]

$$\frac{\partial F}{\partial v} - \frac{d}{d\theta} \frac{\partial F}{\partial v'} = 0$$

$$\frac{\partial F}{\partial w} - \frac{d}{d\theta} \frac{\partial F}{\partial w'} + \frac{d^2}{d\theta^2} \frac{\partial F}{\partial w''} = 0$$

From Eq. (4.16) the partial derivatives are

$$\frac{\partial F}{\partial v} = \frac{1}{a} \left[EA\varepsilon \frac{v - w'}{a} + q(v - \tfrac{1}{2}w') \right]$$

$$\frac{\partial F}{\partial v'} = \frac{1}{a} \left(EA\varepsilon + EI \frac{\kappa}{a} + \tfrac{1}{2}qw \right)$$

$$\frac{\partial F}{\partial w} = \frac{1}{a} [EA\varepsilon + q(a + \tfrac{1}{2}v' + w)]$$

$$\frac{\partial F}{\partial w'} = \frac{1}{a} \left(-EA\varepsilon \frac{v - w'}{a} - \tfrac{1}{2}qv \right)$$

$$\frac{\partial F}{\partial w''} = \frac{1}{a} \left(-EI \frac{\kappa}{a} \right)$$

These expressions can be simplified by introduction of the constitutive equations for the ring. The normal force component N and bending moment M on a cross section are defined by the relations

$$N = \int \bar{\sigma} \, dA \qquad M = \int \bar{\sigma} z \, dA$$

where $\bar{\sigma}$ is the normal stress in the θ direction. Introduction of the expressions $\bar{\sigma} = E\bar{\varepsilon}$ and $\bar{\varepsilon} = \varepsilon + z\kappa$ and integration gives

$$N = EA\varepsilon \qquad M = EI\kappa \qquad (4.17)$$

Equations (4.17) are the constitutive relations for the ring. With these relations and Eq. (4.7) the partial derivatives may be written

$$\frac{\partial F}{\partial v} = \frac{1}{a}[N\beta + q(v - \tfrac{1}{2}w')]$$

$$\frac{\partial F}{\partial v'} = \frac{1}{a}\left(N + \frac{1}{a}M + \tfrac{1}{2}qw\right)$$

$$\frac{\partial F}{\partial w} = \frac{1}{a}[N + q(a + \tfrac{1}{2}v' + w)]$$

$$\frac{\partial F}{\partial w'} = \frac{1}{a}(-N\beta - \tfrac{1}{2}qv)$$

$$\frac{\partial F}{\partial w''} = \frac{1}{a}\left(-\frac{1}{a}M\right)$$

Introduction into the Euler equations and rearrangement gives the final expressions

$$aN' + M' - aN\beta - qa^2\beta = 0$$
$$M'' - aN - a(N\beta)' - qa(v' + w) = qa^2 \qquad (4.18)$$

Equations (4.18) are the equilibrium equations for the ring for the intermediate class of deformations. The equations are nonlinear in the dependent variables. They are the counterpart for the ring of Eq. (1.12) for the straight column and of Eqs. (3.18) for the flat plate. Unlike the column case, no special circumstance transforms them into linear differential equations.

Equations (4.18) are written in terms of the five dependent variables N, M, β, v, and w. They may be expressed in terms of the two variables v and w alone by introduction of the constitutive and kinematic relations. Substitution and rearrangement gives

$$\left[\frac{v'+w}{a} + \frac{1}{2}\left(\frac{v-w'}{a}\right)^2 \right]' + \frac{I}{Aa^2}\left(\frac{v-w'}{a}\right)''$$

$$- \left[\frac{v'+w}{a} + \frac{1}{2}\left(\frac{v-w'}{a}\right)^2 \right] \frac{v-w'}{a} - \frac{qa}{EA}\frac{v-w'}{a} = 0$$

$$\frac{I}{Aa^2}\left(\frac{v-w'}{a}\right)''' - \left[\frac{v'+w}{a} + \frac{1}{2}\left(\frac{v-w'}{a}\right)^2 \right]$$

$$- a\left\{ \left[\frac{v'+w}{a} + \frac{1}{2}\left(\frac{v-w'}{a}\right)^2 \right] \frac{v-w'}{a} \right\}' - \frac{qa}{EA}\frac{v'+w}{a} = \frac{qa}{EA}$$

(4.19)

4.4 LINEAR STABILITY EQUATIONS

Under axisymmetric loading, circular equilibrium configurations of the ring exist for all values of the applied load q. The critical load q_{cr} is the smallest load for which the ring may be maintained in equilibrium in an adjacent noncircular configuration.

4.4a Adjacent-Equilibrium Criterion

According to the adjacent-equilibrium criterion, two infinitesimally adjacent equilibrium configurations exist at $q = q_{cr}$, the circular one and a slightly non-circular one. Both configurations are governed by Eqs. (4.19). Let v_0, w_0 represent the circular configuration and $v_0 + v_1$, $w_0 + w_1$ the noncircular one, where v_1, w_1 is an infinitesimally small increment. For the circular form, v_0 and its derivatives and w_0' and its derivatives equal zero. Thus we let

$$v \to v_1$$
$$w \to w_0 + w_1$$

in Eqs. (4.19). In the resulting equation, the terms containing only q or w_0 may be omitted; the sum of such terms is equal to zero because w_0 represents an equilibrium configuration. Terms that are quadratic or cubic in v_1, w_1 may be omitted because of the smallness of the incremental displacement. The remaining terms are

$$EAa^2(v_1' + w_1)' + EI(v_1 - w_1')'' - EAaw_0(v_1 - w_1') - qa^3(v_1 - w_1') = 0$$
$$EI(v_1 - w_1')''' - EAa^2(v_1' + w_1) - EAaw_0(v_1 - w_1')' - qa^3(v_1' + w_1) = 0$$

Specialization of Eqs. (4.18) for axisymmetric deformation gives for w_0 the expression

$$w_0 = -\frac{a^2}{EA} q$$

Replacement of w_0 by the function of q and simplification gives the expressions

$$EAa^2(v_1' + w_1)' + EI(v_1 - w_1')'' = 0$$
$$EAa^2(v_1' + w_1) - EI(v_1 - w_1')''' + qa^3(w_1'' + w_1) = 0 \tag{4.20}$$

Equations (4.20) are the stability equations for the ring subjected to external fluid pressure. They are seen to be homogeneous linear equations in v_1, w_1. These equations for the ring are the counterpart of Eq. (1.23) for the column and of Eqs. (3.34) for the flat plate.

4.4b   Minimum Potential Energy Criterion

In this section Eqs. (4.20) are rederived by application of the minimum potential energy criterion. From Sec. 4.3, the total potential energy of the loaded ring may be written

$$V = U_m + U_b + \Omega \tag{4.21}$$

where

$$U_m = \frac{EAa}{2} \int_0^{2\pi} \left[\frac{v' + w}{a} + \frac{1}{2}\left(\frac{v - w'}{a}\right)^2 \right]^2 d\theta$$

$$U_b = \frac{EIa}{2} \int_0^{2\pi} \left(\frac{v' - w''}{a^2}\right)^2 d\theta$$

$$\Omega = q \int_0^{2\pi} [wa + \tfrac{1}{2}(v^2 - vw' + v'w + w^2)] \, d\theta$$

To determine the character of the potential energy corresponding to a deformed configuration $v = v_0$, $w = w_0$, we let

$$v \to v_0 + v_1$$
$$w \to w_0 + w_1 \tag{4.22}$$

where the variation v_1, w_1 is arbitrarily small (see Sec. A.4). For the circular form, v_0 and its derivatives and w_0' and its derivatives equal zero. Furthermore, the circular form is an equilibrium configuration for all values of q, so that δV is equal to zero and only the second variation need be investigated. Introduction of Eqs. (4.22) and collection of all terms that are quadratic in v_1, w_1 gives

$$\delta^2 V = \delta^2 U_m + \delta^2 U_b + \delta^2 \Omega \tag{4.23}$$

where

$$\delta^2 U_m = \frac{EA}{a} \int_0^{2\pi} \left[(v_1' + w_1)^2 + \frac{w_0}{a} (v_1 - w_1')^2 \right] d\theta \qquad (4.24a)$$

$$\delta^2 U_b = \frac{EI}{a^3} \int_0^{2\pi} [(v_1' - w_1'')^2] \, d\theta \qquad (4.24b)$$

$$\delta^2 \Omega = q \int_0^{2\pi} (v_1{}^2 - v_1 w_1' + v_1' w_1 + w_1{}^2) \, d\theta \qquad (4.24c)$$

But $w_0/a = -(a/EA)q$. Therefore rearrangement gives

$$\delta^2 V = \frac{1}{a^3} \int_0^{2\pi} [EAa^2(v_1' + w_1)^2 + EI(v_1' - w_1'')^2$$

$$+ qa^3(v_1 w_1' + v_1' w_1 + w_1{}^2 - w_1'^2)] \, d\theta$$

Simplification of the integrand by integration by parts gives, for the final expression, the equation

$$\delta^2 V = \frac{1}{a^3} \int_0^{2\pi} F \, d\theta$$

where

$$F = EAa^2(v_1' + w_1)^2 + EI(v_1' - w_1'')^2 + qa^3(w_1{}^2 - w_1'^2)$$

The expression in Eq. (4.24c) for the second variation of the potential energy of the applied fluid pressure is the same as the corresponding expression in Ref. 4.3, p. 142.

   [It may be noted that the term containing w_0 in Eq. (4.24a) sometimes is rearranged as follows:

$$\frac{EA}{a} \int_0^{2\pi} \frac{w_0}{a} (v_1 - w_1')^2 \, d\theta = \int_0^{2\pi} N \left(\frac{v_1 - w_1'}{a} \right)^2 a \, d\theta$$

where, from Eqs. (4.9) and (4.17), $N = EAw_0/a$. In this form the term (with a change in sign) occasionally is called the work of the hoop force N during buckling.]

   According to the Trefftz criterion (Sec. A.5), the equations for loss of stability are obtained by introduction of F into the Euler equations of the calculus of variations. The Euler equations in this case are

$$\frac{\partial F}{\partial v_1} - \frac{d}{d\theta} \frac{\partial F}{\partial v_1'} = 0$$

$$\frac{\partial F}{\partial w_1} - \frac{d}{d\theta} \frac{\partial F}{\partial w_1'} + \frac{d^2}{d\theta^2} \frac{\partial F}{\partial w_1''} = 0$$

Introduction of the expression for F and rearrangement is found to yield the stability equations in Eqs. (4.20).

4.5 SOLUTION FOR FLUID-PRESSURE LOADING

Equations (4.20) for loss of stability of the ring under fluid-pressure loading are constant-coefficient homogeneous equations. The general solution is readily found for arbitrary boundary conditions. For the complete ring, however, the boundary requirement is simply that v_1, w_1 and their derivatives be periodic in θ. A solution of the form

$$v_1 = B \sin n\theta$$
$$w_1 = C \cos n\theta$$

(4.25)

where B and C are constants and n is a positive integer, is seen to satisfy both the differential equations and the periodicity requirement. Introduction into Eqs. (4.20) and simplification gives

$$\left[n(nB + C) + n^2 \frac{I}{Aa^2}(B + nC) \right] \sin n\theta = 0$$

$$\left[(nB + C) + n^3 \frac{I}{Aa^2}(B + nC) - (n^2 - 1)\frac{qa}{EA}C \right] \cos n\theta = 0$$

The equations must be satisfied for all values of θ, and therefore the sine and cosine factors may be divided out. Rearrangement gives

$$n^2\left(1 + \frac{I}{Aa^2}\right)B + n\left(1 + n^2\frac{I}{Aa^2}\right)C = 0$$

$$n\left(1 + n^2\frac{I}{Aa^2}\right)B + \left[\left(1 + n^4\frac{I}{Aa^2}\right) - (n^2 - 1)\frac{qa}{EA}\right]C = 0$$

(4.26)

For $n = 1$ the first equation becomes

$$\left(1 + \frac{I}{Aa^2}\right)(B + C) = 0$$

Therefore $B = -C$ for $n = 1$ and, from Eqs. (4.25), $v_1 = C \sin \theta$ and $w_1 = C \cos \theta$. But this displacement mode is seen to represent a rigid-body translation of the ring. We postulate that the ring is constrained against such translation, and we consider only modes for which n is greater than unity.

For a nontrivial solution, the determinant of the coefficients of B, C in Eqs. (4.26) must equal zero:

$$\begin{vmatrix} n^2\left(1 + \dfrac{I}{Aa^2}\right) & n\left(1 + n^2\dfrac{I}{Aa^2}\right) \\[3ex] n\left(1 + n^2\dfrac{I}{Aa^2}\right) & \left(1 + n^4\dfrac{I}{Aa^2}\right) - (n^2 - 1)\dfrac{qa}{EA} \end{vmatrix} = 0 \qquad (4.27)$$

Rearrangement gives

$$q = \frac{n^2 - 1}{1 + I/Aa^2}\frac{EI}{a^3} \qquad n = 2, 3, 4, \ldots$$

The term I/Aa^2 is much smaller than unity. For example, for a rectangular cross-sectional ring of thickness h, $I/Aa^2 = (h/a)^2/12$. For $h/a < 0.1$, $I/Aa^2 < 0.0008$. Therefore the term may be neglected to give

$$q = (n^2 - 1)\frac{EI}{a^3} \qquad n = 2, 3, 4, \ldots \qquad (4.28)$$

The smallest eigenvalue is seen to correspond to $n = 2$. For that value Eq. (4.28) gives

$$q_{\mathrm{cr}} = 3\frac{EI}{a^3} \qquad (4.29)$$

This result, due to Lévy (Ref. 4.4), is considered to be the classical solution for a ring subjected to external fluid pressure.

4.6 RING ON AN ELASTIC FOUNDATION

To provide a model for further illustration of the role of mode shape in stability analysis (compare Sec. 2.2d), the analysis in Sec. 4.5 is extended to include the stabilizing influence of a Winkler foundation. A more accurate analysis of the influence of an elastic foundation may be found in, for example, Ref. 4.5. For our purposes, we consider a ring completely filled with a soft elastic medium that is treated as an infinite set of uncoupled radial springs. For such a foundation the pressure q_f, in pounds per inch, between ring and foundation is given by the relation [compare Eq. (2.39)]

$$q_f = -k_f w \qquad (4.30)$$

where k_f, in pounds per square inch, is a known constant representing the foundation modulus.

The strain energy of the foundation may be written

$$U_f = -\frac{a}{2} \int_0^{2\pi} q_f w \, d\theta$$

Introduction of Eq. (4.30) gives

$$U_f = \frac{ak_f}{2} \int_0^{2\pi} w^2 \, d\theta \qquad (4.31)$$

and for the second variation of the foundation strain energy,

$$\delta^2 U_f = ak_f \int_0^{2\pi} w_1{}^2 \, d\theta$$

Addition of this term to the expression for the second variation of the total potential energy in Eq. (4.23) and application of the Trefftz criterion $\delta(\delta^2 V) = 0$ gives, for the Euler equations, the expressions

$$EAa^2(v_1' + w_1)' + EI(v_1 - w_1')'' = 0$$
$$EAa^2(v_1' + w_1) - EI(v_1 - w_1')''' + qa^3(w_1'' + w_1) - k_f a^4 w_1 = 0 \qquad (4.32)$$

Equations (4.32) differ from Eqs. (4.20) only in the addition of the term $k_f a^4 w_1$ in the second equation. Introduction of Eqs. (4.25) leads to a pair of homogeneous algebraic equations. The criterion that the determinant of the coefficients equals zero then gives the expression

$$\begin{vmatrix} n^2\left(1 + \dfrac{I}{Aa^2}\right) & n\left(1 + n^2\dfrac{I}{Aa^2}\right) \\[3mm] n\left(1 + n^2\dfrac{I}{Aa^2}\right) & \left(1 + n^4\dfrac{I}{Aa^2} + \dfrac{k_f a^2}{EA}\right) - (n^2 - 1)\dfrac{qa}{EA} \end{vmatrix} = 0 \qquad (4.33)$$

Consequently,

$$q = \frac{n^2 - 1}{1 + I/Aa^2}\frac{EI}{a^3} + \frac{1}{n^2 - 1}k_f a \qquad (4.34)$$

As previously noted, $I/Aa^2 \ll 1$. Therefore, approximately,

$$q = (n^2 - 1)\frac{EI}{a^3} + \frac{1}{n^2 - 1}k_f a \qquad n = 2, 3, 4, \ldots \qquad (4.35)$$

For $k_f = 0$, Eq. (4.35) reduces to Eq. (4.28). Equation (4.35) is analogous to Eq. (2.42) for a column on a Winkler foundation. For given values of EI, k_f, and a, the value of n may be determined by trial to give the smallest eigenvalue, q_{cr}. Numerical examples are considered in the problems at the end of the chapter.

4.7   CENTRALLY DIRECTED PRESSURE LOADING

In the analysis of Sec. 4.3, the applied pressure was assumed to remain normal to the surface of the ring during the incremental deformation. In this section the pressure is assumed to be centrally directed. An examination of this problem furnishes insight into the role of the Ω term in the expression $V = U + \Omega$ and forms a background for the presentation of the Donnell approximations in the following section.

An analysis of a ring subjected to centrally directed pressure is given in Ref. 4.6. This method of loading also is conservative (Ref. 4.1). A sketch of a ring element before and after deformation is shown in Fig. 4.5. After deformation the force dP^* is directed toward the original center of the ring. Therefore the angle between dP^* and the z direction is approximately v/a. The magnitude of the total force is assumed to remain constant during the deformation; i.e., the change in length of the ring element during deformation is neglected. Then the z and θ components of dP^* are given approximately by the expressions

$$dP_z^* = qa\, d\theta \qquad dP_\theta^* = qv\, d\theta$$

and the potential energy of the applied pressure may be written

$$\Omega = qa \int_0^{2\pi} \left(w + \frac{1}{2a} v^2 \right) d\theta$$

The second variation of Ω is seen to be simply

$$\delta^2 \Omega = q \int_0^{2\pi} v_1{}^2\, d\theta \qquad (4.36)$$

Equation (4.36) is the same as eq. (15) of Ref. 4.6. This expression for centrally directed pressure corresponds to Eq. (4.24c) for normally directed pressure.

The expressions for $\delta^2 U_m$ and $\delta^2 U_b$ are the same as for normally directed pressure. Consequently, the second variation of the total potential energy becomes

$$\delta^2 V = \frac{1}{a^3} \int_0^{2\pi} F\, d\theta$$

where now

$$F = EAa^2(v_1' + w_1)^2 + EI(v_1' - w_1'')^2 + qa^3(2v_1 w_1' - w_1'^2)$$

The Euler equations for this integrand are found to be

$$EAa^2(v_1' + w_1)' + EI(v_1 - w_1')'' - qa^3 w_1' = 0$$
$$EAa^2(v_1' + w_1) - EI(v_1 - w_1')''' + qa^3(v_1 + w_1')' = 0$$

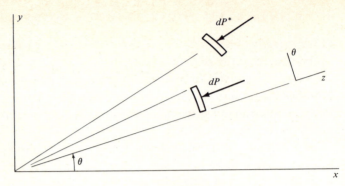

FIGURE 4.5
Centrally directed pressure loading.

These equations correspond to Eqs. (4.20) for normally directed pressure. Introduction of Eqs. (4.25) leads to the stability determinant

$$
\begin{vmatrix}
n^2\left(1 + \dfrac{I}{Aa^2}\right) & n\left(1 + n^2\dfrac{I}{Aa^2} - \dfrac{qa}{EA}\right) \\[2ex]
n\left(1 + n^2\dfrac{I}{Aa^2} - \dfrac{qa}{EA}\right) & \left(1 + n^4\dfrac{I}{Aa^2}\right) - n^2\dfrac{qa}{EA}
\end{vmatrix} = 0 \qquad (4.38)
$$

Again, it may be shown that n is equal to or greater than 2 if the ring is constrained against rigid-body translation. Equation (4.38) gives

$$
(n^2 - 1)^2 \frac{I}{Aa^2} + (2 - n^2)\frac{qa}{EA} + \left(\frac{qa}{EA}\right)^2 = 0 \qquad (4.39)
$$

For thin rings, $I/Aa^2 \ll 1$, as previously noted. Therefore, as may be seen from an examination of the quadratic formula, the two roots of Eq. (4.39) are quite different in size. The larger one is given approximately by neglecting the first term in Eq. (4.39), and the smaller one by neglecting the third term. Therefore, for the smaller root, approximately,

$$
q = \frac{(n^2 - 1)^2}{n^2 - 2}\frac{EI}{a^3} \qquad n = 2, 3, 4, \ldots \qquad (4.40)
$$

Equation (4.40) corresponds to Eq. (4.28). The smallest eigenvalue occurs for $n = 2$, for which

$$
q_{cr} = 4.5\,\frac{EI}{a^3} \qquad (4.41)
$$

This is the result given in Ref. 4.6, eq. (37). The analysis leads to the surprising result that the critical pressure is 50 percent higher if the applied pressure is centrally directed rather than normal to the surface.

The eigenvalues for fluid-pressure loading and for centrally directed pressure loading, respectively, have been found to be

$$q = (n^2 - 1)\frac{EI}{a^3} \quad \text{and} \quad q = \frac{(n^2 - 1)^2}{n^2 - 2}\frac{EI}{a^3}$$

In both cases the critical load corresponds to $n = 2$, and the magnitude of the critical load for the circular ring is seen to be extremely sensitive to the behavior of the applied load during ring deformation.

On the other hand, if n were much larger than unity, as it might be, say, for a ring on an elastic foundation, the critical loads given by the two equations would be approximately the same, and the distinction between methods of loading would be of little consequence. Displacement functions such as

$$w_1 = C \cos n\theta$$

for which n is much larger than unity, are termed *rapidly varying* functions of the circumferential coordinate. An important set of simplifying approximations of wide applicability suitable for situations in which the displacements are rapidly varying functions of the coordinates is discussed in the following section.

4.8 THE DONNELL APPROXIMATIONS

A set of simplifying approximations for circular cylindrical shells was introduced by Donnell in 1933 (Ref. 4.7) and subsequently generalized for shells of arbitrary shape. (The generalization is presented here as the Donnell-Mushtari-Vlasov approximations in Chap. 6.) The additional limitations introduced by the Donnell approximations are as follows. It is required that the wavelength of the deformation pattern be small relative to the radius of curvature, i.e., that the displacement components be rapidly varying functions of the circumferential coordinate, and that the tangential displacement components be small relative to the normal displacement component. The latter requirement ordinarily is met for the present intermediate class of deformations. Koiter shows in Ref. 4.8 for the general case of an arbitrary shell that if these limitations are satisfied, the contribution of the nonlinear displacement terms in the expression for the potential energy of the applied loads [in, for example, Eq. (4.15)] always is negligibly small; i.e., the applied load always may be treated as a dead load.

For circular rings, satisfaction of these requirements means that $w'/a \gg v/a$ in the expression for the rotation β in Eq. (4.7). Then, approximately,

$$\beta = -\frac{w'}{a} \qquad (4.42)$$

and from Eq. (4.10),

$$\kappa = -\frac{w''}{a^2} \qquad (4.43)$$

For dead-loading,

$$\Omega = qa \int_0^{2\pi} w \, d\theta$$

and therefore

$$\delta^2 \Omega = 0$$

Then, for a ring subjected to external pressure, the expression for the second variation of the total potential energy in Eq. (4.23) becomes simply

$$\delta^2 V = \delta^2 U_m + \delta^2 U_b \qquad (4.44)$$

where

$$\delta^2 U_m = \frac{EA}{2} \int_0^{2\pi} \left[(v_1' + w_1)^2 + \frac{w_0}{a} (w_1')^2 \right] d\theta$$

and

$$\delta^2 U_b = \frac{EI}{2} \int_0^{2\pi} [(w_1'')^2] \, d\theta$$

The corresponding Euler equations are found to be [compare Eqs. (4.20)]

$$(v_1' + w_1)' = 0 \qquad (4.45a)$$

$$EAa^2(v_1' + w_1) + EIw_1{}^{iv} + qa^3 w_1'' = 0 \qquad (4.45b)$$

From Eq. (4.45a), however,

$$v_1' + w_1 = c_1$$

where c_1 is a constant. Therefore

$$v_1 = -\int w_1 \, d\theta + c_1\theta + c_2$$

For periodicity, $c_1 = 0$. Consequently,

$$v_1' + w_1 = 0 \qquad (4.46)$$

This condition is referred to as *inextensional buckling*. Equations (4.45b) and (4.46) now give [compare Eq. (1.23)]

$$EIw_1{}^{iv} + qa^3 w_1'' = 0 \qquad (4.47)$$

The solution is of the form

$$w_1 = C \cos n\theta \qquad n = 2, 3, 4, \ldots \qquad (4.48)$$

and the eigenvalues are

$$q = n^2 \frac{EI}{a^3} \qquad n = 2, 3, 4, \ldots \qquad (4.49)$$

In comparison with the more accurate solution for fluid-pressure loading given by Eq. (4.28), the Donnell approximations are seen to lead to eigenvalues that are too high by a factor of $n^2/(n^2 - 1)$. For $n = 2$ the error is 25 percent. For $n = 10$, say (as it might be for a ring on an elastic foundation), the error is only 1 percent.

Equation (4.46) gives some insight into the basis for neglect of the v terms in the kinematic relations. From Eqs. (4.25) and (4.46), $B = C/n$. Therefore

$$v_1 = -\frac{C}{n} \sin n\theta \qquad w_1 = C \cos n\theta$$

If the displacement components are rapidly varying functions of the coordinate θ, that is, if $n \gg 1$, the amplitude of the displacement v_1 is much smaller than the amplitude of w_1, and v_1 is very much smaller than w_1'.

The Donnell approximations are utilized in the analysis of circular cylindrical shells in Chap. 5, and the corresponding Donnell-Mushtari-Vlasov approximations are used for shells of general shape in Chap. 6.

REFERENCES

4.1 BODNER, S. R.: On the Conservativeness of Various Distributed Force Systems, *J. Aeronaut. Sci.*, vol. 25, pp. 132–133, February 1958.

4.2 KAPLAN, W.: "Advanced Calculus," Addison-Wesley, Reading, Mass., 1952.

4.3 PEARSON, C. E.: General Theory of Elastic Stability, *Q. Appl. Math.*, vol. 14, pp. 133–144, 1956.

4.4 LÉVY, M.: *J. Math. Pure Appl.* (*Liouville*), ser. 3, vol. 10, p. 5, 1884.

4.5 SEIDE, P.: The Stability under Axial Compression and Lateral Pressure of Circular Cylindrical Shells with a Soft Elastic Core, *J. Aerospace Sci.*, vol. 29, pp. 851–862, July 1962.

4.6 BORESI, A. P.: A Refinement of the Theory of Buckling of Rings under Uniform Pressure, *J. Appl. Mech.*, vol. 22, no. 1, pp. 95–103, 1955.

4.7 DONNELL, L. H.: Stability of Thin-walled Tubes under Torsion, *NACA Rep.* 479, 1933.

4.8 KOITER, W. T.: General Equations of Elastic Stability for Thin Shells, *Proceedings of the Symposium on the Theory of Shells to Honor Lloyd Hamilton Donnell*, University of Houston, Houston, Tex., 1967, pp. 187–228.

4.9 CHWALLA, E., and C. F. KOLLBRUNNER: Beiträge zum Knickproblem des Bogenträgers und des Rahmens, *Stahlbau*, vol. 11, no. 10, pp. 73–78, 1938.

PROBLEMS

4.1 A thin circular ring subjected to uniform external pressure has an 8-in. radius and a rectangular cross section of width 1.2 in. and thickness 0.2 in. Determine the critical pressure for elastic buckling if $E = 10 \times 10^6$ psi.   *Ans.* $q_{cr} = 46.9$ lb/in.

4.2 If the ring in Prob. 4.1 is filled with a soft elastic medium whose foundation modulus $k_f = 20,000$ psi, determine the critical value of the external pressure for elastic buckling.

4.3 Equation (4.35) for a ring on an elastic foundation gives the external pressure q as a function of the wavelength parameter n. The value of n may be determined by trial to give the minimum eigenvalue q_{cr}. (*a*) By treating n as a continuous variable and utilizing the relationship $dq/dn = 0$, derive an expression, suitable for $n \gg 1$, for the approximate value of n for which q is a minimum. (*b*) For the example in Prob. 4.2, compare the value of n given by this expression with the integral value obtained in Prob. 4.2. (*c*) By introducing into Eq. (4.35) the expression for n in part *a* of this problem, derive an approximate expression for q_{cr} that is independent of n. (Compare Prob. 2.4.)

4.4 Repeat the analysis of a ring on an elastic foundation in terms of the Donnell approximations. Show that Eq. (4.35) is replaced by the expression $q = n^2 EI/a^3 + k_f a/n^2$. Repeat Probs. 4.1 and 4.2 using the Donnell forms of the respective solutions. Compare the results with the numerical values found in those problems, and comment on the accuracy of the Donnell forms for each case.

4.5 Rederive the expression for q_{cr} in Eq. (4.29) by substituting the displacement functions in Eqs. (4.25) directly into the expression for the second variation of the potential energy in Eqs. (4.24), and then applying the criterion for loss of stability of a multiple degree-of-freedom system in Sec. A.2. (This procedure is called a *direct energy method*. See the discussion of the Rayleigh-Ritz procedure in Chap. 8.) Carefully note the difficulty that arises if the fact that $I/Aa^2 \ll 1$ is utilized before the final step in the analysis. This exercise serves as an excellent illustration of the systematic problem of small differences of large numbers in the numerical analysis of shell stability.

4.6 Rederive the Donnell form of the solution in Eq. (4.49) by substituting the displacement functions in Eqs. (4.25) directly into the coupled differential equations in Eqs. (4.45) rather than into Eq. (4.47).

4.7 In the stability analysis of rings, the curvature change $\kappa = -(w'' - v')/a^2$ obtained from Eqs. (4.7) and (4.10) frequently is replaced by the relationship $\kappa = -(w'' + w)/a^2$. (*a*) Rederive the expression for q_{cr} in Eq. (4.29) in terms of this kinematic relationship. Note that in this solution the buckling turns out to be inextensional, that is, $(v_1' + w_1) = 0$. Consequently, $-(w_1'' + w_1) = -(w_1'' - v_1')$, and the two expressions for curvature change are equivalent. Use the adjacent-equilibrium stability criterion in the analysis. (*b*) Repeat the analysis using the minimum potential energy criterion.

4.8 (*a*) Derive ring equilibrium equations corresponding to Eqs. (4.18) in terms of the Donnell approximations. [In Eqs. (4.9) and (4.16), replace Eqs. (4.7) and (4.10) by Eq. (4.42), and omit all quadratic terms in the displacement components in the expression for Ω in Eq. (4.15)]. (*b*) Show that in this form the equations governing equilibrium in the nonlinear regime reduce to a single linear differential equation analogous to Eq. (1.12) for the straight bar.   *Ans.* (*a*) $N' = 0$, $M'' - aN + aNw'' = qa^2$

4.9 An expression for the second variation of the total potential energy of a ring subjected to external fluid pressure is given in Eq. (4.23). The corresponding expression for dead-load pressure may be obtained simply by setting $\delta^2\Omega = 0$ in that equation. (*a*) Show that the Euler equations for the resulting integrand are

$$EAa^2(v_1' + w_1)' + EI(v_1 - w_1')'' + qa^3(v_1 - w_1') = 0$$
$$EAa^2(v_1' + w_1) - EI(v_1 - w_1')''' - qa^3(v_1 - w_1')' = 0$$

(*b*) In Ref. 4.9, eq. 15, Chwalla and Kollbrunner give, for this case, the following equation, derived in terms of the assumption that the buckling is inextensional [i.e., that $(v_1' + w_1) = 0$]:

$$v_1{}^{vi} + (2 + \mu)v_1{}^{iv} + (1 + 2\mu)v_1'' + \mu v_1 = 0$$

where $\mu \equiv qa^3/EI$. Show that, for $(v_1' + w_1) = 0$, the equations in part *a* reduce to the same expression.

5

CIRCULAR CYLINDRICAL SHELLS

5.1   INTRODUCTION

The stability of circular cylindrical shells is treated in a separate chapter for several reasons. Cylinders are common shell configurations in structural applications. Furthermore, the equations governing the stability of cylindrical shells are much simpler than those for shells of general shape, and a separate study of cylindrical shells can serve as an introduction to the more complex topic of general shell stability.

Stability equations for cylindrical shells have been available in the literature since the late 1800s. Earliest solutions for cylinders subjected to axial compression were presented by Lorenz in 1911 (Ref. 5.1). Solutions for buckling under uniform lateral pressure were given by Southwell in 1913 (Ref. 5.2) and by von Mises in 1914 (Ref. 5.3). In 1932 Flügge (Ref. 5.4) presented a comprehensive treatment of cylindrical shell stability, including combined loading and cylinders subjected to bending. Results for cylinders subjected to torsional loading were given by Schwerin in 1925 (Ref. 5.5) and Donnell in 1933 (Ref. 5.6).

In the present development of stability equations for circular cylindrical shells, major emphasis is placed on the relatively simple equations suggested by

Donnell in Ref. 5.6. Although the simplifications somewhat limit their range of applicability, the Donnell equations form the basis for more stability analyses in the literature than any other set of cylindrical shell equations. The equations give accurate results for cylindrical panels that are relatively flat before deformation and for complete cylindrical shells whose displacement components in the deformed configuration are rapidly varying functions of the circumferential coordinate. Such shells are sometimes termed *quasi-shallow*. Most structural applications are in these categories. A somewhat more accurate set of cylindrical shell equations that are not subject to the shallowness limitation also is presented, and numerical results based on the two sets of equations are compared.

In recent years much of the work in shell analysis has been oriented toward computer solutions. In such cases simplicity of the governing equations is of less importance. Consequently, interest in the Donnell equations has diminished. Nevertheless, the relative simplicity of the equations makes them ideally suited for this introductory examination of shell stability.

The circular ring in Chap. 4 was our first example in which elements of the structure undergo lateral displacement prior to loss of stability. Similarly, the cylindrical shell is our first example in which the prebuckling deformation of the ideal structure entails rotation of structural elements. When rotations occur, the prebuckling deformation is nonlinear, as emphasized in Sec. 1.1. In such cases, terms representing prebuckling rotations appear explicitly, not only in the nonlinear equations governing the prebuckling deformation, but also in the linear stability equations. Fortunately, the influence of prebuckling rotations is negligibly small in many instances. Consequently, in this and subsequent chapters, we consider two kinds of stability equations, those in which prebuckling rotation terms are retained and those in which such terms are omitted. Applications of both kinds of equations are examined.

The Donnell forms of the equilibrium and stability equations are derived in Secs. 5.2 and 5.3, respectively. The set of cylindrical shell equations that are not subject to the shallowness limitation is developed in Sec. 5.4. Applications and numerical examples are presented in Sec. 5.5.

5.2   DONNELL FORM OF THE NONLINEAR EQUILIBRIUM EQUATIONS

Many publications have appeared in which equations are derived directly for circular cylindrical shells. In others the equations are derived for shells of general shape and then specialized for circular cylindrical shells. A discussion of the literature on shell theory is included in Chap. 6, General Shells.

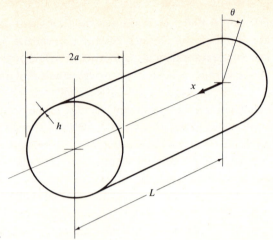

FIGURE 5.1
Circular cylindrical shell.

Let us consider a thin-walled circular cylindrical shell of length L, wall thickness h, and undeformed-middle-surface radius a, with $h \ll a$. The middle surface of the cylinder is referred to cylindrical coordinates x, θ, as shown in Fig. 5.1. Distances from the middle surface are measured by a coordinate z, positive outward. Displacement components in the x, θ, and z directions, respectively, are denoted by u, v, and w. The cylinder is considered to be subjected to edge loading and to a dead-load lateral pressure p, in pounds per square inch, positive outward. In general, $p = p(x,\theta)$.

In shell theory, as in plate theory, it is convenient to express internal forces and moments in terms of forces and moments per unit distance along the edge of a shell element, as shown in Fig. 5.2. The force and moment intensities are related to the internal stresses by the equations

$$N_x = \int_{-h/2}^{h/2} \bar{\sigma}_x \left(1 + \frac{z}{a}\right) dz \qquad N_\theta = \int_{-h/2}^{h/2} \bar{\sigma}_\theta \, dz$$

$$N_{x\theta} = \int_{-h/2}^{h/2} \bar{\tau}_{x\theta} \left(1 + \frac{z}{a}\right) dz \qquad N_{\theta x} = \int_{-h/2}^{h/2} \bar{\tau}_{\theta x} \, dz$$

$$Q_x = \int_{-h/2}^{h/2} \bar{\tau}_{xz} \left(1 + \frac{z}{a}\right) dz \qquad Q_\theta = \int_{-h/2}^{h/2} \bar{\tau}_{\theta z} \, dz$$

$$M_x = a \int_{-h/2}^{h/2} \bar{\sigma}_x \left(1 + \frac{z}{a}\right) z \, dz \qquad M_\theta = a \int_{-h/2}^{h/2} \bar{\sigma}_\theta \, z \, dz$$

$$M_{x\theta} = a \int_{-h/2}^{h/2} \bar{\tau}_{x\theta} \left(1 + \frac{z}{a}\right) z \, dz \qquad M_{\theta x} = a \int_{-h/2}^{h/2} \bar{\tau}_{\theta x} \, z \, dz$$

$$(5.1)$$

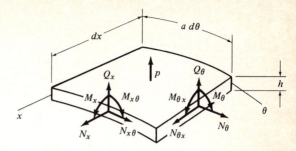

FIGURE 5.2
Cylindrical shell element.

where N_x, $N_{x\theta}$, N_θ, $N_{\theta x}$ are in-plane normal and shearing force intensities, in pounds per inch; Q_x, Q_θ are transverse shearing force intensities, in pounds per inch; M_x, M_θ are bending moment intensities, in inch-pounds per inch; and $M_{x\theta}$, $M_{\theta x}$ are twisting moment intensities, in inch-pounds per inch. As before, the symbols $\bar{\sigma}_x$, $\bar{\tau}_{x\theta}$, etc., denote stress components at any point through the shell wall thickness, as distinguished from σ_x, $\tau_{x\theta}$, etc., which refer to corresponding quantities on the middle surface ($z = 0$) only.

5.2a Summation of Forces and Moments

The nonlinear equilibrium equations may be derived by summation of forces and moments for a cylindrical shell element in a slightly deformed configuration, as shown in Fig. 5.3. As in Chap. 3, force and moment intensities are shown in separate sketches to simplify the diagrams. For the intermediate class of deformations, the angles of rotation β_x and β_θ are assumed to be small, and sines and cosines of the angles are replaced by the angles themselves and by unity, respectively. Furthermore, quadratic terms representing nonlinear interaction between the small transverse shearing forces and the rotations are assumed to be negligibly small, as in Secs. 1.2 and 3.3. Then summation of forces in the x, θ, and z directions, respectively, gives the equations

$$aN_{x,x} + N_{\theta x,\theta} = 0 \qquad (5.2a)$$

$$aN_{x\theta,x} + N_{\theta,\theta} + Q_\theta = 0 \qquad (5.2b)$$

$$Q_{\theta,\theta} + aQ_{x,x} - N_\theta - aN_x\beta_{x,x} - aN_{x\theta}\beta_{\theta,x} - N_{\theta x}\beta_{x,\theta} - N_\theta\beta_{\theta,\theta} = -pa \qquad (5.2c)$$

Equation (5.2b) corresponds to Eq. (3.11) for flat plates. For sufficiently shallow cylinders the Q_θ term in this equation may be expected to make a negligibly small contribution to the equilibrium of forces in the circumferential direction. Consequently, the term is omitted in the Donnell quasi-shallow shell equations.

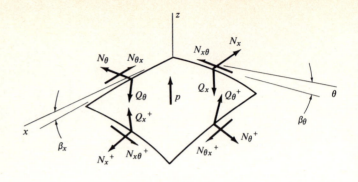

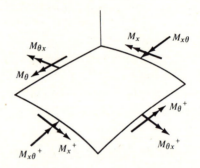

FIGURE 5.3
Cylindrical shell element in deformed configuration.

Summation of moments relative to the x and θ directions gives the equations

$$aQ_\theta = M_{\theta,\theta} + aM_{x\theta,x} \qquad (5.3a)$$

$$aQ_x = aM_{x,x} + M_{\theta x,\theta} \qquad (5.3b)$$

As for plates, the sixth equilibrium equation, $\sum M_z = 0$, leads to an identity and yields no additional information (see Sec. 3.3).

Use of Eqs. (5.3) to eliminate the $Q_{x,x}$ and $Q_{\theta,\theta}$ terms in Eq. (5.2c) now gives the three equilibrium equations

$$aN_{x,x} + N_{\theta x,\theta} = 0$$

$$aN_{x\theta,x} + N_{\theta,\theta} = 0 \qquad (5.4)$$

$$a^2 M_{x,xx} + aM_{x\theta,x\theta} + aM_{\theta x,x\theta} + M_{\theta,\theta\theta} - aN_\theta - a^2 N_x \beta_{x,x}$$
$$- a^2 N_{x\theta}\beta_{\theta,x} - aN_{\theta x}\beta_{x,\theta} - aN_\theta \beta_{\theta,\theta} = -pa^2$$

For sufficiently thin shells, z/a may be neglected relative to unity in Eqs. (5.1). Then $N_{\theta x} = N_{x\theta}$ and $M_{\theta x} = M_{x\theta}$. In terms of this approximation, Eqs. (5.4) may be written in the simpler form

$$aN_{x,x} + N_{x\theta,\theta} = 0$$
$$aN_{x\theta,x} + N_{\theta,\theta} = 0 \tag{5.5}$$
$$a^2 M_{x,xx} + 2aM_{x\theta,x\theta} + M_{\theta,\theta\theta} - aN_\theta - a^2 N_x \beta_{x,x}$$
$$- aN_{x\theta}(a\beta_{\theta,x} + \beta_{x,\theta}) - aN_\theta \beta_{\theta,\theta} = -pa^2$$

The constitutive equations for thin-walled isotropic elastic cylinders are the same as those for flat plates in Eqs. (3.8), namely,

$$N_x = C(\varepsilon_x + v\varepsilon_\theta) \qquad M_x = D(\kappa_x + v\kappa_\theta)$$
$$N_\theta = C(\varepsilon_\theta + v\varepsilon_x) \qquad M_\theta = D(\kappa_\theta + v\kappa_x) \tag{5.6}$$
$$N_{x\theta} = C\frac{1-v}{2}\gamma_{x\theta} \qquad M_{x\theta} = D(1-v)\kappa_{x\theta}$$

The middle-surface kinematic relations on which the Donnell equations are based are (Ref. 5.6)

$$\varepsilon_x = u_{,x} + \tfrac{1}{2}\beta_x^2 \qquad \beta_x = -w_{,x} \qquad \kappa_x = \beta_{x,x}$$
$$\varepsilon_\theta = \frac{v_{,\theta} + w}{a} + \tfrac{1}{2}\beta_\theta^2 \qquad \beta_\theta = -\frac{w_{,\theta}}{a} \qquad \kappa_\theta = \frac{\beta_{\theta,\theta}}{a} \tag{5.7}$$
$$\gamma_{x\theta} = \left(\frac{u_{,\theta}}{a} + v_{,x}\right) + \beta_x \beta_\theta \qquad\qquad \kappa_{x\theta} = \frac{1}{2}\left(\frac{\beta_{x,\theta}}{a} + \beta_{\theta,x}\right)$$

In comparison, the expressions given for flat plates in Eqs. (3.5) are

$$\varepsilon_x = u_{,x} + \tfrac{1}{2}\beta_x^2 \qquad \beta_x = -w_{,x} \qquad \kappa_x = \beta_{x,x}$$
$$\varepsilon_y = v_{,y} + \tfrac{1}{2}\beta_y^2 \qquad \beta_y = -w_{,y} \qquad \kappa_y = \beta_{y,y}$$
$$\gamma_{xy} = (u_{,y} + v_{,x}) + \beta_x \beta_y \qquad \kappa_{xy} = \tfrac{1}{2}(\beta_{x,y} + \beta_{y,x})$$

and, in terms of the present notation for cylindrical shells, the corresponding expressions for quasi-shallow circular rings in Eqs. (4.9), (4.10), and (4.42) are

$$\varepsilon_\theta = \frac{v_{,\theta} + w}{a} + \tfrac{1}{2}\beta_\theta^2 \qquad \beta_\theta = -\frac{w_{,\theta}}{a} \qquad \kappa_\theta = \frac{\beta_{\theta,\theta}}{a}$$

The expressions for ε_θ, β_θ, and κ_θ in Eqs. (5.7) are seen to be the same as the corresponding expressions for quasi-shallow circular rings, and the remainder of the kinematic relations are of the form of the corresponding expressions for flat plates. Equations (5.7) are the expressions given by specialization for cylindrical coordinates of the Donnell-Mushtari-Vlasov kinematic relations in Eqs. (6.6), (6.7a), (6.7c), and (6.8).

Introduction of the constitutive and kinematic relations into the equilibrium equations leads to a coupled set of three nonlinear differential equations in the three variables u, v, w. A more compact set of equations is obtained by introduction of only a portion of the constitutive and kinematic relations, to give

$$aN_{x,x} + N_{x\theta,\theta} = 0$$

$$aN_{x\theta,x} + N_{\theta,\theta} = 0 \qquad (5.8)$$

$$D\nabla^4 w + \frac{1}{a} N_\theta - \left(N_x w_{,xx} + \frac{2}{a} N_{x\theta} w_{,x\theta} + \frac{1}{a^2} N_\theta w_{,\theta\theta} \right) = p$$

where

$$\nabla^4 w \equiv w_{,xxxx} + \frac{2}{a^2} w_{,xx\theta\theta} + \frac{1}{a^4} w_{,\theta\theta\theta\theta}$$

Equations (5.8) are nonlinear equilibrium equations for quasi-shallow cylindrical shells. They have been widely used in the literature for large deflection analyses of cylindrical shells (see, for example, Ref. 5.7).

5.2b Stationary Potential Energy

In this section the nonlinear equilibrium equations in Eqs. (5.8) are rederived on the basis of the stationary potential energy criterion.

The total potential energy is the sum of the strain energy U of the cylindrical shell and the potential energy Ω of the applied pressure:

$$V = U + \Omega \qquad (5.9)$$

An expression for the shell strain energy may be derived in the same manner as for flat plates in Sec. 3.3b. The resulting expressions are [compare Eqs. (3.20)—(3.22)]

$$U = U_m + U_b$$

$$U_m = \frac{aC}{2} \iint \left(\varepsilon_x^2 + \varepsilon_\theta^2 + 2v\varepsilon_x \varepsilon_\theta + \frac{1-v}{2} \gamma_{x\theta}^2 \right) dx \, d\theta \qquad (5.10)$$

$$U_b = \frac{aD}{2} \iint [\kappa_x^2 + \kappa_\theta^2 + 2v\kappa_x \kappa_\theta + 2(1-v)\kappa_{x\theta}^2] \, dx \, d\theta$$

where ε_x, ε_θ, ..., $\kappa_{x\theta}$ are given by Eqs. (5.7). Equations (5.10) are given by specialization for cylindrical coordinates of the strain energy expression for a shell of general shape in Eqs. (6.3).

For a cylindrical shell subjected to dead-load lateral pressure, the potential energy of the applied pressure is simply

$$\Omega = -a \iint pw \, dx \, d\theta \qquad (5.11)$$

The Euler equations for an integrand of the form of that in Eqs. (5.9) to (5.11) are, from Eqs. (A.35),

$$\frac{\partial F}{\partial u} - \frac{\partial}{\partial x}\frac{\partial F}{\partial u_{,x}} - \frac{\partial}{\partial \theta}\frac{\partial F}{\partial u_{,\theta}} = 0$$

$$\frac{\partial F}{\partial v} - \frac{\partial}{\partial x}\frac{\partial F}{\partial v_{,x}} - \frac{\partial}{\partial \theta}\frac{\partial F}{\partial v_{,\theta}} = 0 \qquad (5.12)$$

$$\frac{\partial F}{\partial w} - \frac{\partial}{\partial x}\frac{\partial F}{\partial w_{,x}} - \frac{\partial}{\partial \theta}\frac{\partial F}{\partial w_{,\theta}} + \frac{\partial^2}{\partial x^2}\frac{\partial F}{\partial w_{,xx}} + \frac{\partial^2}{\partial x \, \partial \theta}\frac{\partial F}{\partial w_{,x\theta}} + \frac{\partial^2}{\partial \theta^2}\frac{\partial F}{\partial w_{,\theta\theta}} = 0$$

Introduction of the integrand and rearrangement again leads to the equilibrium equations in Eqs. (5.8). (See Prob. 5.2.)

Equations (5.8) are nonlinear equilibrium equations for quasi-shallow cylindrical shells. They are the counterpart for such shells of Eqs. (3.18) for flat plates. As for plates, the equations form a coupled set of three equations in the four variables N_x, $N_{x\theta}$, N_θ, and w. Three equations in three variables u, v, w may be obtained by introduction of the constitutive and kinematic relations from Eqs. (5.6) and (5.7). A simpler set of two equations in two variables is given by use of a stress function f analogous to that used for plates in Eqs. (3.27). The resulting equations are found to be

$$D\nabla^4 w + af_{,xx} - (f_{,\theta\theta} w_{,xx} - 2f_{,x\theta} w_{,x\theta} + f_{,xx} w_{,\theta\theta}) = p$$

$$\nabla^4 f - \frac{Eh}{u^4}(w_{,x\theta}{}^2 - w_{,xx} w_{,\theta\theta} + aw_{,xx}) = 0 \qquad (5.13)$$

where

$$N_x = f_{,\theta\theta} \qquad N_\theta = a^2 f_{,xx} \qquad N_{x\theta} = -af_{,x\theta}$$

Equations (5.13) were first presented by Donnell (Ref. 5.8).

The *linear* equilibrium equations corresponding to Eqs. (5.8) are obtained by simple omission of all quadratic and higher-order terms in u, v, w from the nonlinear equations. The resulting equations are

$$aN_{x,x} + N_{x\theta,\theta} = 0 \qquad (5.14a)$$

$$aN_{x\theta,x} + N_{\theta,\theta} = 0 \qquad (5.14b)$$

$$D\nabla^4 w + \frac{1}{a}N_\theta = p \qquad (5.14c)$$

where now

$$N_x = C(\varepsilon_x + v\varepsilon_\theta) \qquad \varepsilon_x = u_{,x}$$

$$N_\theta = C(\varepsilon_\theta + v\varepsilon_x) \qquad \varepsilon_\theta = \frac{v_{,\theta} + w}{a}$$

$$N_{x\theta} = C\frac{1-v}{2}\gamma_{x\theta} \qquad \gamma_{x\theta} = \frac{u_{,\theta}}{a} + v_{,x}$$

Unlike the corresponding linear equilibrium equations for plates in Eqs. (3.30), Eq. (5.14c) is not uncoupled from Eqs. (5.14a), and (5.14b). In this respect the analysis of cylindrical shells is more complicated than that of plates.

The linear equilibrium equations in Eqs. (5.14) form a coupled set of three equations in the four variables N_x, N_θ, $N_{x\theta}$, w. A set of three equations in three variables is obtained by introduction of the constitutive and kinematic relations, as follows:

$$a^2 u_{,xx} + \frac{1-v}{2} u_{,\theta\theta} + \frac{1+v}{2} av_{,x\theta} + vaw_{,x} = 0 \qquad (5.15a)$$

$$\frac{1+v}{2} au_{,x\theta} + \frac{1-v}{2} a^2 v_{,xx} + v_{,\theta\theta} + w_{,\theta} = 0 \qquad (5.15b)$$

$$D\nabla^4 w + \frac{1}{a^2} C(v_{,\theta} + w + vau_{,x}) = p \qquad (5.15c)$$

Donnell has shown in Ref. 5.6 (see Prob. 5.3) that these equations may be partially uncoupled to give

$$\nabla^4 u = -\frac{v}{a} w_{,xxx} + \frac{1}{a^3} w_{,x\theta\theta} \qquad (5.16a)$$

$$\nabla^4 v = -\frac{2+v}{a^2} w_{,xx\theta} - \frac{1}{a^4} w_{,\theta\theta\theta} \qquad (5.16b)$$

$$D\nabla^8 w + \frac{1-v^2}{a^2} Cw_{,xxxx} = \nabla^4 p \qquad (5.16c)$$

where $\nabla^8 w \equiv \nabla^4(\nabla^4 w)$. The linear equations also may be written in the form of a coupled pair of equations in w and the stress function f, as in Eqs. (5.13).

In shell theory, equilibrium equations that govern deformation without bending are called *membrane* equations, and those that govern deformation by bending or by combined bending and stretching are called *bending* equations.

Equations (5.14) to (5.16) are linear bending equations for the stress and deformation analysis of quasi-shallow cylindrical shells. Linear membrane equations corresponding to Eqs. (5.14) may be obtained by setting the bending stiffness parameter $D = 0$ in those equations. Then Eqs. (5.14) reduce to the expressions

$$aN_{x,x} + N_{x\theta,\theta} = 0$$
$$aN_{x\theta,x} + N_{\theta,\theta} = 0 \qquad (5.17)$$
$$N_\theta = pa$$

These equations are seen to be statically determinate; i.e., there are three equations in only three variables. In general, the prebuckling deformation of cylindrical shells entails both bending and stretching.

In summary, two equivalent sets of nonlinear bending equations for quasi-shallow cylindrical shells are given in this section, in Eqs. (5.8) and (5.13). Three equivalent sets of the corresponding linear bending equations are given in Eqs. (5.14) to (5.16). The corresponding linear membrane equations are given in Eqs. (5.17).

5.2c Nonlinear Equilibrium Paths

Numerous approximate solutions of the nonlinear equilibrium equations have appeared in the literature. Perhaps the most widely known is the von Kármán–Tsien solution (Ref. 5.7) for long cylindrical shells subjected to axial compression. The von Kármán–Tsien solution is discussed in some detail in Chap. 9 because of its historical importance and its value as an example of the use of numerical solution methods in nonlinear analysis. In Ref. 5.7 the cylinder is assumed, for simplicity, to be infinitely long and the prebuckling deformation analysis is based on the linear membrane equations. Consequently, the primary equilibrium path is a straight line. The analysis in Ref. 5.7 yields secondary equilibrium paths of the general form of that in Fig. 5.4, where axial load P is plotted vs. the axial displacement u. Points on the primary and secondary equilibrium paths denote circular cylindrical and noncylindrical configurations of the shell, respectively. As noted for flat plates, the equilibrium paths show immediately the existence of a bifurcation point and the magnitude of the corresponding load. However, the equilibrium paths are based on approximate numerical solutions of nonlinear differential equations. The purpose of the stability analysis of the following section is to permit determination of the bifurcation-point load for various methods of loading through solution of only linear differential equations.

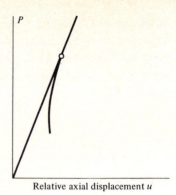

FIGURE 5.4
Equilibrium paths for cylinder subjected
to axial compression.

Relative axial displacement u

5.3 DONNELL FORM OF THE LINEAR STABILITY EQUATIONS

The equations for determination of the bifurcation-point load of a quasi-shallow circular cylindrical shell subjected to edge loading and dead-load lateral pressure are derived in Sec. 5.3a by application of the adjacent-equilibrium criterion. The equations are rederived in Sec. 5.3b by use of the minimum potential energy criterion.

5.3a Adjacent-Equilibrium Criterion

To investigate the possible existence of adjacent-equilibrium configurations, we give small increments to the *displacement* variables and examine the two adjacent configurations represented by the displacements before and after the increment, as follows. Let

$$u \to u_0 + u_1$$
$$v \to v_0 + v_1 \qquad (5.18)$$
$$w \to w_0 + w_1$$

where (u_0,v_0,w_0) and (u,v,w) are adjacent equilibrium configurations corresponding to a single value of applied load (no increment is given to the applied pressure p), and (u_1,v_1,w_1) is an arbitrarily small increment. The mechanics of the linearization procedure are simplest if applied to a relatively compact form of the nonlinear equations such as Eqs. (5.8). Accordingly, we let

$$N_x \to N_{x0} + \Delta N_x$$
$$N_\theta \to N_{\theta 0} + \Delta N_\theta \qquad (5.19)$$
$$N_{x\theta} \to N_{x\theta 0} + \Delta N_{x\theta}$$

where terms with 0 subscripts correspond to the u_0, v_0, w_0 displacement, and the ΔN_x, ΔN_θ, $\Delta N_{x\theta}$ are increments corresponding to u_1, v_1, w_1. We also let N_{x1}, $N_{\theta 1}$, $N_{x\theta 1}$ represent the portions of ΔN_x, ΔN_θ, $\Delta N_{x\theta}$, respectively, that are linear in u_1, v_1, w_1. For example, from Eqs. (5.6) and (5.7),

$$N_x = C\left[(u_{,x} + \tfrac{1}{2}w_{,x}^2) + v\left(\frac{v_{,\theta} + w}{a} + \frac{1}{2}\frac{w_{,\theta}^2}{a^2}\right)\right]$$

Then

$$N_x + \Delta N_x = C\left[(u_{0,x} + u_{1,x} + \tfrac{1}{2}w_{0,x}^2 + w_{0,x}w_{1,x} + \tfrac{1}{2}w_{1,x}^2)\right.$$
$$\left. + v\left(\frac{v_{0,\theta} + w_0}{a} + \frac{v_{1,\theta} + w_1}{a} + \frac{1}{2}\frac{w_{0,\theta}^2}{a^2} + \frac{w_{0,\theta}w_{1,\theta}}{a^2} + \frac{1}{2}\frac{w_{1,\theta}^2}{a^2}\right)\right]$$

Therefore

$$N_{x0} = C\left[(u_{0,x} + \tfrac{1}{2}w_{0,x}^2) + v\left(\frac{v_{0,\theta} + w_0}{a} + \frac{1}{2}\frac{w_{0,\theta}^2}{a^2}\right)\right]$$

$$\Delta N_x = C\left[(u_{1,x} + w_{0,x}w_{1,x} + \tfrac{1}{2}w_{1,x}^2)\right.$$
$$\left. + v\left(\frac{v_{1,\theta} + w_1}{a} + \frac{w_{0,\theta}w_{1,\theta}}{a} + \frac{1}{2}\frac{w_{1,\theta}^2}{a^2}\right)\right]$$

and

$$N_{x1} = C\left[(u_{1,x} + w_{0,x}w_{1,x}) + v\left(\frac{v_{1,\theta} + w_1}{a} + \frac{w_{0,\theta}w_{1,\theta}}{a}\right)\right]$$

In the equations obtained on introduction of Eqs. (5.18) and (5.19) into Eqs. (5.8), all terms in u_0, v_0, w_0 alone (or their counterpart in the form of N_{x0}, etc.) and the p term drop out because u_0, v_0, w_0 is an equilibrium configuration, i.e., a solution to the equilibrium equations. Furthermore, quadratic and higher-order terms in u_1, v_1, w_1 (or their counterpart in the form of N_{x1}, etc.) may be neglected because of the smallness of u_1, v_1, w_1. The resulting equations are

$$aN_{x1,x} + N_{x\theta 1,\theta} = 0$$

$$aN_{x\theta 1,x} + N_{\theta 1,\theta} = 0$$

$$D\nabla^4 w_1 + \frac{1}{a}N_{\theta 1} - \left[(N_{x0}w_{1,xx} + w_{0,xx}N_{x1})\right. \tag{5.20}$$

$$\left. + \frac{2}{a}(N_{x\theta 0}w_{1,x\theta} + w_{0,x\theta}N_{x\theta 1}) + \frac{1}{a^2}(N_{\theta 0}w_{1,\theta\theta} + w_{0,\theta\theta}N_{\theta 1})\right] = 0$$

where

$$N_{x0} = C(\varepsilon_{x0} + v\varepsilon_{\theta0}) \qquad N_{x1} = C(\varepsilon_{x1} + v\varepsilon_{\theta1})$$

$$N_{\theta0} = C(\varepsilon_{\theta0} + v\varepsilon_{x0}) \qquad N_{\theta1} = C(\varepsilon_{\theta1} + v\varepsilon_{x1}) \qquad (5.21)$$

$$N_{x\theta0} = C\frac{1-v}{2}\gamma_{x\theta0} \qquad N_{x\theta1} = C\frac{1-v}{2}\gamma_{x\theta1}$$

and

$$\varepsilon_{x0} = u_{0,x} + \tfrac{1}{2}w_{0,x}{}^2 \qquad\qquad \varepsilon_{x1} = u_{1,x} + w_{0,x}w_{1,x}$$

$$\varepsilon_{\theta0} = \frac{v_{0,\theta} + w_0}{a} + \frac{1}{2}\frac{w_{0,\theta}{}^2}{a^2} \qquad \varepsilon_{\theta1} = \frac{v_{1,\theta} + w_1}{a} + \frac{w_{0,\theta}w_{1,\theta}}{a^2} \qquad (5.22)$$

$$\gamma_{x\theta0} = \left(v_{0,x} + \frac{u_{0,\theta}}{a}\right) + \frac{w_{0,x}w_{0,\theta}}{a} \qquad \gamma_{x\theta1} = \left(v_{1,x} + \frac{u_{1,\theta}}{a}\right) + \frac{w_{0,x}w_{1,\theta}}{a} + \frac{w_{0,\theta}w_{1,x}}{a}$$

Equations (5.20) correspond to Eqs. (3.34) for plates.

In stability applications the displacement (u_0,v_0,w_0) is commonly called the *prebuckling deformation*, and (u_1,v_1,w_1) is called the *buckling mode*. Equations (5.20) to (5.22) differ from corresponding expressions for edge-loaded flat plates in Eqs. (3.34) to (3.35) in that they include terms $w_{0,x}$, $w_{0,\theta}$, representing prebuckling rotations. Unfortunately, the deformation of shells prior to loss of stability is not rotation-free. The presence of these prebuckling rotation terms in the stability equations introduces a substantial complication that did not arise for straight bars, flat plates, or circular rings in earlier chapters. Fortunately, the influence of prebuckling rotations is negligibly small in many instances. Consequently, we neglect the influence of prebuckling rotations in the remainder of this chapter, and defer a discussion of the general influence of prebuckling nonlinearity to later chapters.

Accordingly, omission of terms containing the rotations $w_{0,x}$, $w_{0,\theta}$ from Eqs. (5.20) gives, for the stability equations, the expressions

$$aN_{x1,x} + N_{x\theta1,\theta} = 0$$

$$aN_{x\theta1,x} + N_{\theta1,\theta} = 0 \qquad (5.23)$$

$$D\nabla^4 w_1 + \frac{1}{a}N_{\theta1} - \left(N_{x0}w_{1,xx} + \frac{2}{a}N_{x\theta0}w_{1,x\theta} + \frac{1}{a^2}N_{\theta0}w_{1,\theta\theta}\right) = 0$$

Similar omission of terms containing $w_{0,x}$, $w_{0,\theta}$ from the kinematic relations reduces Eqs. (5.22) to the form:

$$\varepsilon_{x0} = u_{0,x} \qquad\qquad \varepsilon_{x1} = u_{1,x}$$

$$\varepsilon_{\theta 0} = \frac{v_{0,\theta} + w_0}{a} \qquad \varepsilon_{\theta 1} = \frac{v_{1,\theta} + w_1}{a} \qquad (5.24)$$

$$\gamma_{x\theta 0} = v_{0,x} + \frac{u_{0,\theta}}{a} \qquad \gamma_{x\theta 1} = v_{1,x} + \frac{u_{1,\theta}}{a}$$

Equations (5.21), (5.23), and (5.24) lead to a coupled set of three linear homogeneous equations in the variables u_1, v_1, w_1. The equations are shown in explicit form in the following section.

5.3b   Minimum Potential Energy Criterion

In this section Eqs. (5.23) are rederived on the basis of the minimum potential energy criterion.

An expression for the total potential energy of a circular cylindrical shell subjected to dead-load lateral pressure is given by Eqs. (5.9) to (5.11). To obtain the corresponding expression for the second variation of the total potential energy, we let

$$u \to u_0 + u_1$$
$$v \to v_0 + v_1$$
$$w \to w_0 + w_1$$

and collect all terms in the resulting expression that are quadratic in the incremental quantities u_1, v_1, w_1. In this operation we note that the expression for the potential energy Ω of the dead-load lateral pressure is a linear functional of the displacement components, so that $\delta^2\Omega = 0$. Then the second-variation expression is found to be

$$\tfrac{1}{2}\delta^2 V = a\,\frac{C}{2}\iint\left(\varepsilon_{x1}^{\,2} + \varepsilon_{\theta 1}^{\,2} + 2\nu\varepsilon_{x1}\varepsilon_{\theta 1} + \frac{1-\nu}{2}\gamma_{x\theta 1}^{\,2}\right)dx\,d\theta$$

$$+\frac{a}{2}\iint\left(N_{x0}\,w_{1,x}^{\,2} + N_{\theta 0}\,\frac{w_{1,\theta}^{\,2}}{a^2} + 2N_{x\theta 0}\,w_{1,x}\,\frac{w_{1,\theta}}{a}\right)dx\,d\theta$$

$$+a\,\frac{D}{2}\iint\left[w_{1,xx}^{\,2} + \frac{w_{1,\theta\theta}^{\,2}}{a^4} + 2\nu w_{1,xx}\,\frac{w_{1,\theta\theta}}{a^2} + 2(1-\nu)\,\frac{w_{1,x\theta}^{\,2}}{a^2}\right]dx\,d\theta$$

$$(5.25)$$

As noted in Sec. 5.3a, the influence of the prebuckling rotations $w_{0,x}$ and $w_{0,\theta}$ is neglected in the remainder of this chapter. Then the ε_{x1}, $\varepsilon_{\theta1}$, $\gamma_{x\theta1}$, N_{x0}, $N_{\theta0}$, $N_{x\theta0}$ in Eq. (5.25) are given by Eqs. (5.21) and (5.24).

For an integrand F of the form of that in Eq. (5.25), the Euler equations are of the form of those in Eqs. (5.12). Introduction of the integrand in accordance with the Trefftz criterion again yields the cylindrical shell stability equations in Eqs. (5.23).

Introduction into Eqs. (5.23) of a portion of the constitutive and kinematic relations from Eqs. (5.21) and (5.24) and rearrangement now gives

$$a^2 u_{1,xx} + \frac{1-v}{2} u_{1,\theta\theta} + \frac{1+v}{2} av_{1,x\theta} + vaw_{1,x} = 0$$

$$\frac{1+v}{2} au_{1,x\theta} + \frac{1-v}{2} a^2 v_{1,xx} + v_{1,\theta\theta} + w_{1,\theta} = 0$$

$$DV^4 w_1 + \frac{1}{a^2} C(v_{1,\theta} + w_1 + vau_{1,x})$$

$$- \left(N_{x0} w_{1,xx} + \frac{2}{a} N_{x\theta0} w_{1,x\theta} + \frac{1}{a^2} N_{\theta0} w_{1,\theta\theta} \right) = 0$$

$$(5.26)$$

Equations (5.26) are a coupled set of three equations in the variables u_1, v_1, w_1. They are the Donnell stability equations in coupled form. As shown by Donnell in Ref. 5.6, the equations may be partially uncoupled and written in the form [compare Eqs. (5.16)]

$$\nabla^4 u_1 = -\frac{v}{a} w_{1,xxx} + \frac{1}{a^3} w_{1,x\theta\theta} \qquad (5.27a)$$

$$\nabla^4 v_1 = -\frac{2+v}{a^2} w_{1,xx\theta} - \frac{1}{a^4} w_{1,\theta\theta\theta} \qquad (5.27b)$$

$$D\nabla^8 w_1 + \frac{1-v^2}{a^2} Cw_{1,xxxx} - \nabla^4 \left(N_{x0} w_{1,xx} + \frac{2}{a} N_{x\theta0} w_{1,x\theta} + \frac{1}{a^2} N_{\theta0} w_{1,\theta\theta} \right) = 0$$

$$(5.27c)$$

Equations (5.27) are called the *Donnell stability equations in uncoupled form*. Equation (5.27c) is seen to be a homogeneous linear equation in w_1 alone, with variable coefficients in N_{x0}, $N_{\theta0}$, $N_{x\theta0}$. Because the influence of prebuckling rotations has been neglected, the coefficients N_{x0}, $N_{\theta0}$, $N_{x\theta0}$ are governed by the linear equilibrium equations in Eqs. (5.14).

In summary, cylindrical shell stability equations in which prebuckling rotation terms are retained are given in this section in Eqs. (5.20). Corresponding equations in which such terms are omitted are given in mutually equivalent forms in Eqs. (5.23), (5.26), and (5.27).

Applications of the Donnell stability equations are given in Sec. 5.5. Before proceeding to the applications, however, the limitations and advantages of the Donnell stability equations are clarified by examination of a somewhat more accurate set of stability equations for circular cylindrical shells.

5.4 STABILITY EQUATIONS FOR NONSHALLOW CYLINDERS

The Donnell stability equations are accurate only for shallow cylindrical panels or for complete cylinders whose displacement components are rapidly varying functions of the circumferential coordinate, as noted. The limited range of applicability results from two factors: (1) the omission of the displacement component v in the kinematic relation $\beta_\theta = (v - w_{,\theta})/a$, and (2) the omission of quadratic terms in the expression for the potential energy of the applied load. The latter omission is equivalent to the assumption that the applied load acts as a dead load during deformation of the shell. Two sets of stability equations are derived in this section. The first set is suitable for nonshallow cylindrical shells subjected to dead-loading, and the second for nonshallow cylindrical shells subjected to fluid lateral pressure.

5.4a Dead-Loading

The Donnell form of the kinematic relations for cylindrical shells is given in Eqs. (5.7). Reintroduction of the v term in the expression for the rotation β_θ gives

$$\varepsilon_x = u_{,x} + \tfrac{1}{2}\beta_x^2 \qquad \beta_x = -w_{,x} \qquad \kappa_x = \beta_{x,x}$$

$$\varepsilon_\theta = \frac{v_{,\theta} + w}{a} + \tfrac{1}{2}\beta_\theta^2 \qquad \beta_\theta = \frac{v - w_{,\theta}}{a} \qquad \kappa_\theta = \frac{\beta_{\theta,\theta}}{a} \qquad (5.28)$$

$$\gamma_{x\theta} = \left(\frac{u_{,\theta}}{a} + v_{,x}\right) + \beta_x\beta_\theta \qquad \kappa_{x\theta} = \frac{1}{2}\left(\frac{\beta_{x,\theta}}{a} + \beta_{\theta,x}\right)$$

These kinematic relations for ε_θ, β_θ, and κ_θ are the same as those for nonshallow circular rings in Eqs. (4.7), (4.9), and (4.10). Equations (5.28) are also the expressions given by specialization for cylindrical coordinates of the Sanders kinematic relations for shells of general shape in Eqs. (6.6) and (6.7).

(Corresponding linear kinematic relations are obtained by omission of the quadratic terms in Eqs. (5.28). The resulting expressions may be written in the form

$$\varepsilon_x = u_{,x} \qquad \beta_x = -w_{,x} \qquad \kappa_x = -w_{,xx}$$

$$\varepsilon_\theta = \frac{v_{,\theta} + w}{a} \qquad \beta_\theta = \frac{v - w_{,\theta}}{a} \qquad \kappa_\theta = \frac{v_{,\theta} - w_{,\theta\theta}}{a}$$

$$\gamma_{x\theta} = \frac{u_{,\theta}}{a} + v_{,x} \qquad\qquad \kappa_{x\theta} = \frac{1}{2}\frac{v_{,x} - 2w_{,x\theta}}{a}$$

These are the kinematic relations given for cylindrical shells in, for example, eqs. (6.8) of Ref. 5.9. They are widely used in the linear stress and deformation analysis of nonshallow cylindrical shells.)

Stability equations based on the kinematic relations of Eqs. (5.28) may be derived by introduction of the relations into the potential energy expression in Eqs. (5.9) to (5.11). The resulting expression for the second variation of the total potential energy of a circular cylindrical shell subjected to dead-load surface pressure may be written in the form

$$\tfrac{1}{2}\delta^2 V = \frac{aC}{2} \iint \left(e_{xx1}^2 + e_{\theta\theta1}^2 + 2v e_{xx1} e_{\theta\theta1} + \frac{1-v}{2} e_{x\theta1}^2 \right) dx\, d\theta$$

$$+ \frac{a}{2} \iint (N_{x0}\beta_{x1}^2 + N_{\theta\theta}\beta_{\theta1}^2 + 2N_{x\theta0}\beta_{x1}\beta_{\theta1})\, dx\, d\theta$$

$$+ \frac{aD}{2} \iint [\chi_{xx1}^2 + \chi_{\theta\theta1}^2 + 2v\chi_{xx1}\chi_{\theta\theta1} + 2(1-v)\chi_{x\theta1}^2]\, dx\, d\theta \qquad (5.29)$$

where terms representing prebuckling rotations have been omitted, and where $e_{xx1}, e_{\theta\theta1}, \ldots, \chi_{x\theta1}$ are linear-displacement parameters defined by the relations [compare Eqs. (5.28)]

$$e_{xx1} = u_{1,x} \qquad \beta_{x1} = -w_{1,x} \qquad \chi_{xx1} = -w_{1,xx}$$

$$e_{\theta\theta1} = \frac{v_{1,\theta} + w_1}{a} \qquad \beta_{\theta1} = \frac{v_1 - w_{1,\theta}}{a} \qquad \chi_{\theta\theta1} = \frac{v_{1,\theta} - w_{1,\theta\theta}}{a^2} \qquad (5.30)$$

$$e_{x\theta1} = \frac{u_{1,\theta}}{a} + v_{1,x} \qquad\qquad \chi_{x\theta1} = \frac{1}{2}\frac{v_{1,x} - 2w_{1,x\theta}}{a}$$

The stability equations are obtained by application of the Trefftz criterion.

Introduction of the integrand in Eq. (5.29) into the Euler equations and rearrangement gives

$$aN_{x1,x} + N_{x\theta1,\theta} = 0$$

$$aN_{x\theta1,x} + N_{\theta1,\theta} + \frac{1}{a}M_{\theta1,\theta} + M_{x\theta1,x} - (N_{\theta0}\beta_{\theta1} + N_{x\theta0}\beta_{x1}) = 0$$

$$aM_{x1,xx} + 2M_{x\theta1,x\theta} + \frac{1}{a}M_{\theta1,\theta\theta} - N_{\theta1} - [aN_{x0}\beta_{x1,x}$$

$$+ N_{x\theta0}(a\beta_{\theta1,x} + \beta_{x1,\theta}) + N_{\theta0}\beta_{\theta1,\theta}] = 0$$

(5.31)

where

$$N_{x1} = C(e_{xx1} + ve_{\theta\theta1}) \qquad M_{x1} = D(\chi_{xx1} + v\chi_{\theta\theta1})$$

$$N_{\theta1} = C(e_{\theta\theta1} + ve_{xx1}) \qquad M_{\theta1} = D(\chi_{\theta\theta1} + v\chi_{xx1}) \qquad (5.32)$$

$$N_{x\theta1} = C\frac{1-v}{2}e_{x\theta1} \qquad M_{x\theta1} = D(1-v)\chi_{x\theta1}$$

and the linear displacement parameters e_{xx1}, $e_{\theta\theta1}, \ldots$, $\chi_{x\theta1}$ are given by Eqs. (5.30).

These equations are the counterpart for nonshallow cylindrical shells of the expressions for quasi-shallow shells in Eqs. (5.23). Introduction of Eqs. (5.30) and (5.32) into Eqs. (5.31) leads to a coupled set of three linear equations in the three variables u_1, v_1, w_1. Such equations correspond to Eqs. (5.26). Unlike Eqs. (5.26), however, there is no apparent way to uncouple the equations to yield relations corresponding to Eqs. (5.27). Consequently, this form of the equations is less convenient in applications than the Donnell equations.

5.4b Fluid-Pressure Loading

Both the shallow and nonshallow shell equations in the preceding sections have been derived for dead loads only. In some situations the solution is inaccurate unless the action of the applied load during shell deformation is taken into account. Examples were examined in Chap. 4, where the effects of fluid pressure and centrally directed pressure were compared for a circular ring. In this section the nonshallow shell equations of Sec. 5.4a are modified for a circular cylindrical shell subjected to fluid-pressure loading that is uniformly distributed over the lateral surface of the cylinder.

For a complete cylindrical shell, the work done by uniform fluid pressure as the shell deforms is the product of the pressure times the change in the volume enclosed by the shell. The system is conservative (Ref. 5.10), and therefore the change in potential energy is the negative of the work done as the shell deforms. The area enclosed by a cross section of the cylinder after deformation is, from Eq. (4.14),

$$A^* = \pi a^2 + a \int \left[w + \frac{1}{2a} (v^2 - vw_{,\theta} + v_{,\theta} w + w^2) \right] d\theta$$

Consequently, for pressure applied to the lateral surface of the cylinder, the change in the potential energy of the applied pressure during the deformation is approximately

$$\Omega = -pa \iint \left[w + \frac{1}{2a} (v^2 - vw_{,\theta} + v_{,\theta} w + w^2) \right] dx\, d\theta \qquad (5.33)$$

where p is positive outward. The expression for dead-load lateral pressure in Eq. (5.11) is obtained from this equation by omission of quadratic terms in the displacement components.

The second variation of the potential energy of the fluid-pressure loading is seen to be

$$\tfrac{1}{2}\, \delta^2 \Omega = -\tfrac{1}{2} p \iint (v_1{}^2 - v_1 w_{1,\theta} + v_{1,\theta} w_1 + w_1{}^2)\, dx\, d\theta \qquad (5.34)$$

where the $\frac{1}{2}$ is retained on both sides of the equation to avoid ambiguity. Addition of the second-variation expression in Eq. (5.34) to that in Eq. (5.29) gives an equation for the second variation of the total potential energy of a cylindrical shell subjected to lateral fluid pressure loading. Application of the Trefftz criterion to the resulting second-variation expression leads to a set of three stability equations corresponding to Eqs. (5.31). The equations are found to be

$$aN_{x1,x} + N_{x\theta1,\theta} = 0 \qquad (5.35a)$$

$$aN_{x\theta1,x} + N_{\theta1,\theta} + \frac{1}{a} M_{\theta1,\theta} + M_{x\theta1,x} - (N_{\theta0}\beta_{\theta1} + N_{x\theta0}\beta_{x1})$$
$$+ p(v_1 - w_{1,\theta}) = 0 \qquad (5.35b)$$

$$aM_{x1,xx} + 2M_{x\theta1,x\theta} + \frac{1}{a} M_{\theta1,\theta\theta} - N_{\theta1}$$
$$- [aN_{x0}\beta_{x1,x} + N_{x\theta0}(a\beta_{\theta1,x} + \beta_{x1,\theta}) + N_{\theta0}\beta_{\theta1,\theta}] + p(v_{1,\theta} + w_1) = 0 \qquad (5.35c)$$

These stability equations for fluid lateral pressure are the counterpart for the cylinder of Eqs. (4.20) for the circular ring. They differ from Eqs. (5.31) for dead loading only in the addition of the terms containing p in Eqs. (5.35b) and (5.35c).

Numerical results based on Eqs. (5.35) are compared in the following section with those based on the Donnell equations and on the well-known Flügge equations for fluid lateral pressure. [For values of the wavelength parameter $n = 2$, results based on Eqs. (5.35) are found to be in close agreement with the Flügge equations, and those based on the Donnell equations are found to be 25 percent too high.]

As noted in Chap. 4, Koiter shows in Ref. 5.11 that the simplifying assumption that the applied load may be treated as a dead load is always justified if the shell is shallow or quasi-shallow.

5.5 APPLICATIONS OF THE STABILITY EQUATIONS

Applications of the Donnell stability equations are presented in this section. A comparison of numerical results based on the equations for nonshallow cylinders with results based on the Donnell equations is included in Sec. 5.5a. In all the applications in this chapter the prebuckling rotation terms are omitted. Examples in which such terms are retained are treated in Chap. 9.

For notational simplicity, the subscripts 1 are omitted from the incremental quantities (for example, w_1, $N_{x\theta1}$, M_{x1}, etc.) in the equations of Sec. 5.5.

5.5a Uniform Lateral Pressure

As a first example let us consider a cylindrical shell that is simply supported at its ends and subjected to uniform external lateral pressure p_e, in pounds per square inch. Under such loading the prebuckling deformation of the shell is axisymmetric, as illustrated in Fig. 5.5a. The critical pressure p_{cr} is defined as the lowest pressure at which the axisymmetric form loses its stability.

If the cylinder is free to extend longitudinally as the lateral pressure is applied, $N_{x0} = 0$. Furthermore, $N_{x\theta0} = 0$ in the absence of torsional loading. Introduction of these values into Eq. (5.27c) simplifies that expression to the form

$$DV^8 w + \frac{1 - v^2}{a^2} C w_{,xxxx} - \frac{1}{a^2} \nabla^4 (N_{\theta 0} w_{,\theta\theta}) = 0 \qquad (5.36)$$

Equation (5.36) is a linear equation with a variable coefficient $N_{\theta 0}(x)$. Because the prebuckling deformation entails bending, the coefficient is governed

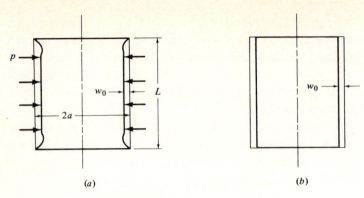

FIGURE 5.5
Cylinder subjected to uniform external lateral pressure.

by the linear bending equations in Eqs. (5.14). Specialization of those equations for axial symmetry gives

$$N'_x = 0 \qquad (5.37a)$$

$$Dw^{iv} + \frac{1}{a} N_\theta = -p_e \qquad (5.37b)$$

[Note that p_e in the example is positive inward, whereas p in Eqs. (5.14) is positive outward.] From the constitutive and kinematic relations, $N_x = C(u_{,x} + vw)$ and $N_\theta = C(w + vu_{,x})$. But $C = Eh/(1 - v^2)$ and, as noted, $N_x = 0$. Therefore

$$N_\theta = Eh \frac{w}{a} \qquad (5.38)$$

Introduction of this relation into Eq. (5.37b) gives

$$Dw^{iv} + \frac{Ehw}{a^2} = -p_e \qquad (5.39)$$

Equations (5.38) and (5.39) furnish the required relationship between the variable coefficient $N_{\theta 0}$ and the applied load p_e.

Equation (5.39) is the well-known equilibrium equation for axisymmetric deformation of a cylindrical shell. It is frequently called the *beam on an elastic foundation equation*, because of its similarity to the equation for displacements of a beam resting on a Winkler foundation. Solution of the equation (see, for example, Ref. 5.9) shows that the function $w(x)$ includes a rapidly decaying exponential factor, so that bending of the shell wall is confined to narrow " boundary" or "discontinuity" zones adjacent to the cylinder ends, as shown in the sketch of the axisymmetric form in Fig. 5.5a. In the interest of simplicity, this localized bending frequently is neglected, and the axisymmetric form is assumed to be cylindrical over its entire length, as illustrated in Fig. 5.5b. Then the coefficient $N_{\theta 0}$ is governed by the linear membrane equations in Eqs. (5.17) rather than by the linear bending equations. This assumption has the major analytical advantage that the governing stability equation then has a constant rather than a variable coefficient. We adopt such a simplification here and defer discussion of variable-coefficient stability equations to Chap. 8.

From the linear membrane equations in Eq. (5.17),

$$N_{\theta 0} = -p_e a \qquad (5.40)$$

Introduction into Eq. (5.36) gives the following stability equation for a cylindrical shell subjected to uniform external lateral pressure:

$$D\nabla^8 w + \frac{1-\nu^2}{a^2}\,Cw_{,xxxx} + \frac{1}{a}\,p_e\nabla^4 w_{,\theta\theta} = 0 \qquad (5.41)$$

For simply supported ends the boundary conditions for Eq. (5.41) are that w be periodic in θ and that

$$w = w_{,xx} = 0 \qquad \text{on } x = 0, L \qquad (5.42)$$

[see Eqs. (3.46)]. Equation (5.41) is a constant-coefficient equation. Solutions of the form

$$w = C_1 \sin \bar{m}x \sin n\theta \qquad (5.43)$$

where C_1 is a constant, $\bar{m} \equiv m\pi a/L$, and $m, n = 1, 2, 3, \ldots$, are seen to satisfy both the differential equation and the boundary conditions. Introduction into Eq. (5.41) and rearrangement gives

$$p_e a = \frac{(\bar{m}^2 + n^2)^2}{n^2}\frac{D}{a^2} + \frac{\bar{m}^4}{n^2(\bar{m}^2 + n^2)^2}(1-\nu^2)C \qquad (5.44)$$

A distinct eigenvalue corresponds to each pair of values of m and n. The smallest eigenvalue in every case is seen to correspond to $m = 1$ (that is, $\bar{m} = \pi a/L$). Accordingly, for the smallest eigenvalues,

$$p_e a = \frac{[(\pi a/L)^2 + n^2]^2}{n^2}\frac{D}{a^2} + \frac{(\pi a/L)^4}{n^2[(\pi a/L)^2 + n^2]^2}(1-\nu^2)C$$

Introduction of the definitions $C = Eh/(1-\nu^2)$ and $D = Eh^3/[12(1-\nu^2)]$ and rearrangement gives

$$\frac{p_e a}{Eh} = \frac{[(\pi a/L)^2 + n^2]^2}{n^2}\frac{(h/a)^2}{12(1-\nu^2)} + \frac{(\pi a/L)^4}{n^2[(\pi a/L)^2 + n^2]^2} \qquad (5.45)$$

For particular values of L/a and a/h, the n corresponding to the smallest eigen-value may be determined by trial.

As an example, let us calculate p_{cr} for a 40-in. diameter cylinder ($a = 20$ in.) that is 20 in. long and has a wall thickness $h = 0.200$ in. The shell material is aluminum, with $E = 10 \times 10^6$ psi and $v = 0.3$. From Eq. (5.45) we find that for $n = 7$, 8, and 9, respectively, $p_e = 117$, 106, and 108 psi, and that p_e is higher for all other values of n. Consequently, $p_{cr} = 106$ psi. The values for the wave-length parameters m and n indicate that, at the bifurcation point, the shell con-figuration on the secondary equilibrium path has one half sine wave in the axial direction and eight complete sine waves in the circumferential direction.

For L/a approaching infinity, Eq. (5.45) reduces to the expression

$$p_e = n^2 \frac{D}{a^3} \qquad (5.46)$$

This result agrees with the value given by the Donnell analysis of the circular ring in Eq. (4.49). For $n = 2$, the eigenvalue is 25 percent higher than the classical value for fluid-pressure loading, as noted.

Although the n in Eq. (5.45) is not a continuous variable, a close estimate of the magnitude of p_{cr} may be obtained through analytical minimization, as discussed in Chap. 2 for the column on an elastic foundation. Results of such a minimization procedure for stability analyses based on the Donnell equations may be presented in a particularly simple form by introduction of the following parameters (see Ref. 5.12). Let

$$\bar{p} \equiv \frac{L^2 a}{\pi^2 D} p_e \qquad \bar{n} \equiv \frac{nL}{\pi a} \qquad Z \equiv \frac{L^2}{ah} (1 - v^2)^{1/2} \qquad (5.47)$$

where $\bar{p}$ = a nondimensional pressure parameter
$\bar{n}$ = a circumferential wavelength parameter
Z = a cylinder geometry variable widely known as the Batdorf parameter.

In terms of these quantities Eq. (5.45) may be rewritten in the simpler form

$$\bar{p} = \frac{(1 + \bar{n}^2)^2}{\bar{n}^2} + \frac{1}{\bar{n}^2(1 + \bar{n}^2)^2} \frac{12}{\pi^4} Z^2 \qquad (5.48)$$

Analytical minimization of $\bar{p}$ with respect to $\bar{n}$ in Eq. (5.48) now gives $\bar{p}_{cr}$ as a function of the single geometric parameter Z instead of the two parameters L/a and a/h. Results of such a procedure are shown in Fig. 5.6 (from Ref. 5.12, Fig. 1). For relatively long cylinders (large values of L or Z) the analytical minimization procedure is excessively inaccurate, because n is small and may not satisfactorily be treated as a continuous variable.

For external lateral pressure, the hoop stress σ_θ is related to the applied

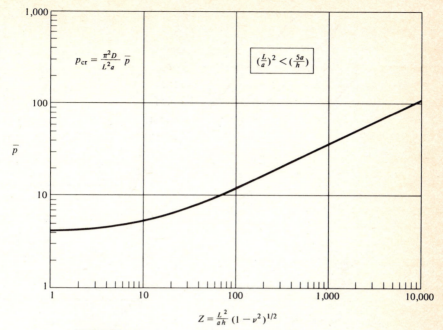

$$Z = \frac{L^2}{ah} (1 - \nu^2)^{1/2}$$

FIGURE 5.6
Critical values of pressure for cylinders subjected to external lateral pressure.

pressure p by the simple equation $\sigma_\theta = pa/h$. Then $\sigma_{\mathrm{cr}} = p_{\mathrm{cr}} a/h$. As the cylinder radius approaches infinity, the value of the parameter Z approaches zero in Fig. 5.6, and the stress σ_{cr} approaches the value given in Fig. 3.8 for long flat plates in compression, namely, $\sigma_{\mathrm{cr}} = 4\pi^2 D/L^2h$.

In Table 5.1 numerical results (from Ref. 5.13) based on Eqs. (5.35) for a

Table 5.1   BUCKLING PRESSURES FOR FLUID
LATERAL PRESSURE

L/a	a/h	Eqs. (5.35)		Flügge	
		$10^6 p_{\mathrm{cr}} a/Eh$	n	$10^6 p_{\mathrm{cr}} a/Eh$	n
32π	1,000	0.2947	2	0.2947	2
	400	1.738	2	1.738	2
	100	27.51	2	27.51	2
8π	1,000	1.111	3	1.111	3
	400	4.971	3	4.972	3
	100	32.79	2	32.80	2
2π	1,000	4.609	6	4.610	6
	400	18.10	5	18.11	5
	100	157.7	4	157.7	4

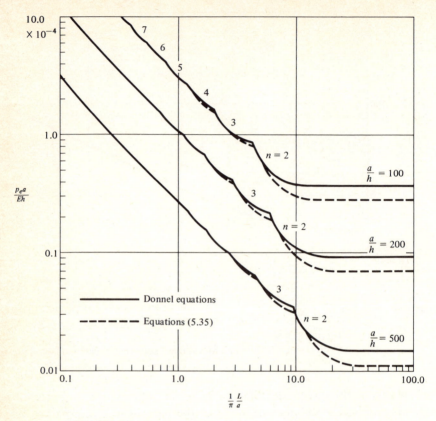

FIGURE 5.7
Critical values of external lateral pressure–comparison of results based on Donnell quasi-shallow-shell equations with those based on the nonshallow-shell equations in Eqs. (5.35).

nonshallow cylindrical shell are compared with those based on the well-known Flügge equations (Ref. 5.14, p. 459) for fluid lateral pressure. [Although written in a somewhat different form, the expression for Ω in Ref. 5.13, eq. (4), can be shown by integration by parts to be the same as that in Eq. (5.33).] Agreement between results based on the two sets of equations is seen to be quite close for nonshallow cylinders, i.e., for small values of the circumferential wavelength parameter n. Numerical results based on the present equations for nonshallow cylindrical shells are compared in Fig. 5.7 with results based on the Donnell equations. For $n = 2$, as noted, the Donnell values are 25 percent too high.

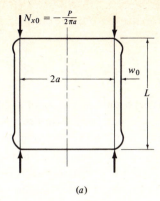 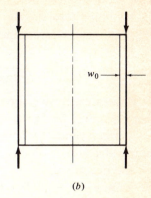

$$N_{x0} = -\frac{P}{2\pi a}$$

(a)

(b)

FIGURE 5.8
Cylinder subjected to axial compression.

5.5b Axial Compression

As a second example we consider a cylinder that is simply supported at its ends and subjected to a uniformly distributed axial compressive load P, in pounds. Under the action of the applied load, the cylinder shortens and, except at the ends, increases in diameter. As in the preceding example, the initial deformation is axisymmetric, and the critical load P_{cr} is the lowest load at which equilibrium in the axisymmetric form ceases to be stable.

A sketch of the axisymmetric form is shown in Fig. 5.8a. We see that w_0 again is a function of x, as for the cylinder subjected to lateral pressure. Again we assume, for simplicity, that the prebuckling deformation may be determined with sufficient accuracy by the linear membrane equations. Thus the axisymmetric form in Fig. 5.8a in effect is replaced by that in Fig. 5.8b.

From a membrane analysis of the unbuckled cylindrical form,

$$N_{x0} = -\frac{P}{2\pi a} \qquad N_{x\theta 0} = N_{\theta 0} = 0$$

Introduction of these values into Eq. (5.27c) gives

$$D\nabla^8 w + \frac{1-v^2}{a^2} C w_{,xxxx} + \frac{P}{2\pi a} \nabla^4 w_{,xx} = 0 \qquad (5.49)$$

Equation (5.49) is a constant-coefficient equation. The boundary conditions and solution form are the same as for the preceding example. Introduction of Eq. (5.43) and rearrangement gives

$$\frac{P}{2\pi a} = \frac{(\overline{m}^2 + n^2)^2}{\overline{m}^2} \frac{D}{a^2} + \frac{\overline{m}^2}{(\overline{m}^2 + n^2)^2} (1 - v^2)C \qquad (5.50)$$

A distinct eigenvalue corresponds to each pair of values of m and n. The pair corresponding to the smallest eigenvalue may be determined by trial.

For cylinders of intermediate length, a close estimate of the smallest eigenvalue may be obtained directly by analytical minimization of P with respect to the quantity $[(\overline{m}^2 + n^2)/\overline{m}]^2$ in Eq. (5.50). Introduction of the definitions $C = Eh/(1 - v^2)$ and $D = Eh^3/[12(1 - v^2)]$ and differentiation leads to the result that P is a minimum for

$$\frac{(\overline{m}^2 + n^2)^2}{\overline{m}^2} = 2[3(1 - v^2)]^{1/2} \frac{a}{h} \qquad (5.51)$$

Introduction into Eq. (5.50) and rearrangement now gives:

$$\frac{P_{cr}}{2\pi a} = \frac{Eh^2/a}{[3(1 - v^2)]^{1/2}}$$

For $\sigma_{cr} \equiv P_{cr}/2\pi ah$, this result may be written

$$\sigma_{cr} = \frac{Eh/a}{[3(1 - v^2)]^{1/2}} \qquad (5.52)$$

Equation (5.52) is considered to be the classical solution for axially compressed cylinders. For $v = 0.3$, Eq. (5.52) becomes

$$\sigma_{cr} = 0.605 \, Eh/a \qquad (5.53)$$

As may be seen from Eq. (5.51), the cylindrical shell subjected to axial compression is a case for which a large number of instability modes correspond to a single bifurcation point. In particular, Eq. (5.52) is the solution for axisymmetric as well as for asymmetric modes (see Prob. 5.10).

Because m and n are positive integers and $\overline{m} \equiv m\pi a/L$, it is impossible to satisfy Eq. (5.51) for short cylinders. Such a difficulty arises for values of the Batdorf parameter Z less than 2.85 (Ref. 5.12, p. 19). In such cases, Eq. (5.50) and the trial-and-error procedure of course may be used to determine the critical load. The calculations are simplified if Eq. (5.50) is first expressed in terms of the parameters Z and $\bar{n}$, as in Sec. 5.5a. Results of such calculations are shown in Fig. 5.9 (Ref. 5.12, fig. 3), where

$$k_a \equiv \frac{L^2 h}{\pi^2 D} \sigma_{cr} \qquad (5.54)$$

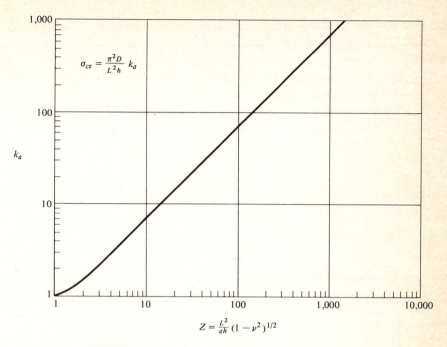

$$\sigma_{cr} = \frac{\pi^2 D}{L^2 h} k_a$$

FIGURE 5.9
Critical values of axial stress for cylinders subjected to axial compression.

For $Z > 2.85$, the values of σ_{cr} given by Fig. 5.9 and Eq. (5.52) are the same.

As the cylinder radius approaches infinity and Z approaches zero, the coefficient k_a in Fig. 5.9 approaches the value 1. Then

$$N_{x0_{cr}} = \sigma_{cr} h = \frac{\pi^2 D}{L^2} \qquad (5.55)$$

This is the equation for the critical load intensity, in pounds per inch, of a "wide column," i.e., a flat plate that is simply supported on the loaded edges and free on the unloaded edges (see Sec. 3.5b).

A very long cylinder can buckle as a column with undeformed cross sections ($m = n = 1$). The present Donnell formulation does not yield the correct results for this case (Ref. 5.15, p. 466), but the critical load is readily determined by setting $I = \pi a^3 h$ in the appropriate column equations in Chaps. 1 and 2. The Donnell formulation also gives somewhat inaccurate results for moderately long cylinders; more accurate values than that in Eq. (5.53) are given by, for example, Ref. 5.15, p. 464, eq. (i).

The sketch of the axially compressed cylindrical shell in Fig. 5.8a shows that, because of the expansion of the shell during loading, the applied load is eccentric relative to the shell wall at midlength. Because of this eccentricity, the prebuckling deformation is nonlinear and the primary equilibrium path is curved from the outset. An analysis of axially compressed cylinders in which this nonlinearity is taken into account is included in Chap. 9. The critical stress is found to be significantly lower than the value given by Eq. (5.52).

5.5c   Torsion

For a cylindrical shell subjected to a twisting moment about its longitudinal axis, Eq. (5.27c) simplifies to the expression

$$DV^8 w + \frac{1 - v^2}{a^2} Cw_{,xxxx} - \frac{2}{a} V^4(N_{x\theta 0}\, w_{,x\theta}) = 0 \qquad (5.56)$$

We assume, for simplicity, that a membrane analysis is adequate for the prebuckling deformation. Then $N_{x\theta 0}$ is a constant, and Eq. (5.56) may be written in the form

$$DV^8 w + \frac{1 - v^2}{a^2} Cw_{,xxxx} - \frac{2}{a} N_{x\theta 0} V^4 w_{,x\theta} = 0 \qquad (5.57)$$

Equation (5.57), like the expression for shear-loaded flat plates in Eq. (3.63), has odd-ordered derivatives with respect to each of the coordinate variables in one term, and even-ordered derivatives in the others. Consequently, a deflection function of the form of that in Eq. (5.43) is not a solution. Under torsional loading, the buckling deformation of a cylindrical shell consists of a number of circumferential waves that spiral around the cylinder from one end to the other. Such waves can be represented by a deflection function of the form

$$w = C_1 \sin (\bar{m}x - n\theta) \qquad (5.58)$$

where C_1 is a constant and $\bar{m} \equiv m\pi a/L$, as before. Equation (5.58) satisfies the differential equation and the requirement of periodicity in the circumferential coordinate, but does not satisfy any of the commonly used boundary conditions at the cylinder ends. Consequently, this simple expression may be used only for long cylinders whose end conditions have little influence on the magnitude of the critical load.

For such cylinders, introduction of Eq. (5.58) into (5.57) and rearrangement gives

$$N_{x\theta 0} = \frac{(\bar{m}^2 + n^2)^2}{2\bar{m}n} \frac{D}{a^2} + \frac{\bar{m}^3}{2(\bar{m}^2 + n^2)^2 n} (1 - v^2)C \qquad (5.59)$$

A distinct eigenvalue corresponds to each pair of values of the wavelength parameters m and n. For long cylinders the shell buckles in two circumferential waves; i.e., the smallest values of $N_{x\theta 0}$ correspond to $n = 2$ (Ref. 5.6). Introduction of that value for n gives

$$N_{x\theta 0} = \frac{(\bar{m}^2 + 4)^2}{4\bar{m}} \frac{D}{a^2} + \frac{\bar{m}^3}{4(\bar{m}^2 + 4)^2} (1 - v^2)C$$

For sufficiently long cylinders, $(m\pi a/L)^2 \ll 4$. Therefore, approximately,

$$N_{x\theta 0} = \frac{4D}{\bar{m}a^2} + \frac{\bar{m}^3(1 - v^2)C}{64}$$

The value of $\bar{m}$ for which $N_{x\theta 0}$ is a minimum may be determined by analytical minimization of $N_{x\theta 0}$ with respect to $\bar{m}$. Introduction of the definitions $C = Eh/(1 - v^2)$ and $D = Eh^3/[12(1 - v^2)]$ and differentiation leads to the result that $N_{x\theta 0}$ is a minimum for

$$\bar{m}^4 = \frac{64}{9(1 - v^2)} \left(\frac{h}{a}\right)^2$$

Introduction into the expression for $N_{x\theta 0}$ and rearrangement now gives

$$\tau_{cr} = \frac{0.272E}{(1 - v^2)^{3/4}} \left(\frac{h}{a}\right)^{3/2} \qquad (5.60)$$

where $\tau_{cr} \equiv N_{x\theta 0}/h$. Equation (5.60) is the result given by Donnell in Ref. 5.6, eq. (2). For $n = 2$, the Donnell equations are inaccurate, as noted; Timoshenko and Gere give a coefficient of 0.236 rather than 0.272 for this case [Ref. 5.15, eq. (11.27)].

For shorter cylinders, a solution can be obtained by use of a deflection function composed of a finite sum of terms of the form of that in Eq. (5.58). Such a procedure is followed by Donnell in Ref. 5.6, where results are given for both simply supported and clamped ends. The analysis in Ref. 5.6, however, is simplified by omission of a few small terms from the equations. Results of a somewhat more accurate analysis based on the Galerkin method (see Chap. 8) are reported in Ref. 5.12 and are shown in Fig. 5.10 in terms of a nondimensional critical stress coefficient k_t defined by the relation

$$k_t = \frac{L^2 h}{\pi^2 D} \tau_{cr} \qquad (5.61)$$

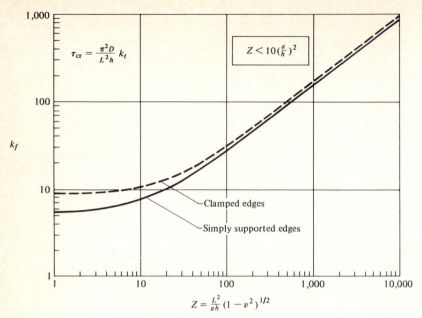

FIGURE 5.10
Critical values of shear stress for cylinders subjected to torsion.

The curves are based on the assumption that the circumferential wavelength parameter n may be treated as a continuous variable, and are limited in applicability to values of the parameter Z less than about $10(a/h)^2$. For longer cylinders, $n = 2$, and Eq. (5.60) rather than Fig. 5.10 is applicable.

As the cylinder radius approaches infinity, the parameter Z approaches zero and the critical stress coefficient k_t for simply supported and clamped edges, respectively, approaches the values 5.35 and 8.98 given in Sec. 3.5c for infinitely long flat plates subjected to shear loading.

5.5d Combined Loading

As an example of combined loading, consider a cylindrical shell with simply supported ends subjected to an axial compressive load P, in pounds, and uniform external lateral pressure p_e, in pounds per square inch. If a membrane analysis again is assumed to be adequate for the axisymmetric deformation, Eq. (5.27c) for such loading specializes to the form

$$D\nabla^8 w + \frac{1 - v^2}{a^2} C w_{,xxxx} + \nabla^4\left(\frac{P}{2\pi a}\, w_{,xx} + \frac{1}{a}\, p_e w_{,\theta\theta}\right) = 0 \qquad (5.62)$$

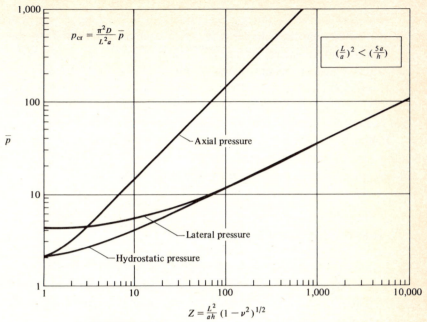

$$Z = \frac{L^2}{ah}(1 - \nu^2)^{1/2}$$

FIGURE 5.11
Comparison of critical values—hydrostatic, lateral, and axial pressure.

Equation (5.62) may be changed to a single-parameter equation by letting

$$\frac{P}{2\pi a} \equiv R p_e a \qquad (5.63)$$

where R is a nondimensional constant. The resulting equation may then be solved for a series of selected values of R (compare Sec. 3.5d).

Introduction into Eq. (5.62) yields the stability equation

$$D\nabla^8 w + \frac{1 - \nu^2}{a^2}Cw_{,xxxx} + p_e\nabla^4\left(\frac{1}{a}w_{,\theta\theta} + aRw_{,xx}\right) = 0 \qquad (5.64)$$

Introduction of the deflection function in Eq. (5.43) and rearrangement now gives

$$p_e a = \frac{(\overline{m}^2 + n^2)^4(D/a^2) + \overline{m}^4(1 - \nu^2)C}{(\overline{m}^2 + n^2)^2(n^2 + R\overline{m}^2)} \qquad (5.65)$$

For any given value of the load ratio R, a distinct eigenvalue corresponds to each pair of values of m and n. The minimum eigenvalue may be determined by trial.

A case of particular interest is represented by the ratio $R = \frac{1}{2}$. For that value $P = \pi a^2 p_e$, and the cylinder is seen to be subjected to a uniform pressure p_e applied to both its lateral surface and its ends. Such loading is termed *hydro-static-pressure loading*. Results based on Eq. (5.65) for the hydrostatic-pressure case are presented in Fig. 5.11 (Ref. 5.12, fig. 5) in terms of the Batdorf parameter Z. Also included for comparison are corresponding results for the pressure p_e applied to the lateral surface only, and to the end surfaces only.

Interaction curves for a wide variety of combinations of compression, bending, and torsional loading are given in Ref. 5.16 in graphical form. A comprehensive list of references also is given in Ref. 5.16.

5.5e   Influence of Boundary Conditions

The uncoupled form of the Donnell stability equations (5.27) is not suited for a general investigation of the influence of end conditions, as may be seen, as follows. The solution given in Eq. (5.43) for the boundary conditions $w = w_{,xx} = 0$ at $x = 0, L$ is of the form

$$w = C_1 \sin \overline{m}x \sin n\theta$$

Introduction of this expression into Eqs. (5.27a) and (5.27b) shows that the corresponding expressions for u and v, respectively, must be of the forms

$$u = A_1 \cos \overline{m}x \sin n\theta$$
$$v = B_1 \sin \overline{m}x \cos n\theta \qquad (5.66)$$

These solution forms are seen to be suitable only for boundary conditions for which $u_{,x} = v = 0$ at $x = 0, L$. The boundary-condition combination $w = w_{,xx} = u = v = 0$, for example, is excluded.

For a general investigation of cylinder end conditions, the *coupled* form of the Donnell equations (5.26) may be used. Those equations are seen to be of second order in u and v and fourth order in w. Therefore, in addition to the periodicity requirement in θ, each set of boundary-condition equations consists of four boundary conditions specified at each end of the cylinder. There are infinitely many such sets of eight boundary conditions each. The conditions need not be the same at the two ends of the cylinder, and one or more of the component conditions may involve elastic restraint against displacement or rotation (cf. the analysis of elastic rotational restraint for columns in Sec. 2.2).

As an example of an analysis of a cylinder whose ends are not simply supported, let us consider a cylindrical shell subjected to external hydrostatic pressure p_e. From a membrane analysis of the prebuckled state under hydrostatic-pressure loading,

$$N_{x0} = -\tfrac{1}{2}p_e a \qquad N_{x\theta 0} = 0 \qquad N_{\theta 0} = -p_e a \qquad (5.67)$$

Introduction of these values into Eqs. (5.26) gives, for the coupled stability equations, the expressions

$$a^2 u_{,xx} + \frac{1-v}{2} u_{,\theta\theta} + \frac{1+v}{2} av_{,x\theta} + vaw_{,x} = 0$$

$$\frac{1+v}{2} au_{,x\theta} + \frac{1-v}{2} a^2 v_{,xx} + v_{,\theta\theta} + w_{,\theta} = 0 \qquad (5.68)$$

$$D\nabla^4 w + \frac{1}{a^2} C(v_{,\theta} + w + vau_{,x}) + p_e a\left(\frac{1}{2} w_{,xx} + \frac{1}{a^2} w_{,\theta\theta}\right) = 0$$

Now if we let

$$u = u_n(x) \cos n\theta$$
$$v = v_n(x) \sin n\theta \qquad (5.69)$$
$$w = w_n(x) \cos n\theta$$

the stability equations reduce to a coupled set of three constant-coefficient, homogeneous, ordinary differential equations in $u_n(x)$, $v_n(x)$, $w_n(x)$, as follows:

$$a^2 u_n'' - \frac{1-v}{2} n^2 u_n + \frac{1+v}{2} anv_n' + vaw_n' = 0$$

$$-\frac{1+v}{2} anu_n' + \frac{1-v}{2} a^2 v_n'' - n^2 v_n - nw_n = 0 \qquad (5.70)$$

$$D\left(a^2 w_n^{iv} - 2n^2 w_n'' + \frac{1}{a^2} n^4 w_n\right) + C(nv_n + w_n + vau_n')$$

$$+ p_e a(\tfrac{1}{2}a^2 w_n'' - n^2 w_n) = 0$$

where primes denote differentiation with respect to x. The general solution of these constant-coefficient equations may be obtained in a straightforward manner, and the critical load then may be determined for arbitrary end conditions.

Such an analysis has been carried out by Sobel (Ref. 5.17), who has shown the solution of the ordinary differential equations to be of the form

$$u_n(x) = \sum_{i=1}^{4} \{\bar{u}_i A_i \sinh [(r_i)^{1/2} x] + \bar{u}_i B_i \cosh [(r_i)^{1/2} x]\}$$

$$v_n(x) = \sum_{i=1}^{4} \{\bar{v}_i A_i \cosh [(r_i)^{1/2} x] + \bar{v}_i B_i \sinh [(r_i)^{1/2} x]\} \qquad (5.71)$$

$$w_n(x) = \sum_{i=1}^{4} \{A_i \cosh [(r_i)^{1/2} x] + B_i \sinh [(r_i)^{1/2} x]\}$$

where

$$\bar{u}_i = \frac{-(r_i)^{1/2}(n^2 + v r_i)}{(r_i - n^2)^2}$$

$$\bar{v}_i = \frac{n[(2 + v)r_i - n^2]}{(r_i - n^2)^2} \qquad i = 1, \ldots, 4$$

and where $r_i (i = 1, \ldots, 4)$ are the roots of the characteristic equation

$$(r - n^2)^4 + 12(1 - v^2)\left(\frac{a}{h}\right)^2 \left[(\tfrac{1}{2}r - n^2)(r - n^2)^2 \frac{p_e a}{Eh} + r^2\right] = 0 \qquad (5.72)$$

Introduction of Eqs. (5.71) into four boundary-condition equations for each end of the cylinder leads to a set of eight linear homogeneous algebraic equations in the A_i, B_i ($i = 1, \ldots, 4$). The equation system has nontrivial solutions only for discrete values of the applied pressure p_e. The eigenvalues are solutions to the equation obtained by setting the determinant of the coefficients of the A_i, B_i equal to zero. The minimum eigenvalue for each particular case (each wave-number) is determined by trial, as before.

Table 5.2　INFLUENCE OF BOUNDARY
CONDITIONS

Cylinder subjected to uniform lateral pressure

	Boundary-condition sets				Critical pressure
1	$w = 0$	$w_{,xx} = 0$	$N_x = 0$	$v = 0$	1.00
2	$w = 0$	$w_{,xx} = 0$	$u = 0$	$v = 0$	1.24
3	$w = 0$	$w_{,xx} = 0$	$N_x = 0$	$N_{x\theta} = 0$	0.96
4	$w = 0$	$w_{,xx} = 0$	$u = 0$	$N_{x\theta} = 0$	1.16
5	$w = 0$	$w_{,x} = 0$	$N_x = 0$	$v = 0$	1.12
6	$w = 0$	$w_{,x} = 0$	$u = 0$	$v = 0$	1.33
7	$w = 0$	$w_{,x} = 0$	$N_x = 0$	$N_{x\theta} = 0$	1.12
8	$w = 0$	$w_{,x} = 0$	$u = 0$	$N_{x\theta} = 0$	1.31

Numerical results from Ref. 5.17 for a cylinder with $a/h = 100$ and $L/a = 1$ are shown in Table 5.2 for eight sets of boundary conditions. For all the sets shown, the boundary conditions at the two ends of the cylinder are the same, and $w = 0$ at $x = 0, L$. For ease of comparison, the results for critical hydrostatic pressure in the table are normalized with respect to the value for the first set, $w = w_{,xx} = N_x = v = 0$. It may be noted that the quantity $N_{x\theta}$ in the table is related to u, v by Eqs. (5.21) and (5.24), as follows: $N_{x\theta} = C(v_{,x} + u_{,\theta})(1 - v)/2$. These results show, surprisingly, that rotational restraint ($w_{,x} = 0$) is less influential in this application than axial restraint ($u = 0$).

For stability equations with variable coefficients, of course, this procedure is not applicable; in such problems a numerical method of analysis must be employed. Indeed, the amount of labor involved in the procedure illustrated in this section is so great that most such applications in the literature utilize numerical methods even for cases in which the stability equations have constant coefficients.

Additional results on the influence of boundary conditions for circular cylindrical shells are included in Chap. 9, where it is shown that the critical load for cylinders subjected to axial compression is strongly affected by cylinder end conditions.

5.6   ORTHOTROPIC CYLINDRICAL SHELLS

Orthotropic cylindrical shells are widely used in structural applications. Examples of shell-wall construction that may be treated as orthotropic include cylindrical sheets stiffened by closely spaced circular rings or longitudinal stringers, fiber-reinforced shells, and corrugated-skin construction. For shell-wall construction that is not symmetrical relative to the shell middle surface, there is a coupling between extensional forces and curvature change and between bending moments and extensional strain. To account for this coupling effect the constitutive equations for orthotropic construction given in Eqs. (3.71) may be generalized to the form

$$
\begin{aligned}
N_x &= C_{11}\varepsilon_x + C_{12}\varepsilon_\theta + C_{14}\kappa_x + C_{15}\kappa_\theta \\
N_\theta &= C_{12}\varepsilon_x + C_{22}\varepsilon_\theta + C_{24}\kappa_x + C_{25}\kappa_\theta \\
N_{x\theta} &= C_{33}\gamma_{x\theta} + C_{36}\kappa_{x\theta} \\
M_x &= C_{14}\varepsilon_x + C_{24}\varepsilon_\theta + C_{44}\kappa_x + C_{45}\kappa_\theta \\
M_\theta &= C_{15}\varepsilon_x + C_{25}\epsilon_\theta + C_{45}\kappa_x + C_{55}\kappa_\theta \\
M_{x\theta} &= C_{36}\gamma_{x\theta} + C_{66}\kappa_{x\theta}
\end{aligned}
\tag{5.73}
$$

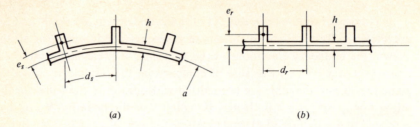

(a)   (b)

FIGURE 5.12
Shell-wall construction for stiffened cylinder.

The magnitudes of the stiffness parameters in Eqs. (5.73) may be determined experimentally, or they may be calculated in terms of the elastic constants E and v of the skin and stiffener materials.

As an illustration let us consider a thin-walled cylindrical shell, reinforced by closely spaced circular rings attached to the inside of the shell skin and with longitudinal stringers attached to the outside, as illustrated in Fig. 5.12. If the stiffeners and skin are made of the same material, the stiffness parameters are given by the following expressions (Ref. 5.18):

$$C_{11} = C + \frac{EA_s}{d_s} \qquad C_{22} = C + \frac{EA_r}{d_r}$$

$$C_{12} = vC \qquad C_{33} = \frac{1-v}{2} C$$

$$C_{14} = e_s \frac{EA_s}{d_s} \qquad C_{15} = 0 \qquad C_{36} = 0$$

$$C_{24} = 0 \qquad C_{25} = -e_r \frac{EA_r}{d_r}$$

$$C_{44} = D + \frac{EI_s}{d_s} \qquad C_{55} = D + \frac{EI_r}{d_r}$$

$$C_{45} = vD \qquad C_{66} = (1-v) D + \frac{1}{2} \left(\frac{GJ_s}{d_s} + \frac{GJ_r}{d_r} \right)$$

(5.74)

where subscripts s and r refer to stringers and rings, respectively; I_s and I_r are moments of inertia of the stiffener cross sections relative to the shell middle surface; and the eccentricities e_s and e_r represent the distance from the shell middle surface to the centroid of the stiffener cross section [compare Eqs. (3.72)]. It is assumed that the effects of moments of inertia about the z axis, warping, and the product of inertia may be neglected. The coupling parameters C_{14} and C_{25} are positive for outside stiffeners and negative for inside ones.

The equilibrium and kinematic relations for cylindrical shells are given by Eqs. (5.5) and (5.7), respectively. Introduction of Eqs. (5.73) into Eqs. (5.5) gives

$$aC_{11}\varepsilon_{x,x} + aC_{12}\varepsilon_{\theta,x} + C_{33}\gamma_{x\theta,\theta} + aC_{14}\kappa_{x,x} = 0$$

$$C_{12}\varepsilon_{x,\theta} + C_{22}\varepsilon_{\theta,\theta} + aC_{33}\gamma_{x\theta,x} + C_{25}\kappa_{\theta,\theta} = 0$$

$$a^2 C_{44}\kappa_{x,xx} + C_{45}(a^2\kappa_{\theta,xx} + \kappa_{x,\theta\theta}) + 2aC_{66}\kappa_{x\theta,x\theta} + C_{55}\kappa_{\theta,\theta\theta}$$
$$+ C_{14}a^2\varepsilon_{x,xx} + C_{25}\varepsilon_{\theta,\theta\theta} - a(C_{12}\varepsilon_x + C_{22}\varepsilon_\theta + C_{25}\kappa_\theta)$$
$$- a[aN_x\beta_{x,x} + N_{x\theta}(a\beta_{\theta,x} + \beta_{x,\theta}) + N_\theta\beta_{\theta,\theta}] = -p$$

where the term in square brackets is retained in its original form in anticipation of subsequent linearization. Introduction of the kinematic relations, linearization, and omission of terms $w_{0,x}$ and $w_{0,\theta}$ representing prebuckling rotations gives

$$C_{11}a^2 u_{1,xx}$$
$$+ C_{33}u_{1,\theta\theta} + (C_{12} + C_{33})av_{1,x\theta} + C_{12}aw_{1,x} - C_{14}a^3 w_{1,xxx} = 0$$

$$(C_{12} + C_{33})au_{1,x\theta} + C_{33}a^2 v_{1,xx} + C_{22}(v_{1,\theta\theta} + w_{1,\theta}) - C_{25}w_{1,\theta\theta\theta} = 0$$

$$C_{44}w_{1,xxxx} + \frac{2}{a^2}(C_{45} + C_{66})w_{1,xx\theta\theta} + \frac{1}{a^4}C_{55}w_{1,\theta\theta\theta\theta}$$

$$+ \frac{1}{a}\left[C_{12}u_{1,x} + \frac{1}{a}C_{22}(v_{1,\theta} + w_1)\right] \qquad (5.75)$$

$$- \left[C_{14}u_{1,xxx} + \frac{1}{a^2}C_{25}(v_{1,\theta\theta\theta} + 2w_{1,\theta\theta})\right]$$

$$- \left(N_{x0}w_{1,xx} + \frac{2}{a}N_{x\theta0}w_{1,x\theta} + \frac{1}{a^2}N_{\theta0}w_{1,\theta\theta}\right) = 0$$

These equations are the same as those given for eccentrically stiffened orthotropic cylindrical shells in Refs. 5.18 and 5.19. For the isotropic case, $A_s = A_r = I_s = I_r = J_s = J_r = 0$, and Eqs. (5.75) reduce to Eqs. (5.26).

As a more specific example, let us consider a stringer-stiffened cylindrical shell subjected to an axisymmetric compressive load P. Then $A_r = I_r = J_r = 0$ in Eqs. (5.74), and $N_{x\theta0} = N_{\theta0} = 0$ and $N_{x0} = -P/2\pi a$ in Eqs. (5.75). For simplicity, we consider only simply supported boundary conditions. Then u_1, v_1, w_1 must be periodic in θ, and

$$w_1 = w_{1,xx} = v_1 = u_{1,x} = 0 \qquad \text{at } x = 0, L$$

The governing equations are constant-coefficient equations. For asymmetric buckling, solutions of the forms of those in Eqs. (5.43) and (5.66) are seen to satisfy both the differential equations and boundary conditions. Introduction into Eqs. (5.75) and rearrangement gives

$$a_{11}A_1 + a_{12}B_1 + a_{13}C_1 = 0$$
$$a_{12}A_1 + a_{22}B_1 + a_{23}C_1 = 0 \qquad (5.76)$$
$$a_{13}A_1 + a_{23}B_1 + (a_{33} - \bar{P})C_1 = 0$$

where

$$a_{11} = C_{11}\bar{m}^2 + C_{33}n^2$$
$$a_{12} = (C_{12} + C_{33})\bar{m}n$$
$$a_{13} = -\left(C_{12}\bar{m} + \frac{C_{14}}{a}\bar{m}^3\right)$$
$$a_{22} = C_{33}\bar{m}^2 + C_{22}n^2 \qquad (5.77)$$
$$a_{23} = -C_{22}n$$
$$a_{33} = C_{22} + \frac{1}{a^2}[C_{44}\bar{m}^4 + 2(C_{45} + C_{66})\bar{m}^2n^2 + C_{55}n^4]$$

$$\bar{P} = \bar{m}^2 \frac{P}{2\pi a}$$

The homogeneous equation system (5.76) has nontrivial solutions only for discrete values of the load parameter $\bar{P}$. For a nontrivial solution, the determinant of the coefficients of the A_1, B_1, C_1 must equal zero:

$$\begin{vmatrix} a_{11} & a_{12} & a_{13} \\ a_{12} & a_{22} & a_{23} \\ a_{13} & a_{23} & (a_{33} - \bar{P}) \end{vmatrix} = 0$$

Expansion by minors yields

$$a_{13}\begin{vmatrix} a_{12} & a_{13} \\ a_{22} & a_{23} \end{vmatrix} - a_{23}\begin{vmatrix} a_{11} & a_{13} \\ a_{12} & a_{23} \end{vmatrix} + (a_{33} - \bar{P})\begin{vmatrix} a_{11} & a_{12} \\ a_{12} & a_{22} \end{vmatrix} = 0$$

Rearrangement gives the final expression

$$\frac{P}{2\pi a} = \frac{a_{33}}{\bar{m}^2} + \frac{2a_{12}a_{23}a_{13} - a_{22}a_{13}^2 - a_{11}a_{23}^2}{(a_{11}a_{22} - a_{12}^2)\bar{m}^2} \qquad (5.78)$$

where, as noted, $\bar{m} \equiv m\pi a/L$, and $m,n = 1, 2, 3, \ldots$. Equations (5.77) and (5.78) determine the load P as a function of the stiffness parameters C_{ij} and wavelength parameters m, n. For given values of the C_{ij}'s, the values of m and n may be chosen by trial to yield the smallest eigenvalue P_{cr}.

For axisymmetric buckling, the solution forms used for asymmetric buckling may be replaced by the expressions

$$u_1 = A_1 \cos mx$$
$$v_1 = 0 \qquad\qquad (5.79)$$
$$w_1 = C_1 \sin mx$$

Introduction into Eqs. (5.75) and rearrangement gives the expression

$$\frac{P}{2\pi a} = \frac{C_{22}}{\overline{m}^2} + \frac{C_{44}}{a^2}\,\overline{m}^2 - \frac{(C_{12} + C_{14}\,\overline{m}^2)^2}{C_{11}\overline{m}^2} \qquad (5.80)$$

For any particular example, the critical load is the smaller of the values given by Eq. (5.78) or (5.80).

As a numerical example, we consider the stringer-stiffened aluminum shell illustrated in Fig. 5.12a. Let $E = 10 \times 10^6$ psi, $v = 0.3$, $h = 0.0283$ in., $a = 9.55$ in., $L = 38.0$ in., $d_s = 1.00$ in., $e_s = 0.165$ in. The stringers are 0.300 in. high and 0.097 in. wide. Then $A_s = (0.300)(0.097) = 0.0291$ in.^{2}, $I_s = (\frac{1}{12})(0.097)$ $(0.300)^3 + (0.0291)(0.165)^2 = 0.001382$ in.^{4}, $J_s = (\frac{1}{3})(0.300)(0.097)^3 = 0.0000913$ in.^{4}, $C = 311 \times 10^3$ lb/in., and $D = 20.7$ lb-in. (cf. the example in Sec. 3.6). Introduction of these values into Eqs. (5.74) gives, for the stiffness parameters, the values $C_{11} = 602 \times 10^3$, $C_{22} = 311 \times 10^3$, $C_{12} = 93.3 \times 10^3$, and $C_{33} = 109 \times 10^3$ lb/in.; $C_{14} = 48.1 \times 10^3$ lb; $C_{25} = 0$; $C_{44} = 13.84 \times 10^3$, $C_{55} = 20.7$, $C_{45} = 6.23$, and $C_{66} = 190.1$ lb-in.

With these values for the cylinder dimensions and stiffness parameters, Eqs. (5.77) and (5.78) express the load P as a function of the positive integers m and n. By trial and error the minimum value of P is found to correspond to $m = 1$ and $n = 6$. For these values, $P_{cr} = 68,600$ lb.

Numerical results based on Eqs. (5.75) for other stiffener configurations and other boundary conditions are given in both Refs. 5.18 and 5.19. Because of the large number of geometric variables in orthotropic construction, it is not practical to provide the designer with a set of charts from which to determine the critical load. In general, he must resort to the use of a computer program. Use of the computer is especially essential for shell configurations more complicated than cylinders. A comprehensive computer program for the buckling of shells of revolution with various wall constructions, including cylindrical shells as a special case, is described in Ref. 5.20. Explicit provision is made in the program for the following wall constructions: (1) shells with circumferential and longitudinal stiffening, (2) shells with skew stiffeners, (3) fiber-reinforced (layered) shells, (4) isotropic and orthotropic layered shells, (5) corrugated ring-stiffened shells, and (6) shells with one corrugated and one smooth skin, with rings. Such comprehensive programs are coming into common use in structural applications.

It should be noted that the results of Refs. 5.18, 5.19, and 5.20 show a surprisingly large effect of stringer eccentricity. This phenomenon was first noted by van der Neut (Ref. 5.21).

5.7   FAILURE OF CYLINDRICAL SHELLS

The equilibrium paths in Fig. 5.4 for an initially perfect cylindrical shell subjected to axial compression are shown again in Fig. 5.13. Also included in Fig. 5.13 are corresponding curves for two slightly imperfect shells. Three characteristics are apparent from the diagram: (1) The buckling load represents the ultimate strength of the structure. (2) The buckling load of the imperfect shell may be substantially lower than the bifurcation-point load of the perfect one. (3) For shells that are nominally alike, the buckling loads may vary widely because of small unintentional differences in the initial shape of the shell.

Although the equilibrium paths for axial compression in Fig. 5.13 represent the most extreme case in differences in behavior between perfect and imperfect models, these three characteristics apply for other methods of loading as well. Only for certain stiffened cylindrical shells does the secondary equilibrium path slope upward rather than downward from the bifurcation point. Theoretical and experimental buckling loads for four kinds of loading, namely, axial compression, torsion, uniform lateral pressure, and hydrostatic pressure, are compared in Figs. 5.14 to 5.17 (Ref. 5.16, figs. 1 and 35b). (The theoretical curve shown for axial compression is for cylinders with clamped ends.) The differences

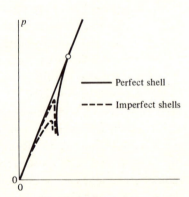

FIGURE 5.13
Equilibrium paths for initially perfect and imperfect cylindrical shells.

Perfect shell

Imperfect shells

Edge displacement u at $x = L$

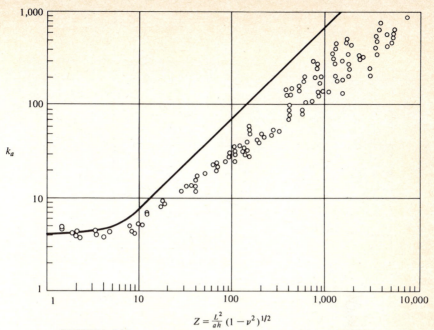

$$Z = \frac{L^2}{ah}(1 - \nu^2)^{1/2}$$

FIGURE 5.14
Comparison of theoretical and experimental values for cylinders subjected to axial compression.

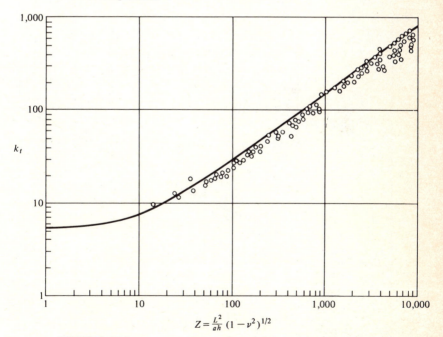

$$Z = \frac{L^2}{ah}(1 - \nu^2)^{1/2}$$

FIGURE 5.15
Comparison of theoretical and experimental values for cylinders subjected to torsion.

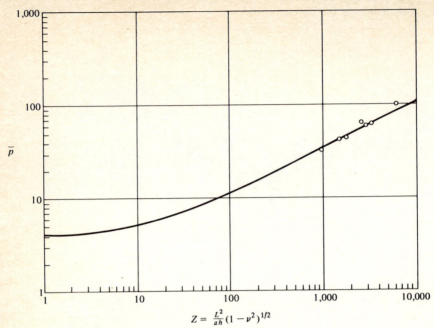

FIGURE 5.16
Comparison of theoretical and experimental values for cylinders subjected to lateral pressure.

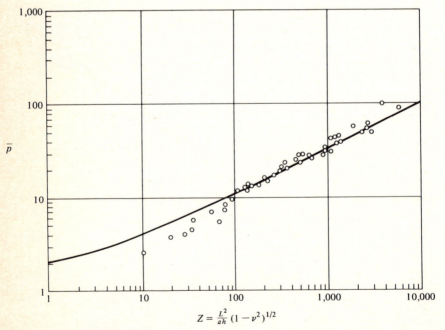

FIGURE 5.17
Comparison of theoretical and experimental values for cylinders subjected to hydrostatic pressure.

between theoretical and experimental values are seen to be substantial in some cases. It should be noted that the graphs are log-log plots that tend to deemphasize the apparent difference between theoretical and experimental loads; the lowest experimental value for $Z = 3,000$ in Fig. 5.14 is only 10 percent of the corresponding theoretical value.

Because of these three characteristics, the design of cylindrical shells is based on the theoretical critical load modified by empirical reduction, or *knock down*, factors for each kind of loading. The magnitude of the reduction factor in each case depends on both the average difference between theoretical and experimental values for the critical load and the severity of scatter of the test data.

As an example of the establishment of reduction factors, let us investigate the establishment of such factors for axially compressed cylindrical shells for values of the Batdorf parameter $Z > 2.85$. For this parameter range the theoretical critical stress is given in Eq. (5.53) as $\sigma_{cr} = 0.605Eh/a$, but the data in Fig. 5.14 show that a coefficient much smaller than 0.605 must be used in design. Furthermore, tests indicate that thinner cylindrical shells require a larger reduction factor than thicker ones because of their greater sensitivity to initial imperfections. Experimental critical stresses from a collection of data in Ref. 5.22 are plotted vs. radius/thickness ratios in Fig. 5.18. Given such information, a reduction factor for each value of the radius/thickness ratio can be obtained by drawing some appropriate lower-bound curve through the test data. The curve shown in the figure is the "90 percent probability" curve recommended for design in Ref. 5.22. The reduction factor for $a/h = 500$, for example, is seen to be 0.24. Multiplication of the coefficient 0.605 by this factor gives, for the design stress, the value $\sigma_{cr} = 0.15Eh/a$.

Cylindrical shell load reductions for other methods of loading are somewhat less severe than those for axial compression. Comprehensive collections of test data and design recommendations for cylindrical shells and curved plates are given in Refs. 5.16 and 5.23. Results for various combinations of compression, torsion, bending, and internal-pressure loading are included. Most industrial organizations establish their own design criteria for shell buckling and issue them in "structural methods" handbooks. They frequently are looseleaf and are continually updated. Such books are in marked contrast to, say, the firmly established, industry-wide Manual of Steel Construction (Ref. 5.24) for the design of columns and beams.

In spite of the wealth of test data for homogeneous isotropic cylinders, this semiempirical basis for design is only partially satisfactory. The number of possible kinds of wall construction (stringer-stiffened, filament-wound, etc.) is

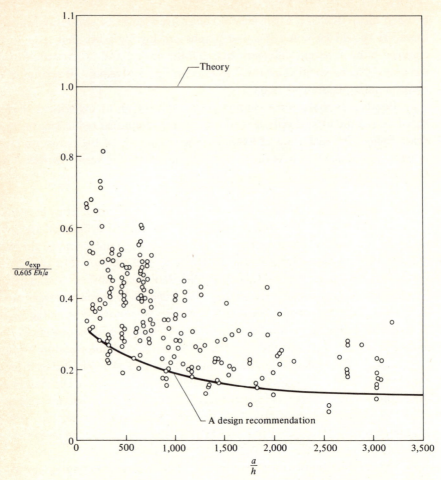

FIGURE 5.18
Distribution of test data for cylinders subjected to axial compression.

limitless. Tests of new configurations are time-consuming and expensive, especially if several specimens of each kind must be tested to compensate for severe scatter of data. As noted, the degree of sensitivity of a particular shell configuration to initial imperfections depends on the shape of the secondary equilibrium path for the initially perfect shell. Consequently, an analysis of the postbuckling behavior of a given shell configuration can provide a useful supplement to the information available to the designer. Analyses of postbuckling behavior are discussed in Chaps. 7 and 9.

REFERENCES

5.1 LORENZ, R.: Die nichtachsensymmetrische Knickung dünnwanger Hohlzylinder, *Phys. Z.*, vol. 13, pp. 241–260, 1911.

5.2 SOUTHWELL, R. V.: On the Collapse of Tubes by External Pressure, *Phil. Mag.*, vol. 25, pp. 687–698, 1913.

5.3 VON MISES, R.: Der kritische Ausendruck zylindrischer Rohre, *Z. Ver. Deutsch. Ing.*, vol. 58, pp. 750–755, 1914.

5.4 FLÜGGE, W.: Die Stabilität der Kreiszylinderschale, *Ing.-Arch.*, vol. 3, pp. 463–506, 1932.

5.5 SCHWERIN, E.: Die Torsions-Stabilität des dünnwandigen Rohres, *Z. Angew. Math. Mech.*, vol. 5, pp. 235–243, 1925.

5.6 DONNELL, L. H.: Stability of Thin-walled Tubes under Torsion, *NACA Rep.* 479, 1933.

5.7 VON KÁRMÁN, T. and H.-S. TSIEN: The Buckling of Thin Cylindrical Shells under Axial Compression, *J. Aeronaut. Sci.*, vol. 8, pp. 303–312, August 1941.

5.8 DONNELL, L. H.: A New Theory for the Buckling of Thin Cylindrical Shells under Axial Compression and Bending, *ASME Trans.*, vol. 56, pp. 795–806, November 1934.

5.9 KRAUS, H.: "Thin Elastic Shells," Wiley, New York, 1967.

5.10 BODNER, S. R.: On the Conservativeness of Various Distributed Force Systems, *J. Aeronaut. Sci.*, vol. 25, pp. 132–133, February 1958.

5.11 KOITER, W. T.: General Equations of Elastic Stability for Thin Shells, *Proc. Symp. Theory of Shells to Honor Lloyd Hamilton Donnell*, University of Houston, Houston, Tex. 1967, pp. 187–223.

5.12 BATDORF, S. B.: A Simplified Method of Elastic Stability Analysis for Thin Cylindrical Shells, *NACA Rep.* 874, 1947.

5.13 BRUSH, D. O., and B. O. ALMROTH: Thin Shell Buckling Analysis: A General Expression for the Second Variation of the Strain Energy, *Proc. Fourth U.S. Nat. Congr. Appl. Mech.*, The American Society of Mechanical Engineers, New York, 1962.

5.14 FLÜGGE, W.: "Stresses in Shells," 2d ed., Springer, Berlin, 1973.

5.15 TIMOSHENKO, S. P., and J. M. GERE: "Theory of Elastic Stability," 2d ed., McGraw-Hill, New York, 1961.

5.16 GERARD, G., and H. BECKER: Handbook of Structural Stability, Part III, Buckling of Curved Plates and Shells, *NACA TN* 3783, 1957.

5.17 SOBEL, L. H.: Effects of Boundary Conditions on the Stability of Cylinders Subject to Lateral and Axial Pressures, *AIAA J.*, vol. 2, pp. 1437–1440, August 1964.

5.18   BARUCH, M., and J. SINGER: Effect of Eccentricity of Stiffeners on the General Instability of Stiffened Cylindrical Shells under Hydrostatic Pressure, *J. Mech. Eng. Sci.*, vol. 5, pp. 23–27, 1963.

5.19   SINGER, J., M. BARUCH, and O. HARARI: On the Stability of Eccentrically Stiffened Cylindrical Shells under Axial Compression, *Intl. J. Solids Structures*, vol. 3, pp. 445–470, 1967.

5.20   ALMROTH, B. O., and D. BUSHNELL: Computer Analysis of Various Shells of Revolution, *AIAA J.*, vol. 6, pp. 1848–1855, October 1968.

5.21   VAN DER NEUT, A.: General Instability of Stiffened Cylindrical Shells under Axial Compression, Nationaal Luchtvaart-Laboratorium, Amsterdam, Report S-314, *Rep. Trans.*, vol. 13, pp. 57–84, 1947.

5.22   HARRIS, L. A., H. S. SUER, W. T. SKENE, and R. J. BENJAMIN: The Stability of Thin-walled Unstiffened Circular Cylinders under Axial Compression, Including the Effects of Internal Pressure, *J. Aeronaut. Sci.*, vol. 24, pp. 587–596, August 1957.

5.23   BAKER, E. H., L. KOVALEVSKY, and F. L. RISH: "Structural Analysis of Shells," McGraw-Hill, New York, 1972.

5.24   Manual of Steel Construction, 7th ed., American Institute of Steel Construction, New York, 1970.

PROBLEMS

5.1   Show that the cylinder equilibrium equations in Eqs. (5.8) may be derived from those in Eqs. (5.5) by introduction of the appropriate constitutive and kinematic relations.

5.2   Show that application of the stationary potential energy criterion to the potential energy expression in Eqs. (5.9) to (5.11) leads to the cylinder equilibrium equations in Eqs. (5.8).

5.3   Following the procedure outlined by Donnell in Ref. 5.6, show that the variable v may be eliminated from Eqs. (5.15a) and (5.15b) by differentiating Eq. (5.15b) with respect to x and θ and differentiating Eq. (5.15a) twice with respect to x and then twice with respect to θ. Following this procedure, derive Eq. (5.16a). Show that the u and v may be eliminated from Eq. (5.15c) by application of the operator ∇^4 and use of Eqs. (5.16a) and (5.16b). In this way, derive Eq. (5.16c).

5.4   Show that application of the Trefftz criterion to the second-variation expression in Eq. (5.25) leads to the cylinder stability equations in Eqs. (5.23).

5.5   Show that introduction of the appropriate constitutive and kinematic relations into Eqs. (5.23) and rearrangement yields the stability equations in Eqs. (5.26).

5.6   An aluminum cylinder with simply supported ends has a diameter of 32 in., a length of 16 in., and a wall thickness of 0.032 in. Compare the critical pressure for external lateral pressure with that for external hydrostatic pressure. Let $E = 10 \times 10^6$ psi and $\nu = 0.3$, and use Eqs. (5.45) and (5.65). Compare these results with values based on Fig. 5.11.

5.7 A stainless-steel cylinder is subjected to external lateral pressure. For $a/h = 500$ and $L/a = 4$, compare values of the critical pressure based on Eq. (5.45) and Figs. 5.7 and 5.11.

5.8 An aluminum cylinder with simply supported ends is subjected to external lateral pressure. If $a/h = 200$ and $L/a = 1.0$, determine the critical pressure based on Eq. (5.45). For the same value of the wavelength parameter n, calculate the corresponding eigenvalue based on the equations for nonshallow shells in Eqs. (5.35). Use the displacement functions in Eqs. (5.43) and (5.66).

5.9 Rederive the expression in Eq. (5.52) for the critical stress of an axially compressed cylinder by introduction of the displacement functions in Eqs. (5.43) and (5.66) into the *coupled* form of the Donnell equations (5.26).

5.10 Determine the value of the wavelength parameter $\bar{m}$ in Sec. 5.6*b* for axisymmetric buckling of an axially compressed cylindrical shell. Show that the critical stress for axisymmetric buckling is given by Eq. (5.52).   *Ans.* $\bar{m} = [3(1 - \nu^2)]^{1/4}(2a/h)^{1/2}$

5.11 Determine the critical load and mode for a simply supported cylinder subjected to axial compression if $a/h = 100$ and L/a is only 0.1. Use Eq. (5.50), and let $E = 10 \times 10^6$ psi and $\nu = 0.3$.

5.12 An aluminum cylinder with simply supported ends has a diameter of 32 in., a length of 16 in., and a wall thickness of 0.032 in. Calculate the critical value of the shearing stress for torsional loading. Compare the result based on Eq. (5.60) with that based on Fig. 5.10.

5.13 An aluminum cylinder with simply supported ends has a diameter of 32 in., a length of 16 in., and a wall thickness of 0.032 in. It is subjected to an external lateral pressure p_e, in pounds per square inch, and an axial tension P, in pounds. For $P/2\pi a = -\frac{1}{2}p_e a$, determine the critical value of the lateral pressure p_e. Calculate the corresponding value of the axial load P.

5.14 A stringer stiffened aluminum cylinder with simply supported ends is subjected to axial compression. The cylinder radius and wall thickness are 22 and 0.100 in., respectively. The stiffeners are 0.60 in. high and 0.20 in. thick and are uniformly spaced a distance 1.57 in. apart around the circumference of the cylinder. Compare the value of the critical load for outside stiffeners with that for inside stiffeners.

6

GENERAL SHELLS

6.1 INTRODUCTION

A structural shell may be defined as a body in which the distance from any point inside the body to some reference surface (usually the shell middle surface) is small in comparison with any typical dimension of the reference surface (such as a radius of curvature). Shell theory is an attempt to use the assumption of small-ness in the dimension normal to the surface to reduce by approximation the three-dimensional problem of deformation of the body to a two-dimensional problem. The essence of the theory is that the displacement of any point inside the shell wall is expressed in terms of the displacement components of a corre-sponding point on the reference surface.

The procedure for reduction to a two-dimensional problem is not unique. A large number of publications on the subject have appeared since the first useful shell theory was presented by Love in 1888 (Ref. 6.1). The difficulties involved in the derivation of a shell theory are reviewed in Refs. 6.2, 6.3, and 6.4. Refer-ence 6.4 includes a discussion of the problem of obtaining estimates of error bounds. Naghdi (Ref. 6.2) points out that many of the theories presented, including Love's theory, contain some inconsistencies. In view of this statement, it may be worthwhile to note that in almost all engineering applications the better-known theories, if applicable in the case, lead to practically identical results. The continuing effort in the field is not motivated by an inability to solve practical problems, but rather by a desire to define a theory that is characterized by simplicity, consistency, and clarity. Also, there are some special shell configurations for which presumably small terms may be of significance, and for which different results are obtained from different theories. No attempt is made here to give the reader a basic knowledge of this complicated subject, nor even to give a complete review of publications in the field. More complete treatments of these matters may be found in the references at the end of the chapter. In the derivation of the present shell equations we sacrifice generality and restrict ourselves to conditions under which the derivations are relatively simple yet sufficiently complete to illustrate the basic principles involved.

All shell theories available today that include shell bending are based on the assumption that the strains in the shell are small enough to be discarded in comparison with unity. Also, in what we refer to as first-order approximations, it is assumed that the shell is thin enough that quantities such as the thickness/radius ratio may be discarded in comparison with unity. Higher-order approximations have been derived for special cases; an example is given in the Flügge equations (Ref. 6.5), in which the square of the thickness/radius ratio is omitted in comparison with unity.

In addition to the assumptions of small strains and small thickness/radius ratios, Love invoked the approximations previously applied by Kirchhoff in thin-plate analysis (Sec. 3.2). That is, Love assumed that (1) normals to the reference surface remain straight and normal during the deformation, and (2) the transverse normal stress is negligibly small. The assumption that normals remain normal to the deformed surface is equivalent to the assumption that the resistance to deformation under transverse shear is infinite. Within the framework of a linear first-order theory, modifications of Love's theory have been presented in order to remove some of its inconsistencies. Among such modifications the papers by Budiansky and Sanders (Ref. 6.6) and by Koiter (Ref. 6.7) are frequently referenced. Higher-order theories, in which transverse shearing deformations are included, also have been presented (for example, Refs. 6.8 and 6.9).

Nonlinear shell equations were first derived for shells of specialized geometries. Donnell presented a particularly simple set of equations for cylindrical shells in Ref. 6.10, as noted in Chap. 5. The Donnell approximations were extended to shallow shells of otherwise general geometry in a form that we refer to as the Donnell-Mushtari-Vlasov equations for quasi-shallow shells (Ref. 6.11). A frequently used set of equations for nonlinear analysis of shells of revolution was presented by Reissner (Ref. 6.12).

For shells of general shape, nonlinear equations were presented by Novozhilov in 1953 (Ref. 6.13). The equations were derived in terms of lines-of-curvature coordinates, i.e., lines that coincide with the lines of principal curvature of the surface (Ref. 6.14, pp. 13–14). More general nonlinear equations were later published by Sanders (Ref. 6.15), Leonard (Ref. 6.16), Budiansky (Ref. 6.17), Naghdi and Nordgren (Ref. 6.18), Koiter (Ref. 6.19), and Marlowe and Flügge (Ref. 6.20). Equations for the stability analysis of shells of general shape may be obtained from these nonlinear equilibrium equations by application of the adjacent-equilibrium criterion. Stability equations are presented in explicit form by Koiter (Ref. 6.21), by Budiansky (Ref. 6.17), and by Simmonds and Danielson (Ref. 6.22).

For nonlinear shell theory, different levels of approximation have been employed. Most of the publications referenced above include equations that are valid for any size of the rotations, as well as somewhat simpler equations that are valid for "moderate" rotations. When specialized for columns, the latter equations are identical with the equations for the intermediate class of deformations discussed in Chap. 1.

All the general shell equations referenced here, except those in Ref. 6.13, were derived by use of tensor analysis, and a serious student of the subject should become familiar with this technique. A suitable introduction is available in, for example, Ref. 6.23. To get a basic feeling for the physical background of the theory, however, the student may be well advised to first familiarize himself with shell theory through one of the textbooks that presents the theory without recourse to tensor analysis, such as Ref. 6.14 or 6.24.

The nonlinear equilibrium and linear stability equations given in this chapter are based on analyses in Refs. 6.15 and 6.21. The present equations, however, are limited to shell coordinates that coincide with the lines of principal curvature, and equations for only the intermediate class of deformations are considered. Tensor notation is not used. Particular emphasis is given to the so-called Donnell-Mushtari-Vlasov equations for shells of general shape. The latter equations are chosen because they are believed to be the simplest possible form of the equations for stability analysis, and hence to be especially well suited for illustration of the underlying principles.

Nonlinear equilibrium equations are derived in Sec. 6.2, and the associated linear stability equations in Sec. 6.3. The stability equations are specialized for shells of revolution in Sec. 6.4, and applications are treated in Sec. 6.5.

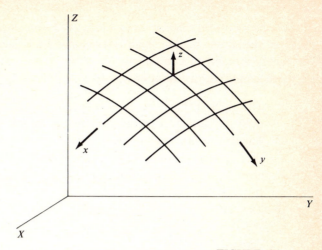

FIGURE 6.1
Coordinate systems.

6.2 NONLINEAR EQUILIBRIUM EQUATIONS

We consider a thin-walled shell of arbitrary shape made of a homogeneous iso-tropic elastic material. For our purpose it is sufficient to consider lines-of-curva-ture coordinates and to use the shell middle surface as the reference surface. The lines-of-curvature coordinates are denoted by the symbols x and y, and the principal radii of curvature by R_x and R_y. Distances ds_x and ds_y along the coordinate lines are given by the relations

$$ds_x = A\,dx \qquad ds_y = B\,dy \qquad (6.1)$$

where A, B are called the Lamé coefficients for the coordinate system; if X, Y, Z denote rectangular cartesian coordinates of a point on the surface (Fig. 6.1), then $X = X(x,y)$, $Y = Y(x,y)$, $Z = Z(x,y)$, and from Ref. 6.13, eq. (I.123),

$$A = \left[\left(\frac{\partial X}{\partial x}\right)^2 + \left(\frac{\partial Y}{\partial x}\right)^2 + \left(\frac{\partial Z}{\partial x}\right)^2\right]^{1/2}$$

$$B = \left[\left(\frac{\partial X}{\partial y}\right)^2 + \left(\frac{\partial Y}{\partial y}\right)^2 + \left(\frac{\partial Z}{\partial y}\right)^2\right]^{1/2} \qquad (6.2)$$

The shell thickness is denoted by h. Points not on the middle surface are located by an additional coordinate z that denotes the distance of a point from the corresponding point on the middle surface. The sign of z is arbitrary; for convex shells it is taken to be positive outward. The coordinates x, y, z form an orthogonal coordinate system.

A theory for thin elastic shells that is valid for deflections of any magnitude provided the strains remain small is developed by Koiter in Refs. 6.7 and 6.19. The theory is based on a strain energy expression derived in terms of the following three simplifying assumptions:

1   The shell is thin, that is, $h/R \ll 1$, where R is the smallest principal radius of curvature of the undeformed middle surface.
2   The strains are small compared with unity, and the strain energy density, i.e., the strain energy per unit volume of the undeformed body, is given by the quadratic function of the strain components for an isotropic elastic solid.
3   The state of stress is approximately plane; i.e., the effect of transverse shearing and normal stresses may be neglected in the strain-energy density.

The third of Koiter's assumptions is essentially equivalent to the Kirchhoff approximations in Sec. 3.2. In terms of these assumptions Koiter gives, for the strain energy of a thin elastic shell, the equations [Ref. 6.7, eqs. (2.11), (2.13), and (2.14)]

$$U = U_m + U_b$$

$$U_m = \frac{C}{2} \iint \left(\varepsilon_x{}^2 + \varepsilon_y{}^2 + 2v\varepsilon_x \varepsilon_y + \frac{1-v}{2} \gamma_{xy}{}^2 \right) AB\, dx\, dy \qquad (6.3)$$

$$U_b = \frac{D}{2} \iint \left[\kappa_x{}^2 + \kappa_y{}^2 + 2v\kappa_x \kappa_y + 2(1-v)\kappa_{xy}{}^2 \right] AB\, dx\, dy$$

where ε_x, ε_y, and γ_{xy} are middle-surface extensional and shearing strain components, and κ_x, κ_y, and κ_{xy} are middle-surface curvature changes and twist.

The total potential energy V of a loaded shell is the sum of the strain energy U and the potential energy Ω of the applied loads:

$$V = U + \Omega \qquad (6.4)$$

To obtain an expression for Ω, let p_x, p_y, p_z denote the x, y, and z components, respectively, of the load distributed over the surface of the shell element, and let u, v, w represent the corresponding components of the displacement of a point on the shell middle surface. Then, for a dead load, the expression for the potential energy of the applied load may be written

$$\Omega = - \iint (p_x u + p_y v + p_z w) AB\, dx\, dy \qquad (6.5)$$

Equations (6.3) to (6.5) plus a suitable set of kinematic relations constitute a general potential energy expression for a shell of arbitrary shape.

The equilibrium and stability equations developed here are based on non-linear middle-surface kinematic relations of the relatively simple form

$$\varepsilon_x = e_{xx} + \tfrac{1}{2}\beta_x{}^2 \qquad \kappa_x = \chi_{xx}$$
$$\varepsilon_y = e_{yy} + \tfrac{1}{2}\beta_y{}^2 \qquad \kappa_y = \chi_{yy} \qquad (6.6)$$
$$\gamma_{xy} = e_{xy} + \beta_x\beta_y \qquad \kappa_{xy} = \chi_{xy}$$

where e_{ij}, β_i, and χ_{ij} are *linear* functions of the middle-surface displacement components u, v, w. Kinematic relations of this form are derived by Sanders (Ref. 6.15) on the basis of the assumptions that (1) the e_{ij} are of the order of magnitude of ε^2, where ε is a number small compared with unity; (2) the rotations β_x, β_y are of the order of magnitude of ε; and (3) the rotation relative to normals to the middle surface is negligibly small compared with the other two rotations. The essence of these assumptions is that the effects of rotations about the normal may be omitted and that, in comparison with unity, strains as well as the squares of the rotations about the shell surface coordinate lines may be discarded. [It should be noted that, in general, all line elements through a given point rotate a different amount about the surface normal. In the relatively simple case that we are considering (lines-of-curvature coordinates), the difference between the rotations of the two intersecting coordinate lines is the shearing strain. The rotation relative to the normal may be interpreted as the average of the rotations of the coordinate lines.] In the notation of Eqs. (6.6), the Sanders kinematic relations may be written [Ref. 6.15, eqs. (A-1) to (A-6)]

$$e_{xx} = \frac{u_{,x}}{A} + \frac{A_{,y}v}{AB} + \frac{w}{R_x} \qquad e_{yy} = \frac{v_{,y}}{B} + \frac{B_{,x}u}{AB} + \frac{w}{R_y}$$

$$e_{xy} = \frac{v_{,x}}{A} + \frac{u_{,y}}{B} - \frac{B_{,x}v + A_{,y}u}{AB} \qquad (6.7a)$$

$$\beta_x = -\frac{w_{,x}}{A} + \frac{u}{R_x} \qquad \beta_y = -\frac{w_{,y}}{B} + \frac{v}{R_y} \qquad (6.7b)$$

$$\chi_{xx} = \frac{\beta_{x,x}}{A} + \frac{A_{,y}\beta_y}{AB} \qquad \chi_{yy} = \frac{\beta_{y,y}}{B} + \frac{B_{,x}\beta_x}{AB}$$

$$2\chi_{xy} = \frac{\beta_{y,x}}{A} + \frac{\beta_{x,y}}{B} - \frac{A_{,y}\beta_x + B_{,x}\beta_y}{AB} \qquad (6.7c)$$

Equations (6.7) may also be obtained from Eqs. (2.7) to (2.9) and (2.11) of Koiter's paper on general shell stability equations in Ref. 6.21 by omission of terms representing rotations relative to middle-surface normals. Both Sanders and Koiter observe that the question of the general importance of such terms is not entirely settled at the present time. They are known not to be negligibly small for helicoidal shells. This matter is discussed in some detail in Ref. 6.7.

Although Eqs. (6.7) are relatively simple, an even simpler set of kinematic relations has come into wide use. Experience with numerical examples has shown that the terms containing u and v in Eqs. (6.7b) are of negligible influence for shell segments (curved plates) that are almost flat and for shells whose displacement components in the deformed configuration are rapidly varying functions of the shell coordinates. Such shells have been termed quasi-shallow, as noted. If terms containing u and v are omitted from the rotation expressions, Eqs. (6.7b) are replaced by the relationships

$$\beta_x = -\frac{w_{,x}}{A} \qquad \beta_y = -\frac{w_{,y}}{B} \qquad (6.8)$$

Introduction of Eqs. (6.8) into Eqs. (6.7c) and simplification gives the following widely used expressions for the curvature changes and twist of quasi-shallow shells:

$$\chi_{xx} = -\frac{w_{,xx}}{A^2} + \frac{A_{,x}w_{,x}}{A^3} - \frac{A_{,y}w_{,y}}{AB^2}$$

$$\chi_{yy} = -\frac{w_{,yy}}{B^2} + \frac{B_{,y}w_{,y}}{B^3} - \frac{B_{,x}w_{,x}}{A^2B} \qquad (6.9)$$

$$\chi_{xy} = -\frac{w_{,xy}}{AB} + \frac{A_{,y}w_{,x}}{A^2B} + \frac{B_{,x}w_{,y}}{AB^2}$$

The simplified expressions in Eqs. (6.7a), (6.8), and (6.9) are the kinematic relations underlying the so-called Donnell-Mushtari-Vlasov (DMV) equations for quasi-shallow shells. They are given by Sanders in Ref. 6.15, p. 35, and by Koiter in Ref. 6.21 eqs. (2.14) and (2.15).

The first set of kinematic relations [Eqs. (6.6) and (6.7)] form the basis for the analysis of nonshallow circular rings and cylindrical shells in Chaps. 4 and 5, Eqs. (4.7), (4.9), (4.10), and (5.28). (See Prob. 6.1.) The second set of kinematic relations [Eqs. (6.6), (6.7a), (6.7c), and (6.8)] yield the Donnell forms of the kinematic relations in Chaps. 4 and 5 [Eqs. (4.9) and (4.42) for rings and Eqs. (5.7) for cylinders].

Kinematic relations that are more accurate (valid for larger rotations) than either of the sets included here are given in most of the references in Sec. 6.1. As noted, the present forms were chosen because of their relative simplicity.

Equations (6.3) to (6.6), (6.7a), (6.7c), and (6.8) constitute the total potential energy expression for the DMV form of the general shell equations. Nonlinear equilibrium equations may be derived from this expression by routine application of the principle of stationary potential energy. The resulting Euler equations of the calculus of variations are found to be (see Prob. 6.2)

$$(BN_x)_{,x} + (AN_{xy})_{,y} - B_{,x}N_y + A_{,y}N_{xy} = -ABp_x$$
$$(AN_y)_{,y} + (BN_{xy})_{,x} - A_{,y}N_x + B_{,x}N_{xy} = -ABp_y$$

$$\left[\frac{1}{A}(BM_x)_{,x}\right]_{,x} - \left(\frac{A_{,y}}{B}M_x\right)_{,y} + \left[\frac{1}{B}(AM_y)_{,y}\right]_{,y} - \left(\frac{B_{,x}}{A}M_y\right)_{,x}$$

$$+ 2\left[M_{xy,xy} + \left(\frac{A_{,y}}{A}M_{xy}\right)_{,x} + \left(\frac{B_{,x}}{B}M_{xy}\right)_{,y}\right] - AB\left(\frac{N_x}{R_x} + \frac{N_y}{R_y}\right)$$

$$- [(BN_x\beta_x + BN_{xy}\beta_y)_{,x} + (AN_y\beta_y + AN_{xy}\beta_x)_{,y}] = -ABp_z$$

(6.10)

where

$$N_x = C(\varepsilon_x + \nu\varepsilon_y) \qquad M_x = D(\kappa_x + \nu\kappa_y)$$
$$N_y = C(\varepsilon_y + \nu\varepsilon_x) \qquad M_y = D(\kappa_y + \nu\kappa_x)$$
$$N_{xy} = C\frac{1-\nu}{2}\gamma_{xy} \qquad M_{xy} = D(1-\nu)\kappa_{xy}$$

(6.11)

These equations, although different in form, are identical with those given by Sanders for the DMV form of the equations in Eqs. (A-7) to (A-9) of Ref. 6.15. They are nonlinear equilibrium equations for thin elastic shells of general shape. More accurate equilibrium equations may be found in, for example, Refs. 6.15 to 6.19.

The corresponding linear equilibrium equations are obtained by simple omission of all terms that would be quadratic or higher-order in u, v, w [that is, omission of all terms in Eqs. (6.6) and (6.10), but not Eq. (6.7c), that contain the rotations β_x or β_y].

Specialization of Eqs. (6.10) for a rectangular flat plate ($A = B = 1$ and $1/R_x = 1/R_y = 0$) gives the von Kármán plate equations in the form expressed in Eqs. (3.18). Specialization for a circular cylindrical shell ($A = 1$, $B = R_y = a$, and $1/R_x = 0$) gives Eqs. (5.8). Similar specialization for a column gives Eq. (1.12).

6.3   LINEAR STABILITY EQUATIONS

In this section the minimum potential energy criterion is used to derive an expression for the second variation of the potential energy of a shell of general shape in terms of the linear displacement parameters e_{ij}, β_i, χ_{ij} of Eqs. (6.6). The DMV kinematic relations are then introduced, and the linear differential equations for loss of stability are obtained by application of the Trefftz criterion. The same stability equations are then rederived by application of the adjacent-equilibrium criterion to the nonlinear equilibrium equations of Sec. 6.2.

6.3a   Minimum Potential Energy Criterion

Equations (6.3) to (6.6) represent a general expression for the potential energy in terms of parameters e_{ij}, β_i, and χ_{ij} that are linear functions of the middle-surface displacement components u, v, w. To obtain an expression for the second variation of the potential energy, we let

$$
\begin{aligned}
u &\to u_0 + u_1 \\
v &\to v_0 + v_1 \\
w &\to w_0 + w_1
\end{aligned}
\qquad (6.12)
$$

and we collect all terms in the resulting expression that are quadratic in the incremental quantities u_1, v_1, w_1. In terms of the linear displacement parameters, e_{xx} is replaced by $e_{xx0} + e_{xx1}$, etc., and terms that are quadratic in the quantities with subscript 1 are collected. The expression for the potential energy of the external forces [Eqs. (6.5)] is a linear functional of the displacement components and makes no contribution to the second-variation expression, that is, $\delta^2 \Omega = 0$. Consequently, $\delta^2 V = \delta^2 U$, or

$$
\delta^2 V = \delta^2 U_m + \delta^2 U_b \qquad (6.13)
$$

Following this procedure, the expressions for the second variation of the membrane and bending strain energy are found to be

$$
\frac{1}{2!}\delta^2 U_m = \frac{C}{2} \iint \left\{ \left[(\delta\varepsilon_x)^2 + (\delta\varepsilon_y)^2 + 2\nu(\delta\varepsilon_x)(\delta\varepsilon_y) + \frac{1-\nu}{2}(\delta\gamma_{xy})^2 \right] \right.
$$

$$
+ \left[(\varepsilon_{x0} + \nu\varepsilon_{y0})(\delta^2\varepsilon_x) + (\varepsilon_{y0} + \nu\varepsilon_{x0})(\delta^2\varepsilon_y) \right.
$$

$$
\left. \left. + \frac{1-\nu}{2}\gamma_{xy0}(\delta^2\gamma_{xy}) \right] \right\} AB \, dx \, dy \qquad (6.14)
$$

$$
\frac{1}{2!}\delta^2 U_b = \frac{D}{2} \iint \left[(\delta\kappa_x)^2 + (\delta\kappa_y)^2 + 2\nu(\delta\kappa_x)(\delta\kappa_y) \right.
$$

$$
\left. + 2(1-\nu)(\delta\kappa_{xy})^2 \right] AB \, dx \, dy
$$

where, from Eqs. (6.6),

$$\delta\varepsilon_x = e_{xx1} + \beta_{x0}\beta_{x1} \qquad\qquad \delta^2\varepsilon_x = \beta_{x1}{}^2 \qquad\qquad \delta\kappa_x = \chi_{xx1}$$
$$\delta\varepsilon_y = e_{yy1} + \beta_{y0}\beta_{y1} \qquad\qquad \delta^2\varepsilon_y = \beta_{y1}{}^2 \qquad\qquad \delta\kappa_y = \chi_{yy1} \qquad (6.15)$$
$$\delta\gamma_{xy} = e_{xy1} + \beta_{y0}\beta_{x1} + \beta_{x0}\beta_{y1} \qquad \delta^2\gamma_{xy} = 2\beta_{x1}\beta_{y1} \qquad \delta\kappa_{xy} = \chi_{xy1}$$

The factor $\frac{1}{2}$ is retained on both sides of Eqs. (6.14) to avoid ambiguity. From Eqs. (6.11),

$$N_{x0} = C(\varepsilon_{x0} + v\varepsilon_{y0})$$
$$N_{y0} = C(\varepsilon_{y0} + v\varepsilon_{x0})$$

$$N_{xy0} = C\frac{1-v}{2}\gamma_{xy0}$$

Therefore introduction of Eqs. (6.14) and (6.15) into (6.13) gives, for the second variation, the expression

$$\frac{1}{2}\delta^2 V = \frac{C}{2}\iint\left[(e_{xx1} + \beta_{x0}\beta_{x1})^2 + (e_{yy1} + \beta_{y0}\beta_{y1})^2\right.$$
$$+ 2v(e_{xx1} + \beta_{x0}\beta_{x1})(e_{yy1} + \beta_{y0}\beta_{y1})$$
$$+ \frac{1-v}{2}(e_{xy1} + \beta_{x0}\beta_{y1} + \beta_{y0}\beta_{x1})^2\Big]AB\,dx\,dy$$
$$+ \frac{1}{2}\iint(N_{x0}\beta_{x1}{}^2 + N_{y0}\beta_{y1}{}^2 + 2N_{xy0}\beta_{x1}\beta_{y1})AB\,dx\,dy$$
$$+ \frac{D}{2}\iint[\chi_{xx1}{}^2 + \chi_{yy1}{}^2 + 2v\chi_{xx1}\chi_{yy1} + 2(1-v)\chi_{xy1}{}^2]AB\,dx\,dy \qquad (6.16)$$

Equation (6.16) is a general expression for the second variation of the total potential energy of a thin shell of general shape (shallow or nonshallow) subjected to arbitrary dead-loading.

For quasi-shallow shells, the incremental displacement parameters given by the DMV approximations are, from Eqs. (6.7a), (6.7c), and (6.8),

$$e_{xx1} = \frac{u_{1,x}}{A} + \frac{A_{,y}v_1}{AB} + \frac{w_1}{R_1}$$

$$\beta_{x1} = -\frac{w_{1,x}}{A} \qquad (6.17)$$

$$\chi_{xx1} = \frac{\beta_{x1,x}}{A} + \frac{A_{,y}\beta_{y1}}{AB}$$

etc. Equations (6.16) and (6.17) constitute the expression for the second variation of the total potential energy for the DMV quasi-shallow shell.

The variables β_{x0} and β_{y0} in Eq. (6.16) represent prebuckling rotations. Unlike the prebuckling deformation of columns, plates, and rings, such deformation of shells is not rotation-free. Consequently, it usually is nonlinear from the outset, and there is no straight-line equilibrium path. On the other hand, the influence of prebuckling nonlinearity often may be neglected in stability applications, as noted in Chap. 5. The state of stress often is such that rotations are small or are confined to a small fraction of the shell surface area. In such cases it can be assumed that the influence of prebuckling nonlinearity may be neglected. The assumption is deferred here primarily because we wish to delineate more clearly the place of stability analysis in the context of general nonlinear analysis, and also because there are applications in which the influence of prebuckling nonlinearity is not insignificant. Some examples are treated in Chap. 9

We note that if the linear membrane or linear bending equations are used for the prebuckling deformation (because the prebuckling deformation is rotation-free or because the nonlinear influence of the rotations is neglected in the pre-buckling analysis), all the prebuckling quantities in Eq. (6.16) can be expressed in terms of an applied load factor λ times their values for a unit load. If the prebuckling rotation terms are omitted in the stability equations, the load factor then appears only to the first power. If prebuckling rotations are retained, however, some terms will contain λ^2. This leads to some complications in numerical analyses based on Eq. (6.16). (Experience has shown that adoption of the intermediate position that first-order terms in prebuckling rotations may be retained and only second-order ones omitted may lead to spurious results. For example, it is shown in Ref. 6.25 that retention of only first-order terms in the prebuckling rotations may give less accurate results than the discarding of pre-buckling rotation terms altogether.)

For a functional of the form of Eqs. (6.16) and (6.17), the Euler equations are given by Eqs. (3.43). Introduction of the integrand of the second-variation expression, in accordance with the Trefftz criterion, leads to the equations

$$(BN_{x1})_{,x} + (AN_{xy1})_{,y} - B_{,x}N_{y1} + A_{,y}N_{xy1} = 0$$

$$(AN_{y1})_{,y} + (BN_{xy1})_{,x} - A_{,y}N_{x1} + B_{,x}N_{xy1} = 0$$

$$\left[\left(\frac{1}{A}(BM_{x1})_{,x}\right)_{,x} - \left(\frac{A_{,y}}{B}M_{x1}\right)_{,y}\right] + \left[\left(\frac{1}{B}(AM_{y1})_{,y}\right)_{,y} - \left(\frac{B_{,x}}{A}M_{y1}\right)_{,x}\right]$$

$$+ 2\left[M_{xy1,xy} + \left(\frac{A_{,y}}{A}M_{xy1}\right)_{,x} + \left(\frac{B_{,x}}{B}M_{xy1}\right)_{,y}\right] - AB\left(\frac{N_{x1}}{R_x} + \frac{N_{y1}}{R_y}\right) \qquad (6.18)$$

$$- [(BN_{x0}\beta_{x1} + BN_{xy0}\beta_{y1})_{,x} + (B\beta_{x0}N_{x1} + B\beta_{y0}N_{xy1})_{,x}$$

$$+ (AN_{y0}\beta_{y1} + AN_{xy0}\beta_{x1})_{,y} + (A\beta_{y0}N_{y1} + A\beta_{x0}N_{xy1})_{,y}] = 0$$

where, from Eqs. (6.6), (6.7a), (6.7c), (6.8), and (6.11),

$$N_{x1} = C[(e_{xx1} + \beta_{x0}\beta_{x1}) + v(e_{yy1} + \beta_{y0}\beta_{y1})]$$

$$N_{y1} = C[(e_{yy1} + \beta_{y0}\beta_{y1}) + v(e_{xx1} + \beta_{x0}\beta_{x1})]$$

$$N_{xy1} = C\frac{1-v}{2}(e_{xy1} + \beta_{x0}\beta_{y1} + \beta_{y0}\beta_{x1})$$

$$M_{x1} = D\left[\left(\frac{\beta_{x1,x}}{A} + \frac{A_{,y}\beta_{y1}}{AB}\right) + v\left(\frac{\beta_{y1,y}}{B} + \frac{B_{,x}\beta_{x1}}{AB}\right)\right]$$

$$M_{y1} = D\left[\left(\frac{\beta_{y1,y}}{B} + \frac{B_{,x}\beta_{x1}}{AB}\right) + v\left(\frac{\beta_{x1,x}}{A} + \frac{A_{,y}\beta_{y1}}{AB}\right)\right]$$

$$M_{xy1} = D\frac{1-v}{2}\left(\frac{\beta_{y1,x}}{A} + \frac{\beta_{x1,y}}{B} - \frac{A_{,y}\beta_x + B_{,x}\beta_y}{AB}\right) \quad (6.19)$$

$$e_{xx1} = \frac{u_{1,x}}{A} + \frac{A_{,y}v_1}{AB} + \frac{w_1}{R_x}$$

$$e_{yy1} = \frac{v_{1,y}}{B} + \frac{B_{,x}u_1}{AB} + \frac{w_1}{R_y}$$

$$e_{xy1} = \frac{v_{1,x}}{A} + \frac{u_{1,y}}{B} - \frac{B_{,x}v_1 + A_{,y}u_1}{AB}$$

$$\beta_{x1} = -\frac{w_{1,x}}{A} + \frac{u_1}{R_x}$$

$$\beta_{y1} = -\frac{w_{1,y}}{B} + \frac{v_1}{R_y}$$

Introduction of Eqs. (6.19) into Eqs. (6.18) gives a set of three linear homogeneous equations in u_1, v_1, w_1, with variable coefficients in N_{x0}, N_{xy0}, N_{y0}, β_{x0}, and β_{y0}. The coefficients are determined by the nonlinear equilibrium equations of Sec. 6.2.

Equations (6.18) and (6.19) are the linear stability equations for the shell of general shape, in terms of the DMV approximations.

6.3b Adjacent-Equilibrium Criterion

The linear stability equations of Eqs. (6.18) alternatively may be obtained from the nonlinear equilibrium equations of Eqs. (6.10) by application of the adjacent-equilibrium criterion. For this purpose we let

$$u \rightarrow u_0 + u_1$$

$$v \rightarrow v_0 + v_1$$

$$w \rightarrow w_0 + w_1$$

where (u_0, v_0, w_0) represents the equilibrium configuration whose stability is under consideration, (u, v, w) is an adjacent equilibrium configuration corresponding to the same value of applied load as the configuration (u_0, v_0, w_0), and (u_1, v_1, w_1) is an arbitrarily small incremental displacement. Furthermore, we let

$$
\begin{aligned}
N_x &\to N_{x0} + \Delta N_x \\
N_y &\to N_{y0} + \Delta N_y \\
N_{xy} &\to N_{xy0} + \Delta N_{xy} \qquad (6.20) \\
\beta_x &\to \beta_{x0} + \beta_{x1} \\
\beta_y &\to \beta_{y0} + \beta_{y1}
\end{aligned}
$$

where ΔN_x, ΔN_y, ΔN_{xy}, β_{x1}, β_{y1} are increments corresponding to (u_1, v_1, w_1). We also let N_{x1}, N_{y1}, N_{xy1} represent the portions of ΔN_x, ΔN_y, ΔN_{xy}, respectively, that are linear in u_1, v_1, w_1. No increment is given to the applied load (p_x, p_y, p_z). Introduction into Eqs. (6.10) and omission, as in earlier chapters, both of terms that add to zero and terms that are quadratic or of higher-order in the incremental quantities again leads to Eqs. (6.18) and (6.19).

Equations (6.18) and (6.19) correspond to the expressions given by Koiter for shallow buckling modes in Ref. 6.21, eqs. (4.14) and (4.15). Reference 6.21 also presents explicit stability equations for nonshallow shells subjected to fluid-pressure loading.

Specialization of Eqs. (6.18) and (6.19) for rectangular coordinates gives the stability equations for rectangular flat plates in Eqs. (3.34). Specialization for cylindrical coordinates and neglect of prebuckling rotations gives the Donnell stability equations for circular cylindrical shells in Eqs. (5.23). Specialization for a circular ring gives the Donnell form of the ring equations in Eqs. (4.45), and for a column gives Eq. (1.23). Specialization for cylindrical coordinates and *retention* of the prebuckling rotation terms leads to the equations used in the analysis of asymmetric buckling from a nonlinear axisymmetric form in Chap. 9 [Eqs. (9.23)].

6.4 SHELLS OF REVOLUTION

Many structural shell configurations are shells of revolution. The middle surface of a shell of revolution may be formed by rotation of a plane curve about an axis in the plane of the curve (Fig. 6.2). Planes normal to the axis of revolution intersect the surface in curves called *parallels*, and planes that contain the axis intersect the surface in curves called *meridians*. The parallels and meridians are lines of principal curvature on the surface. Points on the surface may be referred to coordinates ϕ and θ, where ϕ denotes the angle between the axis of revolution and a normal to the surface, and θ is a circumferential coordinate

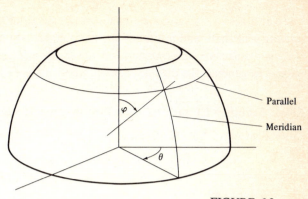

FIGURE 6.2
Shell of revolution.

(Fig. 6.2). The principal radii of curvature of the surface in the ϕ and θ directions may be denoted by r_ϕ and r_θ, respectively. It is convenient to introduce an additional variable r defined by the relation (see Fig. 6.3)

$$r = r_\theta \sin \phi$$

Then distances ds_ϕ and ds_θ along the coordinate lines are given by the relations

$$ds_\phi = r_\phi \, d\phi \qquad ds_\theta = r \, d\theta \qquad (6.21)$$

Thus, if ϕ and θ are taken to correspond to x and y, respectively, the Lamé coefficients are seen to be

$$A = r_\phi \qquad B = r \qquad (6.22)$$

Furthermore,

$$R_x = r_\phi \qquad R_y = r_\theta \qquad (6.23)$$

and from Fig. (6.3),

$$dr = ds_\phi \cos \phi \qquad (6.24)$$

Equations (6.21) and (6.24) yield the additional relationship

$$\frac{dr}{d\phi} = r_\phi \cos \phi$$

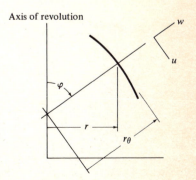

FIGURE 6.3
Meridian of a shell of revolution.

The variables r_ϕ, r_θ, and r characterize the shape of the middle surface of the undeformed shell, and are functions of ϕ alone. We let u, v, w denote middle-surface displacement components in the ϕ, θ, and normal directions, respectively. In general, the displacement components are functions of both ϕ and θ.

6.4a Stability Equations with Prebuckling Rotations Retained

Specialization of Eqs. (6.18) and (6.19) for a shell of revolution now gives

$$(rN_{\phi 1})_{,\phi} + r_\phi N_{\phi\theta 1,\theta} - r_\phi N_{\theta 1} \cos\phi = 0$$

$$(rN_{\phi\theta 1})_{,\phi} + r_\phi N_{\theta 1,\theta} + r_\phi N_{\phi\theta 1} \cos\phi = 0$$

$$\left[\frac{1}{r_\phi}(rM_{\phi 1})_{,\phi}\right]_{,\phi} + 2\left(M_{\phi\theta 1,\phi\theta} + \frac{r_\phi}{r} M_{\phi\theta 1,\theta} \cos\phi\right)$$

$$+ \left[\frac{r_\phi}{r} M_{\theta 1,\theta\theta} - (M_{\theta 1} \cos\phi)_{,\phi}\right] - (rN_{\phi 1} + r_\phi N_{\theta 1} \sin\phi)$$

$$- [(rN_{\phi 0}\beta_{\phi 1} + rN_{\phi\theta 0}\beta_{\theta 1})_{,\phi} + (r\beta_{\phi 0}N_{\phi 1} + r\beta_{\theta 0}N_{\phi\theta 1})_{,\phi}$$

$$+ r_\phi(N_{\theta 0}\beta_{\theta 1} + N_{\phi\theta 0}\beta_{\phi 1})_{,\theta} + r_\phi(\beta_{\theta 0}N_{\theta 1} + \beta_{\phi 0}N_{\phi\theta 1})_{,\theta}] = 0$$

$$(6.25)$$

where the force and moment intensities and the displacement variables are given by the equations

$$N_{\phi 1} = C[(e_{\phi\phi 1} + \beta_{\phi 0}\beta_{\phi 1}) + v(e_{\theta\theta 1} + \beta_{\theta 0}\beta_{\theta 1})]$$

$$N_{\theta 1} = C[(e_{\theta\theta 1} + \beta_{\theta 0}\beta_{\theta 1}) + v(e_{\phi\phi 1} + \beta_{\phi 0}\beta_{\phi 1})]$$

$$N_{\phi\theta 1} = C\frac{1-v}{2}(e_{\phi\theta 1} + \beta_{\phi 0}\beta_{\theta 1} + \beta_{\theta 0}\beta_{\phi 1})$$

$$M_{\phi 1} = D\left[\frac{\beta_{\phi 1,\phi}}{r_\phi} + \frac{v}{r}(\beta_{\theta 1,\theta} + \beta_{\phi 1}\cos\phi)\right]$$

$$M_{\theta 1} = D\left[\frac{1}{r}(\beta_{\theta 1,\theta} + \beta_{\phi 1}\cos\phi) + \frac{v}{r_\phi}(\beta_{\phi 1,\phi})\right]$$

$$M_{\phi\theta 1} = D\frac{1-v}{2}\left[\frac{r}{r_\phi}\left(\frac{\beta_{\theta 1}}{r}\right)_{,\phi} + \frac{\beta_{\phi 1,\theta}}{r}\right]$$

$$(6.26)$$

$$e_{\phi\phi 1} = \frac{1}{r_\phi}(u_{1,\phi} + w_1)$$

$$e_{\theta\theta 1} = \frac{r}{r_\phi}\left(\frac{v_1}{r}\right)_{,\phi} + \frac{u_{1,\theta}}{r}$$

$$e_{\phi\theta 1} = \frac{1}{r}(v_{1,\theta} + u_1\cos\phi + w_1\sin\phi)$$

$$\beta_{\phi 1} = -\frac{w_{1,\phi}}{r_\phi} \qquad \beta_{\theta 1} = -\frac{w_{1,\theta}}{r}$$

The nonlinear equations of equilibrium governing the coefficients $N_{\phi 0}$, $N_{\phi \theta 0}$, $N_{\theta 0}$, $\beta_{\phi 0}$, and $\beta_{\theta 0}$ in Eqs. (6.25) may be obtained by specialization of Eqs. (6.10) for a shell of revolution. If the applied load is axisymmetric, the deformation also is axisymmetric prior to loss of stability. Then $\beta_{\theta 0} = 0$, and $N_{\phi 0}$, $N_{\phi \theta 0}$, $N_{\theta 0}$, and $\beta_{\phi 0}$ are functions of ϕ alone. Specialization of the nonlinear equilibrium equations in Eqs. (6.10) for axisymmetric deformation of a shell of revolution gives

$$\frac{d}{d\phi}(rN_\phi) - r_\phi N_\theta \cos \phi = -rr_\phi p_\phi$$

$$\frac{d}{d\phi}(rN_{\phi \theta}) + r_\phi N_{\phi \theta} \cos \phi = -rr_\phi p_\theta$$

$$\frac{d}{d\phi}\left[\frac{1}{r_\phi}\frac{d}{d\phi}(rM_\phi)\right] - \frac{d}{d\phi}(M_\theta \cos \phi)$$

$$- (rN_\phi + r_\phi N_\theta \sin \phi) - \frac{d}{d\phi}(rN_\phi \beta_\phi) = -rr_\phi p$$

$$(6.27)$$

where p_ϕ, p_θ, and p are the components of the surface load in the ϕ, θ, and normal directions, respectively. From Eqs. (6.11) the constitutive relations are

$$N_\phi = C(\varepsilon_\phi + \nu \varepsilon_\theta) \qquad M_\phi = D(\kappa_\phi + \nu \kappa_\theta)$$
$$N_\theta = C(\varepsilon_\theta + \nu \varepsilon_\phi) \qquad M_\theta = D(\kappa_\theta + \nu \kappa_\phi) \qquad (6.28)$$
$$N_{\phi \theta} = C\frac{1-\nu}{2}\gamma_{\phi \theta}$$

and from Eqs. (6.6), (6.7a), (6.7c), and (6.8), the kinematic relations are

$$\varepsilon_\phi = e_{\phi \phi} + \tfrac{1}{2}\beta_\phi^2 \qquad \varepsilon_\theta = e_{\theta \theta} \qquad \gamma_{\phi \theta} = e_{\phi \theta}$$

$$e_{\phi \phi} = \frac{1}{r_\phi}\left(\frac{du}{d\phi} + w\right) \qquad e_{\theta \theta} = \frac{1}{r}(u \cos \phi + w \sin \phi)$$

$$e_{\phi \theta} = \frac{r}{r_\phi}\frac{d}{d\phi}\left(\frac{v}{r}\right) \qquad \beta_\phi = -\frac{1}{r_\phi}\frac{dw}{d\phi}$$

$$(6.29)$$

$$\kappa_\phi = \frac{1}{r_\phi}\frac{d\beta_\phi}{d\phi} \qquad \kappa_\theta = \frac{1}{r}\beta_\phi \cos \phi$$

If the shell is not subjected to torsional loading, the coefficient $N_{\phi \theta 0} = 0$ in Eqs. (6.25), and the second of Eqs. (6.27) is discarded. (A different form of Eqs. (6.27) that is widely used in the literature is given in Ref. 6.12.)

6.4b Stability Equations with Prebuckling Rotations Omitted

For analyses in which prebuckling rotations may be neglected, the stability equations may be obtained from Eqs. (6.25) by omission of terms containing $\beta_{\phi 0}$ or $\beta_{\theta 0}$. The resulting equations are

$$(rN_{\phi 1})_{,\phi} + r_\phi N_{\phi\theta 1,\theta} - r_\phi N_{\theta 1}\cos\phi = 0$$

$$(rN_{\phi\theta 1})_{,\phi} + r_\phi N_{\theta 1,\theta} + r_\phi N_{\phi\theta 1}\cos\phi = 0$$

$$\left[\frac{1}{r_\phi}(rM_{\phi 1})_{,\phi}\right]_{,\phi} + 2\left(M_{\phi\theta 1,\phi\theta} + \frac{r_\phi}{r}M_{\phi\theta 1,\theta}\cos\phi\right) \tag{6.30}$$

$$+ \left[\frac{r_\phi}{r}M_{\theta 1,\theta\theta} - (M_{\theta 1}\cos\phi)_{,\phi}\right] - (rN_{\phi 1} + r_\phi N_{\theta 1}\sin\phi)$$

$$- [(rN_{\phi 0}\beta_{\phi 1} + rN_{\phi\theta 0}\beta_{\theta 1})_{,\phi} + (r_\phi N_{\theta 0}\beta_{\theta 1} + r_\phi N_{\phi\theta 0}\beta_{\phi 1})_{,\theta}] = 0$$

where

$$N_{\phi 1} = C\left[\frac{1}{r_\phi}(u_{1,\phi} + w_1) + \frac{v}{r}(v_{1,\theta} + u_1\cos\phi + w_1\sin\phi)\right]$$

$$N_{\theta 1} = C\left[\frac{1}{r}(v_{1,\theta} + u_1\cos\phi + w_1\sin\phi) + \frac{v}{r_\phi}(u_{1,\phi} + w_1)\right]$$

$$N_{\phi\theta 1} = C\frac{1-v}{2}\left[\frac{r}{r_\phi}\left(\frac{v_1}{r}\right)_{,\phi} + \frac{u_{1,\theta}}{r}\right]$$

$$M_{\phi 1} = D\left[\frac{\beta_{\phi 1,\phi}}{r_\phi} + \frac{v}{r}(\beta_{\theta 1,\theta} + \beta_{\phi 1}\cos\phi)\right] \tag{6.31}$$

$$M_{\theta 1} = D\left[\frac{1}{r}(\beta_{\theta 1,\theta} + \beta_{\phi 1}\cos\phi) + \frac{v}{r_\phi}\beta_{\phi 1,\phi}\right]$$

$$M_{\phi\theta 1} = D\frac{1-v}{2}\left[\frac{r}{r_\phi}\left(\frac{\beta_{\theta 1}}{r}\right)_{,\phi} + \frac{\beta_{\phi 1,\theta}}{r}\right]$$

$$\beta_{\phi 1} = -\frac{w_{1,\phi}}{r_\phi} \qquad \beta_{\theta 1} = -\frac{w_{1,\theta}}{r}$$

Equations (6.30) are the counterpart, for the symmetrically loaded quasi-shallow shell of revolution, of the Donnell stability equations for cylindrical shells in Eqs. (5.23).

The coefficients $N_{\phi 0}$, $N_{\phi\theta 0}$, $N_{\theta 0}$ in Eqs. (6.30) are determined by the linear equilibrium equations; omission of nonlinear terms from Eqs. (6.27) gives the equations

$$\frac{d}{d\phi}(rN_\phi) - r_\phi N_\theta \cos\phi = -rr_\phi p_\phi$$

$$\frac{d}{d\phi}(rN_{\phi\theta}) + r_\phi N_{\phi\theta} \cos\phi = -rr_\phi p_\theta \qquad (6.32)$$

$$\frac{d}{d\phi}\left[\frac{1}{r_\phi}\frac{d}{d\phi}(rM_\phi)\right] - \frac{d}{d\phi}(M_\theta \cos\phi) - (rN_\phi + r_\phi N_\theta \sin\phi) = -rr_\phi p$$

where the constitutive and kinematic relations are given by Eqs. (6.28) and (6.29), except that now

$$\varepsilon_\phi = e_{\phi\phi} \qquad (6.33)$$

As a simplifying approximation in the determination of the coefficients in the stability equations, the linear bending equations frequently are replaced by the corresponding linear membrane equations, as noted in Chap. 5. From Eqs. (6.32) the latter equations are

$$\frac{d}{d\phi}(rN_\phi) - r_\phi N_\theta \cos\phi = -rr_\phi p_\phi$$

$$\frac{d}{d\phi}(rN_{\phi\theta}) + r_\phi N_{\phi\theta} \cos\phi = -rr_\phi p_\theta \qquad (6.34)$$

$$rN_\phi + r_\phi N_\theta \sin\phi = rr_\phi p$$

Equations (6.34) are seen to be statically determinate, so that solutions can be obtained without use of constitutive and kinematic relations.

If the shell is not subjected to torsional loading, the coefficient $N_{\phi\theta 0} = 0$ in Eqs. (6.30), and the second of Eqs. (6.32) and (6.34) is discarded. In such cases the stability equations obtained by substitution of Eqs. (6.31) into (6.30) may be reduced to ordinary differential equations by introduction of solutions of the form [compare Eqs. (5.69)]

$$u_1 = u_n(\phi) \cos n\theta$$
$$v_1 = v_n(\phi) \sin n\theta$$
$$w_1 = w_n(\phi) \cos n\theta$$

In summary, stability equations for shells of revolution are given in Eqs. (6.25) and (6.30). Prebuckling rotation terms are retained in Eqs. (6.25) and are omitted in Eqs. (6.30). Nonlinear equilibrium equations for symmetrically loaded shells of revolution are given in Eqs. (6.27). Corresponding linear bending equations are given in Eqs. (6.32), and corresponding linear membrane equations in Eqs. (6.34).

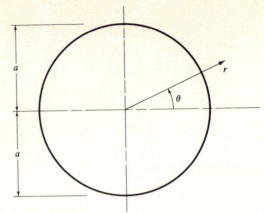

FIGURE 6.4
Circular flat plate.

6.5 APPLICATIONS OF THE STABILITY EQUATIONS

In this section the stability equations for symmetrically loaded shells of revolution in Eqs. (6.30) are specialized for four common structural configurations: circular plates, shallow spherical caps, conical shells, and segments of toroidal shells. Solutions of the equations are discussed.

For notational simplicity the subscript 1 is omitted from the incremental quantities (w_1, N_{r1}, etc.) in the remainder of the chapter.

6.5a Circular Flat Plates

The middle plane of a circular flat plate may be defined by polar coordinates r and θ, as illustrated in Fig. 6.4. In specialization of the shell-of-revolution equations for the plate, r_ϕ and r_θ go to infinity, the angle ϕ goes to zero, and

$$\lim_{r_\phi \to \infty} (r_\phi \, d\phi) = dr$$

Then $\sin \phi = 0$ and $\cos \phi = 1$. Introduction of these values into Eqs. (6.30) gives

$$(rN_r)_{,r} + N_{r\theta,\theta} - N_\theta = 0 \qquad (6.35a)$$

$$(rN_{r\theta})_{,r} + N_{\theta,\theta} + N_{r\theta} = 0 \qquad (6.35b)$$

$$(rM_r)_{,rr} + 2\left(M_{r\theta,r\theta} + \frac{1}{r} M_{r\theta,\theta}\right) + \left(\frac{1}{r} M_{\theta,\theta\theta} - M_{\theta,r}\right)$$
$$- [(rN_{r0}\beta_r + rN_{r\theta 0}\beta_\theta)_{,r} + (N_{r\theta 0}\beta_r + N_{\theta 0}\beta_\theta)_{,\theta}] = 0 \qquad (6.35c)$$

where the subscript ϕ has been replaced by r and where, from Eqs. (6.31),

$$N_r = C\left[u_{,r} + \frac{v}{r}(v_{,\theta} + u)\right]$$

$$N_\theta = C\left[\frac{1}{r}(v_{,\theta} + u) + vu_{,r}\right]$$

$$N_{r\theta} = C\frac{1-v}{2}\left[r\left(\frac{v}{r}\right)_{,r} + \frac{u_{,\theta}}{r}\right]$$

$$M_r = D\left[\beta_{r,r} + \frac{v}{r}(\beta_{\theta,\theta} + \beta_r)\right] \qquad (6.36)$$

$$M_\theta = D\left[\frac{1}{r}(\beta_{\theta,\theta} + \beta_r) + v\beta_{r,r}\right]$$

$$M_{r\theta} = D\frac{1-v}{2}\left[r\left(\frac{\beta_\theta}{r}\right)_{,r} + \frac{\beta_{r,\theta}}{r}\right]$$

$$\beta_r = -w_{,r} \qquad \beta_\theta = -\frac{w_{,\theta}}{r}$$

Introduction of Eqs. (6.36) into (6.35) is seen to yield a set of three homogeneous equations in u, v, w in which the third equation is uncoupled from the first two, as for rectangular plates [compare Eqs. (3.34)].

As a specific example, let us consider the axisymmetric buckling of a plate subjected to uniform compressive loading $N_{r0} = -N$ lb/in. Then $\beta_\theta = N_{r\theta 0} = 0$. To simplify the notation, let $\beta_r \equiv \beta$. Then Eq. (6.35c) specializes to the expression

$$\frac{d^2}{dr^2}(rM_r) - \frac{d}{dr}(M_\theta) + \frac{d}{dr}(rN\beta) = 0 \qquad (6.37)$$

where

$$M_r = D\left(\frac{d\beta}{dr} + \frac{v}{r}\beta\right) \qquad M_\theta = D\left(\frac{1}{r}\beta + v\frac{d\beta}{dr}\right)$$

Integration of Eq. (6.37) gives

$$\frac{d}{dr}(rM_r) - M_\theta + rN\beta = C_1$$

where C_1 is a constant. But $M_r \equiv M_\theta \equiv 0$ for $N = 0$. Therefore $C_1 = 0$. Introduction of the expressions for M_r and M_θ and rearrangement now gives

$$r^2\frac{d^2\beta}{dr^2} + r\frac{d\beta}{dr} - \left(1 - \frac{N}{D}r^2\right)\beta = 0 \qquad (6.38)$$

This equation is the same as eqs. (*a*) and (*b*) in Ref. 6.26, p. 389. As a homogeneous equation, it has nontrivial solutions only for discrete values of the applied load N. The smallest such value is the critical load N_{cr}.

Although Eq. (6.38) is a variable-coefficient equation, its general solution is readily expressed (see Ref. 6.26, p. 389). Let $\alpha^2 \equiv N/D$ and $u \equiv \alpha r$. Then Eq. (6.38) may be rewritten in the form

$$u^2 \frac{d^2\beta}{du^2} + u\frac{d\beta}{du} + (u^2 - 1)\beta = 0$$

The general solution is

$$\beta = C_2 J_1(u) + C_3 Y_1(u)$$

where J_1 and Y_1 are Bessel functions of first order of the first and second kinds, respectively, and C_2 and C_3 are integration constants. But $\beta = 0$ at $r = u = 0$, and $Y_1(0) \to \infty$ (Ref. 6.27, p. 191). Therefore $C_3 = 0$ and

$$\beta = C_2 J_1(u)$$

Solutions for two boundary conditions are given in Ref. 6.26: the plate edge clamped and the plate edge simply supported. For the edge clamped, $\beta = 0$ at $r = a$, where a is the plate radius. Then $J_1(\alpha a) = 0$. From an examination of a table of Bessel functions, the smallest root for which $J_1 = 0$ is $\alpha a = 3.83$ (Ref. 6.27, p. 167). But $\alpha^2 = N/D$. Consequently, the critical load is

$$N_{cr} = 14.68 \frac{D}{a^2} \qquad (6.39)$$

For the edge simply supported, $M_{r1} = D[(d\beta/dr + v\beta/r] = 0$ at $r = a$. Therefore

$$\left[\frac{dJ_1(u)}{dr} + v\frac{J_1(u)}{r}\right]_{r=a} = 0$$

or

$$\left[u\frac{dJ_1(u)}{du} + vJ_1(u)\right]_{u=\alpha a} = 0$$

But $dJ_1/du = J_0 - J_1/u$, where J_0 is the Bessel function of order zero (Ref. 6.27, p. 145). Therefore

$$\alpha a J_0(\alpha a) - (1 - v)J_1(\alpha a) = 0$$

or, for $v = 0.3$,

$$\alpha a J_0(\alpha a) - 0.7J_1(\alpha a) = 0$$

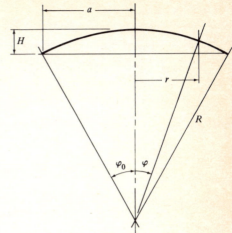

FIGURE 6.5
Shallow spherical cap.

The smallest value of αa for which this relationship is satisfied is $\alpha a = 2.05$ (Ref. 6.27, pp. 156 and 157). Accordingly, the critical load is

$$N_{cr} = 4.20 \frac{D}{a^2} \qquad (6.40)$$

This value is seen to be less than one-third the value in Eq. (6.39) for a clamped edge.

6.5b Shallow Spherical Caps

A segment of a spherical shell is shown in Fig. 6.5. Points on the middle surface are referred to coordinates r and θ. The rise H of the shell is much smaller than the base radius a. Such a shell is called a shallow spherical cap.

From the figure, $r_\phi = R$, a constant, and $\sin \phi = r/R$. Furthermore, approximately, $\cos \phi = 1$ and $r_\phi \, d\phi = dr$. Introduction of these values into Eqs. (6.30) gives

$$(rN_r)_{,r} + N_{r\theta,\theta} - N_\theta = 0 \qquad (6.41a)$$

$$(rN_{r\theta})_{,r} + N_{\theta,\theta} + N_{r\theta} = 0 \qquad (6.41b)$$

$$(rM_r)_{,rr} + 2\left(M_{r\theta,r\theta} + \frac{1}{r} M_{r\theta,\theta}\right) + \left(\frac{1}{r} M_{\theta,\theta\theta} - M_{\theta,r}\right)$$

$$- \frac{r}{R}(N_r + N_\theta) - [(rN_{r0}\beta_r + rN_{r\theta 0}\beta_\theta)_{,r} + (N_{r\theta 0}\beta_r + N_{\theta 0}\beta_\theta)_{,\theta}] = 0 \qquad (6.41c)$$

where, from Eqs. (6.31),

$$N_r = C(e_{rr} + v e_{\theta\theta}) \qquad N_\theta = C(e_{\theta\theta} + v e_{rr}) \qquad N_{r\theta} = C\frac{1-v}{2} e_{r\theta}$$

$$e_{rr} = u_{,r} + \frac{w}{R} \qquad e_{\theta\theta} = \frac{v_{,\theta} + u}{r} + \frac{w}{R} \qquad e_{r\theta} = r\left(\frac{v}{r}\right)_{,r} + \frac{u_{,\theta}}{r} \qquad (6.42)$$

and

$$M_r = D\left[\beta_{r,r} + \frac{v}{r}(\beta_{\theta,\theta} + \beta_r)\right]$$

$$M_\theta = D\left[\frac{1}{r}(\beta_{\theta,\theta} + \beta_r) + v\beta_{r,r}\right]$$

$$M_{r\theta} = D\frac{1-v}{2}\left[r\left(\frac{\beta_\theta}{r}\right)_{,r} + \frac{\beta_{r,\theta}}{r}\right] \qquad (6.43)$$

$$\beta_r = -w_{,r} \qquad \beta_\theta = -\frac{w_{,\theta}}{r}$$

For $R \to \infty$, Eqs. (6.41) to (6.43) reduce to Eqs. (6.35) and (6.36) for the circular flat plate.

Introduction of the constitutive and kinematic relations for M_{r1}, $M_{r\theta1}$, and $M_{\theta1}$ from Eqs. (6.43) into (6.41c) and rearrangement gives

$$D\nabla^4 w + \frac{N_r + N_\theta}{R} - \frac{1}{r}\left[(rN_{r0} w_{,r} + N_{r\theta0} w_{,\theta})_{,r}\right.$$

$$\left. + \left(N_{r\theta0} w_{,r} + N_{\theta0}\frac{w_{,\theta}}{r}\right)_{,\theta}\right] = 0 \qquad (6.44)$$

where

$$\nabla^2(\ ) \equiv \left[(\ )_{,rr} + \frac{1}{r}(\ )_{,r} + \frac{1}{r^2}(\ )_{,\theta\theta}\right]$$

$$\nabla^4(\ ) \equiv \nabla^2\nabla^2(\ )$$

Equations (6.41a), (6.41b), (6.42), and (6.44) lead to a coupled set of three homogeneous equations in u, v, w.

As a specific example let us consider a spherical cap subjected to uniform external pressure p_e, and let us assume that the prebuckling state may be approximated by a membrane analysis. Then $N_{r0} = N_{\theta0} = -p_e R/2$, and $N_{r\theta0} = 0$. Introduction into Eq. (6.44) and rearrangement gives

$$D\nabla^4 w + \frac{N_r + N_\theta}{R} + \tfrac{1}{2}p_e R\nabla^2 w = 0 \qquad (6.45)$$

The set of three equations in u, v, w can be reduced to two equations in w and a stress function f, as follows (compare Ref. 6.28). Let f be defined by the relations

$$N_r = \frac{1}{r} f_{,r} + \frac{1}{r^2} f_{,\theta\theta} \qquad N_\theta = f_{,rr} \qquad N_{r\theta} = -\left(\frac{f_{,\theta}}{r}\right)_{,r} \tag{6.46}$$

Then Eq. (6.45) can be written

$$D\nabla^4 w + \frac{1}{R}\nabla^2 f + \tfrac{1}{2} p_e R \nabla^2 w = 0 \tag{6.47}$$

But from Eqs. (6.42),

$$\frac{1}{r^2} e_{rr,\theta\theta} - \frac{1}{r} e_{rr,r} + \frac{1}{r^2}(r^2 e_{\theta\theta,r})_{,r} - \frac{1}{r^2}(re_{r\theta})_{,r\theta} = \frac{\nabla^2 w}{R} \tag{6.48}$$

$$e_{rr} = \frac{1}{C}(N_r - \nu N_\theta) \qquad e_{\theta\theta} = \frac{1}{C}(N_\theta - \nu N_r) \qquad e_{r\theta} = \frac{2}{1-\nu}\, C N_{r\theta}$$

Therefore, since $C = Eh/(1 - \nu^2)$, the stress function f must satisfy the compatibility condition (see Prob. 6.12)

$$\nabla^4 f = \frac{Eh}{R}\nabla^2 w \tag{6.49}$$

Equations (6.47) and (6.49) reduce the problem to a coupled set of two homogeneous equations in w and f. These equations are the same as eqs. (4) and (5) of Ref. 6.28. The equations have nontrivial solutions only for discrete values of the pressure p_e. The smallest such value is called P_{cr}.

A particularly simple solution is given by Hutchinson in Ref. 6.28 by use of the following coordinate transformations. Let

$$x = r\cos\theta \qquad y = r\sin\theta$$

Then

$$\nabla^2(\ ) = \frac{\partial^2}{\partial r^2}(\ ) + \frac{1}{r}\frac{\partial}{\partial r}(\ ) + \frac{1}{r^2}\frac{\partial^2}{\partial\theta^2}(\ )$$

$$= \frac{\partial^2}{\partial x^2}(\ ) + \frac{\partial^2}{\partial y^2}(\ )$$

(For a general discussion of coordinate transformations and the laplacian operator, see, for example, Ref. 6.29, sec. 2-13.) The differential equations are seen to be satisfied by solutions of the form

$$w = \cos\left(k_x \frac{x}{R}\right)\cos\left(k_y \frac{y}{R}\right)$$

$$f = B_1 \cos\left(k_x \frac{x}{R}\right)\cos\left(k_y \frac{y}{R}\right) \tag{6.50}$$

where k_x, k_y are wavelength parameters and B_1 is a constant. Introduction into Eq. (6.49) gives

$$B_1 = -EhR(k_x^2 + k_y^2)^{-1}$$

Introduction of this expression and the definition $D = Eh^3/[12(1 - v^2)]$ into Eq. (6.47) now yields

$$p_e = \frac{2Eh}{R}\left[(k_x^2 + k_y^2)^{-1} + \frac{(h/r)^2}{12(1 - v^2)}(k_x^2 + k_y^2)\right] \qquad (6.51)$$

An approximate expression for p_{cr} may be obtained by direct analytical minimization of p_e with respect to the quantity $k_x^2 + k_y^2$ (cf. the analysis in Sec. 5.5*b* of long cylindrical shells subjected to axial compression). The smallest p_e is found to correspond to the value

$$k_x^2 + k_y^2 = 2[3(1 - v^2)]^{1/2}\frac{R}{h}$$

Introduction of this value into Eq. (6.51) gives

$$p_{cr} = \frac{2E}{[3(1 - v^2)]^{1/2}}\left(\frac{h}{R}\right)^2 \qquad (6.52)$$

This is the result given in Ref. 6.28, eq. (8).

The expression for the critical pressure in Eq. (6.52) is the same as that given for a complete spherical shell in Ref. 6.26, eq. (11-31). The solution functions in Eqs. (6.50), however, do not satisfy the boundary conditions at the edge of a spherical cap. Consequently, the present simplified analysis is limited to buckle-pattern wavelengths that are small compared with the radius of the shell. Even for such cases, the critical pressure given by Eq. (6.52) is in poor agreement with test data. The discrepancy is due to two factors, the neglect of nonlinearity in the prebuckling analysis and the influence of initial imperfections. The pressurized spherical cap is discussed in further detail in Chap. 9.

6.5*c* Conical Shells

A truncated conical shell with a vertex angle of 2α is illustrated in Fig. 6.6. Points on the middle surface of the shell are defined by a longitudinal coordinate s, as shown, and a circumferential coordinate θ. In the equations for shells of revolution, $r_\phi \to \infty$ for a cone, and

$$\lim_{r_\phi \to \infty} (r_\phi \, d\phi) = ds$$

Furthermore, $r = s \sin \alpha$ and $\phi = \pi/2 - \alpha$. Then $\sin \phi = \cos \alpha$ and $\cos \phi = \sin \alpha$. Introduction of these values into Eqs. (6.30) and rearrangement gives

$$(sN_s)_{,s} + \frac{1}{\sin \alpha} N_{s\theta,\theta} - N_\theta = 0 \qquad (6.53a)$$

$$\frac{1}{\sin \alpha} N_{\theta,\theta} + \frac{1}{s} (s^2 N_{s\theta})_{,s} = 0 \qquad (6.53b)$$

$$(sM_s)_{,ss} + \frac{2}{\sin \alpha} \left(M_{s\theta,s\theta} + \frac{1}{s} M_{s\theta,\theta} \right)$$

$$+ \frac{1}{s \sin^2 \alpha} M_{\theta,\theta\theta} - M_{\theta,s} - N_\theta \cot \alpha$$

$$- \left[(sN_{s0} \beta_s + sN_{s\theta 0} \beta_\theta)_{,s} + \frac{1}{\sin \alpha} (N_{s\theta 0} \beta_s + N_{\theta 0} \beta_\theta)_{,\theta} \right] = 0 \qquad (6.53c)$$

where the subscript ϕ has been replaced by s, and where, from Eqs. (6.31),

$$N_s = C \left[u_{,s} + v \left(\frac{v_{,\theta}}{s \sin \alpha} + \frac{u}{s} + \frac{w}{s} \cot \alpha \right) \right]$$

$$N_\theta = C \left[\left(\frac{v_{,\theta}}{s \sin \alpha} + \frac{u}{s} + \frac{w}{s} \cot \alpha \right) + v u_{,s} \right]$$

$$N_{s\theta} = C \frac{1-v}{2} \left(v_{,s} - \frac{v}{s} + \frac{u_{,\theta}}{s \sin \alpha} \right)$$

$$M_s = D \left[\beta_{s,s} + v \left(\frac{\beta_{\theta,\theta}}{s \sin \alpha} + \frac{\beta_s}{s} \right) \right] \qquad (6.54)$$

$$M_\theta = D \left[\left(\frac{\beta_{\theta,\theta}}{s \sin \alpha} + \frac{\beta_s}{s} \right) + v \beta_{s,s} \right]$$

$$M_{s\theta} = D \frac{1-v}{2} \left[\frac{\beta_{s,\theta}}{s \sin \alpha} + s \left(\frac{\beta_\theta}{s} \right)_{,s} \right]$$

$$\beta_s = - w_{,s} \qquad \beta_\theta = - \frac{w_{,\theta}}{s \sin \alpha}$$

For $\alpha = \pi/2$, these equations reduce to the corresponding expressions for circular flat plates in Eqs. (6.35) and (6.36). At the other extreme, replacement of $s \sin \alpha$ by the radius a and then specialization for $\alpha = 0$ gives the Donnell equations for cylindrical shells in Chap. 5.

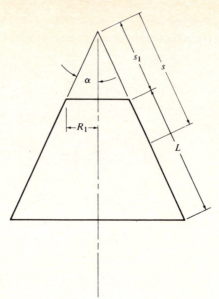

FIGURE 6.6
Conical shell.

Introduction of Eqs. (6.54) into Eqs. (6.53) and rearrangement gives

$$su_{,ss} + u_{,s} - \frac{u}{s} + \frac{1-v}{2}\frac{u_{,\theta\theta}}{s\sin^2\alpha} + \frac{1+v}{2}\frac{v_{,s\theta}}{\sin\alpha}$$

$$-\frac{3-v}{2}\frac{v_{,\theta}}{s\sin\alpha} + \left(vw_{,s} - \frac{w}{s}\right)\cot\alpha = 0$$

$$\frac{1+v}{2}\frac{u_{,s\theta}}{\sin\alpha} + \frac{3-v}{2}\frac{u_{,\theta}}{s\sin\alpha} + \frac{1-v}{2}sv_{,ss}$$

$$+\frac{1-v}{2}\left(v_{,s} - \frac{v}{s}\right) + \frac{v_{,\theta\theta}}{s\sin^2\alpha} + \frac{w_{,\theta}\cot\alpha}{s\sin\alpha} = 0$$

$$Ds\left(w_{,ssss} + 2\frac{w_{,sss}}{s} - \frac{w_{,ss}}{s^2} + \frac{w_{,s}}{s^3} - 2\frac{w_{,s\theta\theta}}{s^3\sin^2\alpha}\right. \tag{6.55}$$

$$\left.+ 2\frac{w_{,ss\theta\theta}}{s^2\sin^2\alpha} + 4\frac{w_{,\theta\theta}}{s^4\sin^2\alpha} + \frac{w_{,\theta\theta\theta\theta}}{s^4\sin^4\alpha}\right)$$

$$+ C\left(\frac{v_{,\theta}}{s\sin\alpha} + \frac{u}{s} + \frac{w\cot\alpha}{s} + vu_{,s}\right)\cot\alpha$$

$$-\left[\left(N_{so}sw_{,s} + N_{s\theta o}\frac{w_{,\theta}}{\sin\alpha}\right)_{,s}\right.$$

$$\left.+ \frac{1}{\sin\alpha}\left(N_{s\theta o}w_{,s} + N_{\theta o}\frac{w_{,\theta}}{s\sin\alpha}\right)_{,\theta}\right] = 0$$

These equations form a coupled set of three homogeneous equations in u, v, w.

As an example, consider a conical shell subjected to uniform external hydrostatic pressure p_e, in pounds per square inch, and an axial compressive load P, in pounds. If we assume that a membrane analysis is adequate for the prebuckling deformation, the coefficients N_{s0}, $N_{s\theta0}$, and $N_{\theta0}$ are given by the simple relations

$$N_{s0} = -\tfrac{1}{2} p_e s \tan \alpha - \frac{P}{\pi s \sin 2\alpha} \qquad (6.56)$$

$$N_{\theta0} = -p_e s \tan \alpha \qquad N_{s\theta0} = 0$$

For this problem the stability equations are variable-coefficient equations. A solution for critical values of the applied load may be obtained by use of numerical methods such as those discussed in Chap. 8. Such an analysis is reported by Baruch, Harari, and Singer in Ref. 6.30, and Eqs. (6.55) and (6.56) are the same as eqs. (4), (28), and (29) of that reference. Extensive numerical results, which were obtained by use of the Galerkin procedure, are given in Ref. 6.30 for hydrostatic-pressure loading ($P = 0$). Results from Ref. 6.30 for truncated conical shells with simply supported ends ($w = w_{,ss} = N_s = v = 0$) are shown in Table 6.1. The results cover a wide range of cone angles from $\alpha = 0°$ (a cylinder) to $\alpha = 85°$ (a shallow cone) and of length/radius ratios from $L/R_1 = \tfrac{1}{2}$ to $L/R_1 = 10$. All the values in Table 6.1 are for the radius/thickness ratio $R_1/h = 100$ and for Poisson's ratio $v = 0.3$. Results for other values of the cone geometry parameters and for other boundary conditions are included in Ref. 6.30.

6.5d Segments of Toroidal Shells

As noted in Sec. 6.4, the middle surface of a shell of revolution may be formed by rotation of a plane curve about an axis in the plane of the curve (Fig. 6.2). If the plane curve is a circular arc of radius b, as shown in Fig. 6.7, the surface

Table 6.1 CRITICAL HYDROSTATIC PRESSURES FOR A CONICAL SHELL

α,	$10^4 p_{cr} R_1/Eh$				
deg	$L/R_1 = \tfrac{1}{2}$	$L/R_1 = 1$	$L/R_1 = 2$	$L/R_1 = 5$	$L/R_1 = 10$
0	21.1	9.84	4.74	1.878	0.968
10	19.40	8.57	3.74	1.126	0.413
30	14.55	5.84	2.24	0.532	0.1386
50	8.81	3.28	1.164	0.233	0.0541
70	3.50	1.201	0.399	0.0726	0.01622
85	0.591	1.710	0.0498	0.00860	0.001917

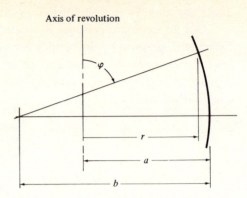

FIGURE 6.7
Meridian of a toroidal shell.

formed is a segment of a torus. For the middle surface of a segment of a toroidal shell, from the figure, $r_\phi = b$ and $r = a - b(1 - \sin \phi)$.

Analyses of shallow segments of toroidal shells are presented by Stein and McElman in Ref. 6.31 and by Hutchinson in Ref. 6.32. For a sufficiently shallow segment in the region of the equator of the torus, the angle ϕ is approximately equal to $\pi/2$. Then, approximately, $\sin \phi = 1$, $\cos \phi = 0$, and $r = a$. Introduction of these values into Eqs. (6.30) yields the stability equations for the toroidal-shell segment. The form of the equations is simplified if we let $dx = b\, d\phi$ and $dy = a\, d\theta$, where x and y are axial and circumferential coordinates, respectively. Then, with the aid of the constitutive and kinematic relations in Eqs. (6.31), the stability equations may be expressed in the form

$$N_{x,x} + N_{xy,y} = 0 \qquad (6.57a)$$

$$N_{xy,x} + N_{y,y} = 0 \qquad (6.57b)$$

$$D\nabla^4 w + \frac{N_x}{b} + \frac{N_y}{a} - (N_{x0}\, w_{,xx} + 2N_{xy0}\, w_{,xy} + N_{y0}\, w_{,yy}) = 0 \qquad (6.57c)$$

where

$$N_x = C\left[\left(u_{,x} + \frac{w}{b}\right) + v\left(v_{,y} + \frac{w}{a}\right)\right]$$

$$N_y = C\left[\left(v_{,y} + \frac{w}{a}\right) + v\left(u_{,x} + \frac{w}{b}\right)\right] \qquad (6.58)$$

$$N_{xy} = C\frac{1 - v}{2}(v_{,x} + u_{,y})$$

These equations are the same as eqs. (17) in Ref. 6.31.

The analyses in Refs. 6.31 and 6.32 are carried out for both the bowed-out segment illustrated in Fig. 6.8a and the bowed-in segment in Fig. 6.8b. For the

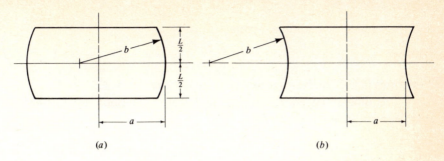

(a)   *(b)*

FIGURE 6.8
Toroidal-shell segments.

latter case the radius of curvature b in Eqs. (6.57) and (6.58) is negative. For $b \to \infty$ the equations reduce to the Donnell equations for cylindrical shells in Chap. 5.

As a specific example we consider a toroidal segment that is simply supported at $x = 0, L$ and subjected to uniform external lateral pressure p_e, in pounds per square inch. We assume, for simplicity, that the prebuckling state may be approximated by a membrane analysis. Then $N_{y0} = -p_e a$ and $N_{x0} = N_{xy0} = 0$. Introduction of these values into Eq. (6.57c) simplifies that expression to the form

$$D\nabla^4 w + \frac{N_x}{b} + \frac{N_y}{a} + p_e a w_{,yy} = 0 \qquad (6.59)$$

For simply supported ends the boundary conditions at $x = 0, L$ are

$$w = w_{,xx} = N_x = v = 0$$

Equations (6.57a), (6.57b), and (6.59) lead to a coupled set of constant-coefficient equations in u, v, w. Although the equations are easily solved, the analysis is simplified somewhat by introduction of a stress function f as in Sec. 6.5b. Let

$$N_x = f_{,yy} \qquad N_{xy} = -f_{,xy} \qquad N_y = f_{,xx}$$

Then the stability equations reduce to the two equations

$$D\nabla^4 w + \frac{f_{,xx}}{a} + \frac{f_{,yy}}{b} + p_e a w_{,yy} = 0 \qquad (6.60a)$$

$$\frac{\nabla^4 f}{Eh} - \frac{w_{,xx}}{a} - \frac{w_{,yy}}{b} = 0 \qquad (6.60b)$$

and the boundary conditions become

$$w = w_{,xx} = f_{,xx} = f = 0 \qquad (6.61)$$

Equations (6.60) are the same as eqs. (1) and (2) in Ref. 6.32.

The differential equations and boundary conditions are satisfied by a displacement function of the form

$$w = C_1 \sin \frac{m\pi x}{L} \sin \frac{ny}{a} \qquad (6.62)$$

where C_1 is a constant and m, n are positive integers. In terms of this displacement function, Eq. (6.60b) gives, for the stress function, the expression

$$f = -\frac{EhL^2}{\pi^2 a} \frac{m^2 + \bar{n}^2 a/b}{(m^2 + \bar{n}^2)^2} C_1 \sin \frac{m\pi x}{L} \sin \frac{ny}{a} \qquad (6.63)$$

where $\bar{n} \equiv nL/\pi a$. Introduction into Eq. (6.60a) now gives, for the eigenvalues, the equation

$$\bar{p} = \frac{(m^2 + \bar{n}^2)^2}{\bar{n}^2} + \frac{(m^2 + \bar{n}^2 a/b)^2}{\bar{n}^2(m^2 + \bar{n}^2)^2} \frac{12}{\pi^4} Z^2 \qquad (6.64)$$

where, as in Sec. 5.6a,

$$\bar{p} \equiv \frac{L^2 a}{\pi^2 D} p_e \qquad Z \equiv \frac{L^2}{ah}(1 - v^2)^{1/2}$$

Equation (6.64) is the expression for the eigenvalues given by Hutchinson in Ref. 6.32.

The smallest eigenvalues occur for $m = 1$ in Eq. (6.64). Accordingly,

$$\bar{p} = \frac{(1 + \bar{n}^2)^2}{\bar{n}^2} + \frac{(1 + \bar{n}^2 a/b)^2}{\bar{n}^2(1 + \bar{n}^2)^2} \frac{12}{\pi^4} Z^2 \qquad (6.65)$$

For $b \to \infty$, this equation reduces to the corresponding expression for a cylindrical shell in Eq. (5.48). For particular values of the shell geometry parameters a, b, and Z, the value of the integer n corresponding to the critical pressure may be determined by trial or by analytical minimization, as in earlier examples.

Results given in Refs. 6.31 and 6.32 for this method of loading are shown in Fig. 6.9 (Ref. 6.31, fig. 3). For given values of a, b, and Z the critical pressures are seen to be substantially higher for bowed-out segments than for bowed-in ones. For $a/b = 0$ the curve is the same as that for a cylindrical shell in Fig. 5.6.

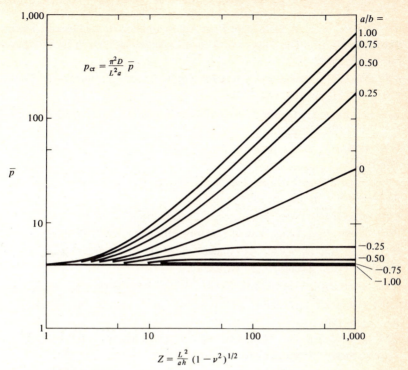

$$p_{cr} = \frac{\pi^2 D}{L^2 a} \bar{p}$$

$$Z = \frac{L^2}{ah} (1 - \nu^2)^{1/2}$$

FIGURE 6.9
Critical values of external pressure for toroidal-shell segments subjected to external lateral pressure.

REFERENCES

6.1 LOVE, A. E. H.: "A Treatise on the Mathematical Theory of Elasticity," 4th ed., Dover, New York, 1927.

6.2 NAGHDI, P. M.: Foundations of Elastic Shell Theory, in "Progress in Solid Mechanics," vol. 4, Wiley, New York, 1963.

6.3 GOLDENVEIZER, A. L.: "Problems in the Rigorous Deduction of the Theory of Thin Elastic Shells," *Proc. 2d IUTAM Symp. Theory Thin Shells*, Springer, Copenhagen, 1967.

6.4 KOITER, W. T., and J. G. SIMMONDS: Foundations of Shell Theory, *WTHD* 40, Delft University of Technology, August 1972.

6.5 FLÜGGE, W.: "Statik und Dynamik der Schalen," Springer, Berlin, 1934.

6.6 BUDIANSKY, B., and J. L. SANDERS: On the "Best" First Order Linear Shell Theory, in "Progress in Applied Mechanics" (Prager Anniversary Volume), Macmillan, New York, 1963.

6.7 KOITER, W. T.: A Consistent First Approximation in the General Theory of Thin Elastic Shells, in "The Theory of Thin Elastic Shells," North-Holland, Amsterdam, pp. 12–33, 1960.

6.8 REISSNER, E.: Stress Strain Relations in the Theory of Thin Elastic Shells, *J. Math. Phys.*, vol. 31, pp. 109–119, 1952.

6.9 NAGHDI, P. M.: On the Theory of Thin Elastic Shells, *Q. Appl. Math.*, vol. 14, pp. 369–380, 1957.

6.10 DONNELL, L. H.: Stability of Thin-walled Tubes under Torsion, *NACA Rep.* 479, 1933.

6.11 VLASOV, V. Z.: "Allgemeine schalen Theorie und ihre Anwending in der Technik," Akademie-Verlag, Berlin, 1958.

6.12 REISSNER, E.: On Symmetrical Deformations of Thin Shells of Revolution, *Proc. Symp. Appl. Math.*, vol. 3, McGraw-Hill, New York, 1950, pp. 27–52.

6.13 NOVOZHILOV, V. V.: "Foundations of the Nonlinear Theory of Elasticity," Greylock Press, Rochester, N.Y., 1953.

6.14 KRAUS, H.: "Thin Elastic Shells," Wiley, New York, 1967.

6.15 SANDERS, J. L.: Nonlinear Theories for Thin Shells, *Q. Appl. Math.*, vol. 21, no. 1, pp. 21–36, 1963.

6.16 LEONARD, R. W.: Nonlinear First Approximation Thin Shell and Membrane Theory, thesis, Virginia Polytechnic Institute, Blacksburg, Va., 1961.

6.17 BUDIANSKY, B.: Notes on Nonlinear Shell Theory, *J. Appl. Mech.*, vol. 35, pp. 392–401, 1968.

6.18 NAGHDI, P. M., and R. P. NORDGREN: Nonlinear Theory of Elastic Shells, *Q. Appl. Math.*, vol. 21, pp. 19–59, 1963.

6.19 KOITER, W. T.: On the Nonlinear Theory of Thin Elastic Shells, *Proc. K. Ned. Akad. Wet.*, ser. B69, 1966.

6.20 MARLOWE, M. B., and W. FLÜGGE: Some New Developments in the Foundations of Shell Theory, thesis, Stanford University, Stanford, Calif., 1968.

6.21 KOITER, W. T.: General Equations of Elastic Stability for Thin Shells, *Proc. Symp. Theory of Shells to Honor Lloyd Hamilton Donnell*, University of Houston, Houston, Tex., 1967, pp. 187–223.

6.22 SIMMONDS, J. G., and D. A. DANIELSON: Nonlinear Shell Theory with Finite Rotation and Stress Function Vectors, *J. Appl. Mech.*, vol. 39, 1972.

6.23 FLÜGGE, W.: "Tensor Analysis in Continuum Mechanics," Springer, Berlin, 1972.

6.24 FLÜGGE, W.: "Stresses in Shells," 2d ed., Springer, Berlin, 1973.

6.25 BUSHNELL, D.: Stress, Stability, and Vibration of Complex, Branched Shells of Revolution, *Proc. AIAA/ASME/SAE 14th Structures, Struct. Dyn., Mater. Conf.*, Williamsburg, Va., 1973.

6.26 TIMOSHENKO, S. P., and J. M. GERE: "Theory of Elastic Stability," 2d ed., McGraw-Hill, New York, 1961.

6.27 JANKE, E., and F. EMDE: Tables of Functions, 4th ed., Dover, New York, 1945.

6.28 HUTCHINSON, J. W.: Imperfection Sensitivity of Externally Pressurized Spherical Shells, *J. Appl. Mech.*, vol. 34, pp. 49–55, 1967.

6.29 KAPLAN, W.: "Advanced Calculus," Addison-Wesley, Reading, Mass., 1952.

6.30 BARUCH, M., O. HARARI, and J. SINGER: Influence of In-Plane Boundary Conditions on the Stability of Conical Shells under Hydrostatic Pressure, *Israel J. Technol.* vol. 5, no. 1–2, pp. 12–24, 1967.

6.31 STEIN, M., and J. A. MCELMAN: Buckling of Segments of Toroidal Shells, *AIAA J.*, vol. 3, pp. 1704–1709, September 1965.

6.32 HUTCHINSON, J. W.: Initial Post-buckling Behavior of Toroidal Shell Segments, *Intl. J. Solids Structures*, vol. 3, pp. 97–115, 1967.

PROBLEMS

6.1 (*a*) Show that specialization for polar coordinates of the kinematic relations in Eqs. (6.6) and (6.7) gives the kinematic relations for the ring in Eqs. (4.7), (4.9), and (4.10). Note that $B = a$. (*b*) Show that similar specialization for cylindrical coordinates gives the kinematic relations for nonshallow cylinders in Eqs. (5.28). Note that, for the cylindrical coordinates x and θ, $A = 1$ and $B = a$.

6.2 Show that application of the stationary potential energy criterion to the potential energy expression in Eqs. (6.3) to (6.6), (6.7a), (6.7c), and (6.8) yields the nonlinear equilibrium equations for the shell of general shape in Eqs. (6.10).

6.3 (*a*) Show that specialization of Eqs. (6.10) for a rectangular flat plate gives the von Kármán plate equations in Eqs. (3.18). (*b*) Show that similar specialization for a circular cylindrical shell gives the equilibrium equations in Eqs. (5.8).

6.4 Show that application of the adjacent-equilibrium criterion to the equilibrium equations in Eqs. (6.10) gives the stability equations for the shell of general shape in Eqs. (6.18).

6.5 Show that specialization of Eqs. (6.16) and (6.17) for rectangular coordinates gives the expression for the second variation of the potential energy for rectangular flat plates in Eqs. (3.41) and (3.42).

6.6 Show that specialization for circular cylindrical shells of the second-variation expression in Eq. (6.16) and the kinematic relations in Eqs. (6.7) leads to the expression in Eqs. (5.29) and (5.30) for the second variation of the total potential energy of a nonshallow cylindrical shell.

6.7 Show that specialization for cylindrical shells of the stability equations in Eqs. (6.18) yields the coupled form of the Donnell equations (5.23).

6.8 Stability equations for shells of revolution subjected to asymmetric loading are given in Eqs. (6.25). Specialize the equations for loads and deformations that are axisymmetric prior to loss of stability. Compare the results with the stability equations in Eqs. (6.30), in which prebuckling rotations have been neglected.

6.9 If a shell of revolution is subjected to axisymmetric loading and $N_{\phi\theta0} = 0$, the stability equations may be reduced to ordinary differential equations. Using the procedure discussed at the end of Sec. 6.4, derive the ordinary differential equations.

6.10 Show that specialization of Eqs. (6.30) for a circular flat plate gives the stability equations in Eqs. (6.35).

6.11 A circular flat plate with a diameter of 40 in. and a thickness of $\frac{1}{8}$ in. is subjected to uniform radial compressive loading. If $E = 29 \times 10^6$ psi and $v = 0.3$, determine the critical load for simply supported and for clamped edges.

6.12 Derive the expression in Eq. (6.49) for the relationship between the stress function f and the displacement component w for a shallow spherical cap. Use Eqs. (6.46) and (6.48).

6.13 A shallow spherical cap has a thickness $h = 0.020$ in. and radius of curvature $R = 8$ in. If $E = 10 \times 10^6$ psi and $v = 0.3$, calculate the theoretical critical pressure p_{cr}.

6.14 A toroidal-shell segment with simply supported ends is subjected to uniform external lateral pressure p_e. (*a*) If $a = 10$ in., $a/b = 0.50$, $L = 2.5$ in., $h = 0.020$ in., $E = 10 \times 10^6$ psi, and $v = 0.3$, determine the critical pressure. Compare results based on Eq. (6.65) with those based on Fig. 6.9. (*b*) Repeat for $a/b = -0.50$.

INFLUENCE OF INITIAL IMPERFECTIONS

7.1 INTRODUCTION

As emphasized in earlier chapters, theoretical bifurcation-point loads are in close agreement with experimental buckling loads in some cases and in sharp disagreement in others. Experimental loads for axially compressed cylindrical shells, for example, may be as low as 10 to 20 percent of corresponding theoretical values, as noted in Chap. 5. An early attempt to establish a correlation between theoretical and experimental results for axially compressed cylindrical shells was reported by Flügge in 1932 (Ref. 7.1). The magnitudes of his experimental buckling loads were approximately one-half the theoretical values. Later Lundquist (Ref. 7.2) and Donnell (Ref. 7.3) tested very thin aluminum cylinders and found even larger discrepancies between theoretical and experimental loads. Their results also revealed the wide scatter in experimental values and the relationship between critical stress and radius/thickness ratio that were discussed in Sec. 5.6 and illustrated in Fig. 5.18.

The reasons for the occasional discrepancies between theoretical and experimental results were at one time the subject of widespread controversy. Now there is general agreement that initial imperfections, i.e., small accidental devia-

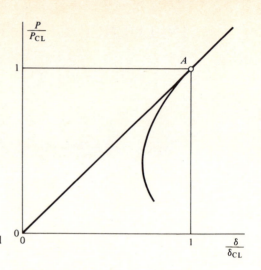

FIGURE 7.1
Equilibrium paths for axially compressed
cylindrical shell.

tions from the assumed initial shape of the structure, are the principal cause of
the disagreement.

A major contribution to the present understanding of the role of initial
imperfections was made by von Kármán and Tsien in 1941 (Ref. 7.4). Their
landmark analysis of the postbuckling equilibrium of axially compressed cylindrical shells showed that the secondary equilibrium path drops sharply downward
from the bifurcation point, as illustrated in Fig. 7.1. (The von Kármán–Tsien
analysis is examined in greater detail in Chap. 9.) Although the authors did not
analyze initially imperfect shells, their results suggest that the corresponding
equilibrium path for such a shell might have the form illustrated in Fig. 7.2,
where buckling occurs at the markedly lower value of the load corresponding
to the limit point B rather than at the bifurcation point A.

A well-known analysis of initially imperfect cylindrical shells presented by
Donnell and Wan in 1950 (Ref. 7.5) yielded equilibrium paths of the form of that
in Fig. 7.2. The Donnell-Wan analysis is based on the nonlinear equilibrium
equations in Eqs. (5.8) and treats a thin-walled shell whose initial shape deviates
slightly from cylindrical. The analysis has the limitation, however, that the
function representing the initial deviation from a perfectly cylindrical shape is
assumed for simplicity to be of the form of the constantly changing displacement
mode. Consequently, it does not represent a particular initial shape.

Rigorous confirmation of the influence of initial imperfections was given
by Koiter (Ref. 7.6) in 1945. His paper received little attention until a summary

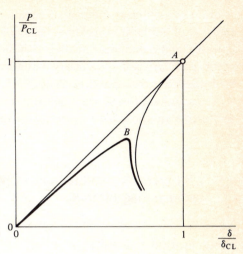

FIGURE 7.2
Equilibrium path for imperfect cylindrical shell.

(Ref. 7.7) appeared in 1963. The Koiter analysis focuses attention on *initial*-postbuckling behavior and provides a theory that is exact in the asymptotic sense, i.e., exact at the bifurcation point itself and a close approximation for postbuckling configurations near the bifurcation point. The central significance of the shape of the secondary equilibrium path in determining the influence of initial imperfections was emphasized in Chap. 1. When the initial portion of the secondary path has a positive slope, considerable postbuckling strength can be developed by the structure, and loss of stability on the primary path does not result in structural collapse. When the initial portion of the secondary path has a negative slope, on the other hand, the buckling is precipitous and the magnitude of the critical load is subject to the influence of initial imperfections. The analysis in Ref. 7.6 showed that a limited amount of information about the secondary path can be obtained by examination of the state of equilibrium at the bifurcation point. Along the primary path, equilibrium always is stable below the bifurcation point and unstable above it. At the bifurcation point itself, however, equilibrium is stable in some cases and unstable in others. When it is stable, the initial portion of the secondary path has a positive slope. When it is unstable, the secondary path has a negative slope and the structure is sensitive to imperfections.

In this chapter we examine the state of equilibrium in the immediate vicinity of the bifurcation point and the role of initial imperfections in the reduction of critical loads. The simplified analyses in this chapter do not require the solution of nonlinear differential equations. In many cases such a linearized analysis of initial-postbuckling behavior is inadequate or even misleading, however, and a nonlinear analysis that more fully traces the postbuckling shape of the secondary path is necessary for understanding the behavior of the structure. Such cases are considered in Chap. 9. Numerical methods that are necessary for the solution of nonlinear equations as well as linear equations with variable coefficients are the subject matter of Chap. 8.

7.2   CYLINDER WITH AXISYMMETRIC IMPERFECTIONS

As a first example of initial-imperfection analysis, we consider an analysis presented by Koiter in Ref. 7.8 of the buckling strength of axially compressed cylinders. In this example the critical load is substantially reduced by imperfections whose magnitude is only a small fraction of the shell-wall thickness. In contrast to the general case, it is possible in this example to investigate the effect of small initial imperfections within the framework of bifurcation analysis. The method is limited in applicability to axisymmetric imperfections, but unlike the general Koiter theory in Ref. 7.6, which is valid only in an asymptotic sense, it is valid for finite imperfections. The more general Koiter theory of initial-postbuckling behavior is discussed in Sec. 7.3.

Nonlinear equilibrium equations for circular cylindrical shells are given in Eqs. (5.8). The form of the equations can be simplified substantially by introduction of a circumferential coordinate s defined by the relationship

$$ds = a\, d\theta$$

In terms of this variable, Eqs. (5.8) may be written as follows for lateral pressure $p = 0$:

$$N_{x,x} + N_{xs,s} = 0 \qquad (7.1a)$$

$$N_{xs,x} + N_{s,s} = 0 \qquad (7.1b)$$

$$D\nabla^4 w + \frac{1}{a}N_s - (N_x w_{,xx} + 2N_{xs} w_{,xs} + N_s w_{,ss}) = 0 \qquad (7.1c)$$

For a slightly imperfect shell, let $w^*(x)$ denote a known small imperfection, i.e., a small deviation of the shell middle surface from a circular cylindrical shape. The unloaded shell in the form including w^* is assumed to be stress-free. Then small angles of rotation $w_{,x}$ in the equations for an initially perfect cylinder are replaced by $(w + w^*)_{,x}$. (Compare the analysis of a slightly crooked column in Sec. 1.4.) From Eqs. (5.7), the strain-displacement relations for the imperfect cylinder now become

$$\varepsilon_x = u_{,x} + \tfrac{1}{2}[(w + w^*)_{,x}]^2 - \tfrac{1}{2}(w^*_{,x})^2$$

$$= u_{,x} + \tfrac{1}{2}w_{,x}{}^2 + w^*_{,x}w_{,x}$$

$$\varepsilon_s = v_{,s} + \frac{w}{a} + \tfrac{1}{2}w_{,s}{}^2 \qquad\qquad (7.2)$$

$$\gamma_{xs} = u_{,s} + v_{,x} + w_{,x}w_{,s} + w^*_{,x}w_{,s}$$

Similarly, Eq. (7.1c) for an initially perfect cylinder is replaced by the expression

$$D\nabla^4 w + \frac{1}{a}N_s - [N_x(w + w^*)_{,xx} + 2N_{xs}w_{,xs} + N_s w_{,ss}] = 0$$

Because both the shell itself and the applied load are axisymmetric, shell configurations on the primary equilibrium path also are axisymmetric. The equilibrium path for axisymmetric deformation can be shown to be of the general form of the path in Fig. 1.10 for an initially crooked column; as the magnitude of the applied load approaches the classical critical load given by Eq. (5.52), the displacement amplitude in the axisymmetric form begins to grow without bounds (Ref. 7.9, sec. 8.2.5). At lower loads there are bifurcation points, however, and at those points equilibrium in an asymmetric form is possible as well. The object of the analysis is to locate the first bifurcation point along the nonlinear equilibrium path.

Following the procedure in earlier chapters, the equilibrium equations may be linearized by letting

$$u \to u_0 + u_1$$

$$v \to v_0 + v_1$$

$$w \to w_0 + w_1$$

where u, v, w and u_0, v_0, w_0 are equilibrium configurations and u_1, v_1, w_1 is an infinitesimally small increment. For the axisymmetric configurations on the primary path, $u_0 = u_0(x)$, $v_0 \equiv 0$, and $w_0 = w_0(x)$. Introduction of these expressions into the equilibrium equations and simplification in accordance with the procedure in earlier chapters gives

$$N_{x1,x} + N_{xs1,s} = 0 \qquad (7.3a)$$

$$N_{xs1,x} + N_{s1,s} = 0 \qquad (7.3b)$$

$$D\nabla^4 w_1 + \frac{1}{a}N_{s1} - [N_{x0}w_{1,xx} + N_{s0}w_{1,ss} + w_0'' N_{x1} + (w^*)'' N_{x1}] = 0 \qquad (7.3c)$$

where primes denote differentiation with respect to x and where, from Eqs. (5.21) and (5.22),

$$N_{x1} = C(e_{xx1} + ve_{ss1})$$
$$N_{s1} = C(e_{ss1} + ve_{xx1})$$
$$N_{xs1} = C\frac{1-v}{2}e_{xs1}$$
$$e_{xx1} = u_{1,x} + w_0' w_{1,x} + (w^*)' w_{1,x}$$
$$e_{ss1} = v_{1,s} + \frac{1}{a}w_1$$
$$e_{xs1} = v_{1,x} + u_{1,s} + w_0' w_{1,s} + (w^*)' w_{1,s}$$

(7.4)

Equations (7.3) and (7.4) can be reduced to three linear homogeneous equations in u_1, v_1, w_1, with the variable coefficients N_{x0}, N_{s0}, w_0, and w^*. The three equations can be further reduced to two by introduction of a stress function f for which

$$N_{x1} = f_{,ss} \qquad N_{s1} = f_{,xx} \qquad N_{xs1} = -f_{,xs} \tag{7.5}$$

For geometric compatibility, from Eqs. (7.4),

$$e_{xx1,ss} + e_{ss1,xx} - e_{xs1,xs} = \frac{1}{a}w_{1,xx} - w_0'' w_{1,ss} - (w^*)'' w_{1,ss}$$

Introduction of the definition $C = Eh/(1 - v^2)$ and rearrangement gives

$$e_{xx1} = \frac{1}{Eh}(N_{x1} - vN_{s1}) \qquad e_{ss1} = \frac{1}{Eh}(N_{s1} - vN_{x1})$$

$$e_{xs1} = 2\frac{1+v}{Eh}N_{xs1}$$

Therefore

$$\nabla^4 f - \left[Eh\frac{1}{a}w_{1,xx} - w_0''w_{1,ss} - (w^*)''w_{1,ss}\right] = 0 \tag{7.6}$$

Introduction of Eqs. (7.5) into (7.3c) now gives

$$D\nabla^4 w_1 + \frac{1}{a}f_{,xx} - [N_{x0}w_{1,xx} + N_{s0}w_{1,ss} + w_0''f_{,ss} + (w^*)''f_{,ss}] = 0 \tag{7.7}$$

Equations (7.6) and (7.7) form two equations in the unknown functions f and w_1.

The variable coefficients N_{x0}, N_{s0}, and w_0 are governed by the original nonlinear equilibrium equations. The form of the equations to be used here is obtained by specialization of the nonlinear equations for axial symmetry and for $N_{xs} \equiv 0$. The resulting expressions are

$$N_x' = 0 \qquad (7.8a)$$

$$Dw^{iv} + \frac{1}{a} N_s - N_x(w'' + w^{*''}) = 0 \qquad (7.8b)$$

Equation (7.8b) is nonlinear in the dependent variables. From Eq. (7.8a), however, N_x is seen to be independent of x. From a consideration of the conditions at the cylinder ends, furthermore, we see that $N_x = -P/2\pi a$ [compare Eqs. (1.11)]. Consequently, Eqs. (7.8) are replaced by the *linear* equation

$$Dw^{iv} + \frac{1}{a} N_s + \frac{P}{2\pi a}(w'' + w^{*''}) = 0 \qquad (7.9)$$

Now let the applied stress σ be defined by the relation

$$\sigma h \equiv \frac{P}{2\pi a} \qquad (7.10)$$

Then specialization of the constitutive and kinematic relations for axial symmetry leads to the expression

$$N_s = \frac{Ehw}{a} - v\sigma h \qquad (7.11)$$

The prebuckling equilibrium equation now may be written in the form

$$Dw^{iv} + \sigma h w'' + \frac{Eh}{a^2} w = \sigma h \left(\frac{v}{a} - w^{*''}\right) \qquad (7.12)$$

The analysis has been reduced to the solution of two homogeneous linear partial differential equations with variable coefficients [Eqs. (7.6) and (7.7)] and an inhomogeneous linear ordinary differential equation with constant coefficients [Eq. (7.12)]. It should be emphasized that, although Eq. (7.12) is a linear differential equation, its solution determines a nonlinear equilibrium path. In this respect it is similar to Eq. (1.26) for a slightly crooked column. Nontrivial solutions to Eqs. (7.6) and (7.7) represent bifurcation points along this nonlinear path.

The preceding equations are applicable for arbitrary axisymmetric imperfections $w^*(x)$. Koiter restricts his analysis to imperfections of the form

$$w^* = -\mu h \cos \frac{2mx}{a} \qquad (7.13)$$

where m is a wavelength parameter, μh denotes the amplitude of the deviation of the middle surface from a cylinder of radius a, and $-L/2 \le x \le L/2$. The minus sign in Eq. (7.13) signifies that the imperfect shell bulges inward at $x = 0$. Introduction into Eq. (7.12) gives

$$Dw^{iv} + \sigma h w'' + \frac{Eh}{a^2} w = \sigma \frac{h}{a} \left(v + 4\mu m^2 \frac{h}{a} \cos \frac{2mx}{a} \right) \qquad (7.14)$$

A solution of the form

$$w = va \frac{\sigma}{E} + B \cos \frac{2mx}{a} \qquad (7.15)$$

where B is a constant, is suitable for sufficiently long cylinders (see the discussion of boundary condition effects for axially compressed cylindrical shells in Sec. 5.5b). Introduction into Eq. (7.14) and rearrangement gives

$$B = -\mu h \frac{4\lambda \rho^2}{4\rho^4 + 1 - 4\lambda m^2} \qquad (7.16)$$

where λ and ρ, respectively, are nondimensional load and wavelength parameters defined by the expressions

$$\lambda \equiv \frac{\sigma}{\sigma_{CL}} = \frac{ca}{Eh} \sigma \qquad \rho^2 \equiv \frac{1}{c} \frac{h}{a} m^2 \qquad c \equiv [3(1 - v^2)]^{1/2} \qquad (7.17)$$

The quantity σ_{CL} is the classical value of the critical stress for the corresponding perfect cylinder, as given in Eq. (5.52).

With this solution for the axisymmetric form, the four coefficients in Eqs. (7.6) and (7.7) are

$$N_{x0} = -\sigma h$$

$$N_{s0} = \frac{EhB}{a} \cos \frac{2mx}{a}$$

$$w_0 = va \frac{\sigma}{E} + B \cos \frac{2mx}{a}$$

$$w^* = -\mu h \cos \frac{2mx}{a}$$

Following Koiter, we let

$$t \equiv \frac{4\rho^4 + 1}{4\rho^4 + 1 - 4\lambda\rho^2}$$

$$t - 1 = \frac{4\lambda\rho^2}{4\rho^4 + 1 - 4\lambda\rho^2} \qquad (7.18)$$

$$\phi \equiv \frac{c}{Eh^3} f$$

Then Eqs. (7.6) and (7.7) may be written

$$\nabla^4 \phi + \frac{C}{ah^2} \left[-w_{1,xx} + 4c\mu t\rho^2 \cos\left(\frac{2mx}{a}\right) w_{1,ss} \right] = 0 \qquad (7.19a)$$

$$\nabla^4 w_1 + \frac{4c}{a} \left[\phi_{,xx} - 4c\mu t\rho^2 \cos\left(\frac{2mx}{a}\right) \phi_{,ss} + \frac{\lambda}{h} w_{1,xx} \right.$$

$$\left. + \frac{c\mu}{h}(t - 1) \cos\left(\frac{2mx}{a}\right) w_{1,ss} \right] = 0 \qquad (7.19b)$$

These equations are the same as eqs. (4.4) and (4.5) in Ref. 7.8. Nontrivial solutions of the homogeneous equations exist only for discrete values of the load parameter λ. The smallest such value represents the critical load.

Equations (7.19) are variable-coefficient equations. A solution can be obtained by use of numerical methods such as those discussed in Chap. 8. We may note that only even-ordered derivatives appear in the equations and that the coefficients are functions of the axial coordinate x alone. Therefore, before application of a numerical method, the equations may be reduced to ordinary differential equations by introduction of solutions of the form

$$w_1 = W(x) \cos\frac{ns}{a} \qquad \phi = F(x) \cos\frac{ns}{a}$$

where n is an integer. Koiter follows this procedure and employs the Galerkin method (Sec. 8.3) in solving the resulting equations. In such an analysis the compatibility condition [Eq. (7.19a)] must be satisfied rigorously to assure that the displacement function associated with the approximate solution of the equilibrium equation is kinematically admissible. The resulting minimum eigenvalue then is an *upper bound* for the actual critical-load parameter.

The imperfection wavelength parameter ρ in Koiter's analysis is taken to have the value

$$\rho = \frac{1}{\sqrt{2}}$$

For this value the axisymmetric deviation $w^*(x)$ from a cylindrical form coincides with the axisymmetric buckling mode of an initially perfect cylindrical shell (see Prob. 5.10). For the perfect cylinder, the displacement amplitude of the axisymmetric buckling mode approaches infinity as the applied load approaches the classical load. Consequently, the compressive hoop stress in regions of maximum inward displacement also approaches infinity, and it seems reasonable to assume that such an imperfection mode should have an especially severe effect.

The displacement function $W(x)$ employed by Koiter is of the form

$$W(x) = \sum a_m \sin \frac{mx}{a}$$

This form is chosen so that the buckling pattern has a node at each point of maximum outward displacement (and therefore maximum hoop tension) in the prebuckling displacement pattern. The circumferential wavelength parameter n, on the other hand, is treated as a continuous variable, as in Sec. 5.5, and its value is determined by minimization to yield the smallest eigenvalue for each combination of parameters.

Results of the Koiter analysis are shown in Fig. 7.3, where P_{cr}/P_{CL} is the ratio of the critical value of the applied load for the imperfect cylinder to the classical critical load for the corresponding perfect cylinder. As noted, μ is the ratio of imperfection amplitude to shell-wall thickness. For $\mu = 0$, the results coincide with results of the classical stability analysis of initially perfect cylindrical shells in Sec. 5.5b. For initially imperfect cylindrical shells, the critical loads are seen to be markedly reduced. For an imperfection amplitude that is only half as large as the shell-wall thickness, for example, the critical load is reduced to less than 30 percent of the value for the corresponding perfect shell. A notable feature of Fig. 7.3 is that the curve has a vertical tangent at $\mu = 0$, indicating that a reduction in critical load occurs even for extremely small imperfections.

The results in Fig. 7.3 represent upper-bound values for a particular imperfection form and axial wavelength. An extension of the analysis that took into consideration other axisymmetric imperfection forms and a more general buckle pattern (Ref. 7.10) did not substantially lower the upper-bound values given in Fig. 7.3.

The Koiter analysis in Ref. 7.8 provides a rigorous demonstration of the fact that small initial imperfections can substantially reduce the buckling load.

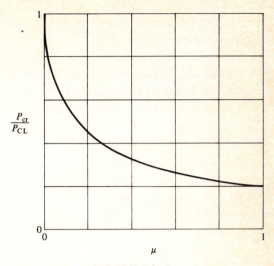

FIGURE 7.3
Influence of imperfection magnitude.

7.3 INITIAL-POSTBUCKLING ANALYSIS

7.3*a* The Koiter Theory

The analysis of imperfect cylindrical shells in the preceding section is limited in applicability to axisymmetric imperfections. Loss of stability occurs at a bifurcation point because the prebuckling deformation is axisymmetric and all asymmetric instability modes are orthogonal to the displacement pattern on the primary path. The differential equations for asymmetric imperfections are easily formulated, but they are inherently nonlinear. The initial-postbuckling theory presented by Koiter in Ref. 7.6 is applicable to asymmetric imperfections, and it avoids the need for solution of nonlinear equations. Furthermore, as noted, it is exact in the asymptotic sense, and it is applicable to shells of arbitrary shape.

For asymmetric imperfections, buckling occurs at a limit point, as illustrated in Fig. 7.2. As previously noted, the influence of imperfections depends on the shape of the secondary equilibrium path for the corresponding perfect structure. If the path drops downward, equilibrium is unstable on the secondary path, buckling is abrupt, and the buckling load of the imperfect structure is reduced. If the path curves upward, the structure may, instead, have considerable postbuckling strength. The solution of the linearized equations of classical stability theory gives no information about the shape of the secondary path. The Koiter initial-postbuckling theory is a higher-order linearization that permits determination of the slope and the curvature of the secondary path at the bifurcation point.

The Koiter formulation is based on an analysis of the potential energy of the loaded structure in buckled equilibrium configurations. The potential energy expressions in earlier chapters are of the general form

$$V = \int_A F(u,v,w,\lambda)\, dA \qquad (7.20)$$

where u, v, w are middle-surface displacement components and λ is an applied load parameter. Application of the stationary potential energy criterion $\delta V = 0$ to such an expression yields the nonlinear differential equations governing all equilibrium paths, both primary and secondary. Corresponding equations for the initial portions of the secondary paths only may be obtained from Eq. (7.20) as follows. Let

$$u \to u_0 + \bar{u}$$
$$v \to v_0 + \bar{v}$$
$$w \to w_0 + \bar{w}$$

where (u_0,v_0,w_0) is an equilibrium configuration on the primary path corresponding to a load parameter λ in the neighborhood of the bifurcation-point load; (u,v,w) is a nearby equilibrium configuration corresponding to the same value of λ; and $(\bar{u},\bar{v},\bar{w})$ is a small finite incremental displacement. The change in potential energy corresponding to the incremental displacement may be written (see Sec. A.2)

$$\Delta V = \frac{1}{2!}\delta^2 V + \frac{1}{3!}\delta^3 V + \frac{1}{4!}\delta^4 V \qquad (7.21)$$

where terms on the right are quadratic, cubic, and quartic, respectively, in the small incremental displacement components. The first-order term is missing because (u_0,v_0,w_0) is an equilibrium configuration. Equations governing the secondary equilibrium paths may be obtained by application of the stationary potential energy criterion to this expression for ΔV. Solution of the resulting differential equations defines the initial portions of the secondary paths. These differential equations also are nonlinear, however. The Koiter theory provides an approximate expression for Eq. (7.21) that makes use of the fact that, in the neighborhood of the bifurcation point, the incremental displacement is approximately of the form of the classical buckle mode. Application of the stationary potential energy criterion to the approximate expression for ΔV then leads to the desired equations governing initial-postbuckling behavior. Development of the theory and justification of the approximations are given in Ref. 7.6. An example is analyzed in detail in Sec. 7.3b, below.

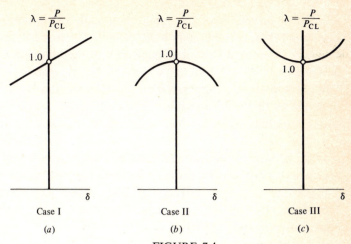

Case I
(a)

Case II
(b)

Case III
(c)

FIGURE 7.4
Equilibrium paths for perfect structures.

When there is only a single buckling mode associated with each bifurcation point, the equation for the secondary equilibrium path obtained from such an analysis may be expressed in the simple form

$$\lambda \equiv \frac{P}{P_{\mathrm{CL}}} = 1 + a_1\,\delta + a_2\,\delta^2 + \cdots \qquad (7.22)$$

where P = applied load

P_{CL} = classical bifurcation-point load

$a_1, a_2, \ldots$ = constants

δ = a measure of lateral displacement amplitude

For sufficiently small values of δ, each term in the expansion is much larger than the following terms. Three cases have been distinguished. In Case I, $a_1 \neq 0$. Then, for sufficiently small values of δ (i.e., for $a_1\,\delta \gg a_2\,\delta^2$), the secondary equilibrium path is approximately a sloping straight line, as illustrated in Fig. 7.4a. In Case II, $a_1 = 0$ and $a_2 < 0$. Then, for small values of δ, the secondary path has the parabolic shape shown in Fig. 7.4b. In Case III, $a_1 = 0$ and $a_2 > 0$. Then the path has the form in Fig. 7.4c. Cases I and II characterize structures that are sensitive to initial imperfections, and Case III those that are not. The detailed analysis in Sec. 7.3b is of a structure of the Case I type.

Since the analysis is exact at the bifurcation point but only approximate for nearby configurations, it is important to emphasize that the question of whether a structure is imperfection-sensitive is answered completely by the stability or instability of the equilibrium state at the bifurcation point itself. The equilibrium is stable only if the potential energy increment ΔV is positive definite. The second variation equals zero at the bifurcation point in all cases. In Case I, the third variation is nonzero and the equilibrium is unstable. In Cases II and III, the third variation vanishes identically and the sign of ΔV is governed by the fourth variation. For Case II, the fourth variation is negative and the equilibrium is unstable. For Case III, it is positive and the equilibrium is stable.

In addition to equations for the secondary equilibrium paths in Fig. 7.4, the theory in Ref. 7.6 also provides approximate expressions for the equilibrium paths of slightly imperfect structures. The equations are obtained by augmentation of Eq. (7.21) by an approximate expression for the potential energy increment caused by the presence of small imperfections. The correction term that is added to the expression for ΔV to account for the effect of small imperfections is of the form

$$\mu \lambda Q_1$$

where μ is a parameter that represents the magnitude of the imperfections, as in Sec. 7.2, and Q_1 is a linear functional of the incremental-displacement components. Then Eq. (7.21) is replaced by the expression

$$\Delta V = \frac{1}{2!} \delta^2 V + \frac{1}{3!} \delta^3 V + \frac{1}{4!} \delta^4 V + \mu \lambda Q_1 \qquad (7.23)$$

Application of the stationary potential energy criterion to this expression again leads to approximate equations for the equilibrium paths. The results are illustrated in Fig. 7.5. A detailed derivation of the equation for the equilibrium path of a slightly imperfect structure of the Case I type is included in Sec. 7.3b.

For Cases I and II there are seen to be limit-point loads λ_L on the paths for the imperfect structures in Fig. 7.5. The magnitude of the limit-point load may be obtained by application of the criterion $d\lambda/d\delta = 0$ to the equation for the equilibrium path. Koiter shows that for both Case I and Case II, the equation for the limit-point load given by this procedure is of the general form [Ref. 7.11, eq. (7.7)]

$$(1 - \lambda_L)^m = \alpha \lambda_L |\mu|$$

where α is a positive constant. For Case I, the exponent $m = 2$, and for Case II, $m = 3/2$. At the bifurcation point, $P = P_{CL}$ and $\lambda = 1$. For limit-point loads sufficiently close to the bifurcation-point load in magnitude, $\lambda_L \approx 1$ and λ_L may

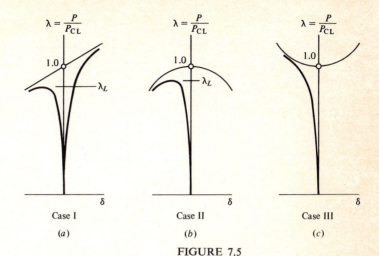

FIGURE 7.5
Paths for slightly imperfect structures.

be replaced by unity on the right side of the equation. Then the equations for the respective cases become, approximately,

Case I: $(1 - \lambda_L)^2 = \alpha |\mu|$

Case II: $(1 - \lambda_L)^{3/2} = \alpha |\mu|$ (7.24)

In both cases the equations are seen to represent parabolas with vertical tangents at $\mu = 0$. Thus the equations have the general form of the P_{cr}/P_{CL} versus μ plot for the imperfect cylindrical shell in Fig. 7.3. The vertical tangent at $\mu = 0$ accounts for the sensitivity of such structures to small imperfections. It should be noted that the expressions for the reduced critical load in Eqs. (7.24) are derived under the assumption that the slope (in Case I) and the curvature (in Case II) remain unchanged as we move away from the bifurcation point. Thus they are accurate only when the limit point occurs close to the bifurcation point, i.e., for very small imperfections.

In most applications there is only a single buckling mode corresponding to each bifurcation-point load. In all such cases the secondary equilibrium paths have the form of one of the three examples in Fig. 7.4. In problems with *multiple* buckling modes at a single bifurcation point, the Koiter analysis is much more complicated. Application of the Koiter theory in such cases is discussed in Ref. 7.6.

The Koiter formulation is based on an analysis of the potential energy expression. In recent years initial-postbuckling analyses based on application

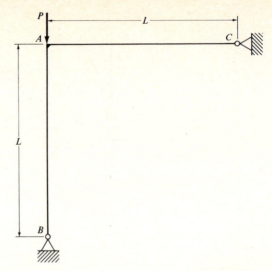

FIGURE 7.6
Two-bar frame.

of a perturbation technique to the nonlinear differential equations of equilibrium have appeared in the literature. An example is given in the report by Roorda and Chilver in Ref. 7.12.

7.3b Two-Bar Frame

An especially simple application of the initial-postbuckling theory was reported by Koiter in Ref. 7.13. Because of its relative simplicity, the analysis is suitable for use here as an illustration of the Koiter initial-postbuckling theory. The model examined in Ref. 7.13 is the two-bar frame shown in Fig. 7.6. The bars are of equal lengths L and bending stiffnesses EI. They are rigidly joined to one another and are simply supported at their far ends. Koiter undertook an initial-postbuckling analysis of this frame because results of a carefully conducted experimental study were available in a report by Roorda in Ref. 7.14. Roorda tested a small frame made of steel straps 1 in. wide, 1/16 in. thick, and 23 in. long. A classical stability analysis of this frame in Sec. 2.2 of this book gave, for the critical load, the expression

$$P_{CL} = 1.406 \frac{\pi^2 EI}{L^2}$$

For $E = 29 \times 10^6$ psi, the theoretical critical load for the test model is $P_{CL} = 15.4$ lb.

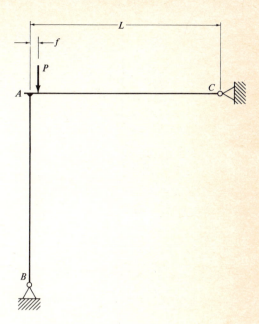

FIGURE 7.7
Frame subjected to an eccentric load.

Of course, a real model is never totally free of imperfections. For the most carefully made frame, the vertical leg will immediately bend to the right or to the left as the load is applied. For the Roorda model, the tendencies to bend to the right or to the left were found by trial to be in approximate balance for a load applied at a distance $f_0 = 0.0013L$ (i.e., 0.03 in.) to the right of the centerline of the vertical leg. Provision was made for applying the load at any distance f to the right or left of the centerline of the vertical leg, as illustrated in Fig. 7.7. In the test program the frame subjected to a slightly eccentric load applied at $f \neq f_0$ represented an equivalent slightly imperfect frame.

In the tests the rotation β_A of joint A was measured optically and was used as the displacement parameter in plots of the equilibrium paths. Results of two of the tests for the smallest eccentricity $|f - f_0|$ that could be achieved are shown in Fig. 7.8 (Ref. 7.14, fig. 5), where $\lambda \equiv P/P_{CL}$. Negative values of β_A represent counterclockwise rotations of joint A. For small values of β_A the equilibrium paths are seen to be of the kind represented by Case I in Fig. 7.4a and 7.5a.

For counterclockwise rotations the equilibrium path exhibits a limit point at a load P approximately equal to $0.99P_{CL}$. Experimental limit-point loads λ_L for values of $f < f_0$ were obtained by applying the load at other locations.

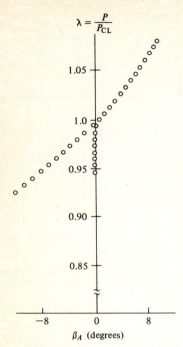

FIGURE 7.8
Experimental equilibrium paths.

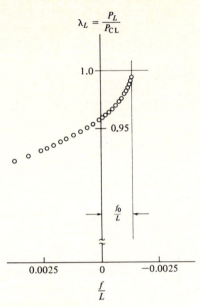

FIGURE 7.9
Experimental limit-point load vs. imperfection amplitude.

A plot of λ_L versus load eccentricity ratio f/L is shown in Fig. 7.9. In this figure the ratio $(f - f_0)/L$ represents an equivalent imperfection parameter. The curve is seen to have the characteristic parabolic form with a vertical tangent at $(f - f_0)/L = 0$.

In the following theoretical analysis, approximate equations are determined for (1) the secondary equilibrium path of an initially perfect frame, (2) the corresponding equilibrium path of a slightly imperfect frame, and (3) the relationship between limit-point load and imperfection magnitude. Theoretical and experimental results are compared at the end of Sec. 7.3b.

Analysis of an initially perfect frame The total potential energy V of the loaded frame is the sum of the strain energy of the frame and the potential energy of the applied load P. A sketch of the frame in a deformed configuration (with exaggerated displacements) is shown in Fig. 7.10. In this analysis we let x and y denote axial coordinates for members BA and CA, respectively, as indicated.

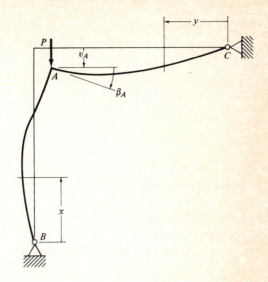

FIGURE 7.10
Frame in deformed configuration.

We also let w, η denote lateral and axial displacement components, respectively, of points on the centroidal surface of member BA, and v, ξ denote corresponding quantities for member CA. Then, from Eq. (A.40), the total potential energy may be written

$$V = \int_0^L \left\{ \frac{EA}{2} \left[\frac{d\xi}{dx} + \frac{1}{2} \left(\frac{dw}{dx} \right)^2 \right]^2 + \frac{EI}{2} \left(\frac{d^2 w}{dx^2} \right)^2 \right\} dx$$

$$+ \int_0^L \left\{ \frac{EA}{2} \left[\frac{d\eta}{dy} + \frac{1}{2} \left(\frac{dv}{dy} \right)^2 \right]^2 + \frac{EI}{2} \left(\frac{d^2 v}{dy^2} \right)^2 \right\} dy - P v_A \qquad (7.25)$$

where v_A is the vertical component of the displacement of joint A, as indicated in the sketch. [Additional terms included by Koiter in Ref. 7.13, eq. (5.2), have a negligibly small effect on the final results.] The associated forced boundary conditions are

$$\xi = w = 0 \text{ (at } x = 0) \qquad \frac{dw}{dx} = -\beta_A \text{ (at } x = L)$$

$$\eta = v = 0 \text{ (at } y = 0) \qquad \frac{dv}{dy} = -\beta_A \text{ (at } y = L) \qquad (7.26)$$

$$\xi \text{ (at } x = L) = -v \text{ (at } y = L) \qquad \eta \text{ (at } y = L) = w \text{ (at } x = L)$$

An expression for the change in potential energy, ΔV, corresponding to a small incremental displacement is obtained by letting

$$\xi \to \xi_0 + \bar{\xi}$$
$$\eta \to \eta_0 + \bar{\eta}$$
$$v \to v_0 + \bar{v}$$
$$w \to w_0 + \bar{w}$$

where $(\xi_0, \eta_0, v_0, w_0)$ represents an equilibrium configuration on the primary equilibrium path in the neighborhood of the bifurcation point; (ξ, η, v, w) is a nearby equilibrium configuration corresponding to the same value of applied load; and $(\bar{\xi}, \bar{\eta}, \bar{v}, \bar{w})$ is a small finite incremental displacement. We assume that at the bifurcation point the vertical member supports a load P and the horizontal member is unstressed and undeformed. Then $\eta_0 = v_0 = w_0 = 0$, and

$$\xi_0 = -\frac{Px}{EA}$$

Then, from Eq. (7.25), the change in potential energy becomes

$$\Delta V = \frac{1}{2!}\delta^2 V + \frac{1}{3!}\delta^3 V + \frac{1}{4!}\delta^4 V \qquad (7.27)$$

where

$$\frac{1}{2!}\delta^2 V = \frac{1}{2}\int_0^L \left[EA\left(\frac{d\bar{\xi}}{dx}\right)^2 - P\left(\frac{d\bar{w}}{dx}\right)^2 + EI\left(\frac{d^2\bar{w}}{dx^2}\right)^2 \right] dx$$

$$+ \frac{1}{2}\int_0^L \left[EA\left(\frac{d\bar{\eta}}{dy}\right)^2 + EI\left(\frac{d^2\bar{v}}{dy^2}\right)^2 \right] dy \qquad (7.28)$$

$$\frac{1}{3!}\delta^3 V = \frac{EA}{2}\int_0^L \frac{d\bar{\xi}}{dx}\left(\frac{d\bar{w}}{dx}\right)^2 dx + \frac{EA}{2}\int_0^L \frac{d\bar{\eta}}{dy}\left(\frac{d\bar{v}}{dy}\right)^2 dy \qquad (7.29)$$

$$\frac{1}{4!}\delta^4 V = \frac{EA}{8}\int_0^L \left(\frac{d\bar{w}}{dx}\right)^4 dx + \frac{EA}{8}\int_0^L \left(\frac{d\bar{v}}{dy}\right)^4 dy \qquad (7.30)$$

An equation for the secondary equilibrium path may be obtained from this expression for ΔV by application of the stationary potential energy criterion. We see, however, that the resulting differential equations would be nonlinear in the finite incremental displacement components $\bar{\xi}, \bar{\eta}, \bar{v}, \bar{w}$. For points on the secondary path sufficiently close to the bifurcation point, on the other hand, the incremental-displacement components are of the form of the classical buckling mode. That form is readily determined from a classical stability analysis. Accordingly, we limit the range of validity of the analysis by assuming that the small finite displacement components $\bar{\xi}, \bar{\eta}, \bar{v}, \bar{w}$ are of the form of the classical buckling-mode components. For this purpose it is convenient to normalize the classical buckling mode by taking $\beta_A = 1$ for the normalized mode. Then the rotation β_A may be considered to be the amplitude parameter for the displacement components. Thus we let

$$
\begin{aligned}
\bar{\xi} &= \beta_A \xi_1 \\
\bar{\eta} &= \beta_A \eta_1 \\
\bar{v} &= \beta_A v_1 \\
\bar{w} &= \beta_A w_1
\end{aligned}
\tag{7.31}
$$

where ξ_1, η_1, v_1, w_1 are the components of the normalized classical buckling mode.

Stability equations for determination of the classical buckling mode may be obtained from the second-variation expression. Introduction of Eqs. (7.31) into Eq. (7.28) gives, for the second variation, the expression

$$
\begin{aligned}
\tfrac{1}{2}\delta^2 V = \frac{1}{2}\int_0^L &\left[EA\left(\frac{d\xi_1}{dx}\right)^2 - P\left(\frac{dw_1}{dx}\right)^2 + EI\left(\frac{d^2w_1}{dx^2}\right)^2 \right] dx\,\beta_A{}^2 \\
&+ \frac{1}{2}\int_0^L \left[EA\left(\frac{d\eta_1}{dy}\right)^2 + EI\left(\frac{d^2v_1}{dy^2}\right)^2 \right] dy\,\beta_A{}^2
\end{aligned}
$$

The stability equations may be obtained from this expression in the usual way by introduction of the integrands into the Euler equations. To derive the natural boundary conditions, however, it is convenient in this example to use the integration-by-parts procedure of the calculus of variations (Sec. A.3). In that procedure we give small variations to each of the four displacement components in the second-variation expression, and we collect all terms in the corresponding increment in $\delta^2 V$ that are linear in the variational displacements. The resulting expression is

$$
\begin{aligned}
\delta(\tfrac{1}{2}\delta^2 V) = \int_0^L &\left(EA\,\frac{d\xi_1}{dx}\frac{d\,\delta\xi_1}{dx} - P\,\frac{dw_1}{dx}\frac{d\,\delta w_1}{dx} \right.\\
&\left. + EI\,\frac{d^2 w_1}{dx^2}\frac{d^2\,\delta w_1}{dx^2} \right) dx\,\beta_A{}^2 \\
&+ \int_0^L \left(EA\,\frac{d\eta_1}{dy}\frac{d\,\delta\eta_1}{dy} + EI\,\frac{d^2 v_1}{dy^2}\frac{d^2\,\delta v_1}{dy^2} \right) dy\,\beta_A{}^2
\end{aligned}
$$

where the variational displacements $\delta\xi_1$, $\delta\eta_1$, δv_1, δw_1 must be admissible; i.e., they must be continuous and twice-differentiable and they must satisfy the forced boundary conditions in Eqs. (7.26). According to the Trefftz criterion, $\delta(\delta^2 V/2) = 0$ at the critical load (Sec. A.5). Thus integration by parts and rearrangement gives

$$\left[EA\frac{d\xi_1}{dx}\delta\xi_1\right]_0^L + \left[EA\frac{d\eta_1}{dy}\delta\eta_1\right]_0^L - \left[\left(EI\frac{d^3w_1}{dx^3} + P\frac{dw_1}{dx}\right)\delta w_1\right]_0^L$$

$$- \left[EI\frac{d^3v_1}{dy^3}\delta v_1\right]_0^L + \left[EI\frac{d^2w_1}{dx^2}\frac{d\,\delta w_1}{dx}\right]_0^L + \left[EI\frac{d^2v_1}{dy^2}\frac{d\,\delta v_1}{dy}\right]_0^L$$

$$- \int_0^L EA\frac{d^2\xi_1}{dx^2}\delta\xi_1\,dx - \int_0^L EA\frac{d^2\eta_1}{dy^2}\delta\eta_1\,dy$$

$$+ \int_0^L \left(EI\frac{d^4w_1}{dx^4} + P\frac{d^2w_1}{dx^2}\right)\delta w_1\,dx + \int_0^L EI\frac{d^4v_1}{dy^4}\delta v_1\,dy = 0$$

Since the variational displacements satisfy the forced boundary conditions but are otherwise arbitrary, this equation equals zero only if (see the Appendix)

$$\frac{d^2w_1}{dx^2} = 0 \text{ at } x = 0 \qquad \frac{d^2v_1}{dy^2} = 0 \text{ at } y = 0$$

$$EA\frac{d\xi_1}{dx} + EI\frac{d^3v_1}{dy^3} = \frac{d^2w_1}{dx^2} + \frac{d^2v_1}{dy^2} = 0 \text{ at } x = y = L \qquad (7.32)$$

$$EA\frac{d\eta_1}{dy} - EI\frac{d^3v_1}{dx^3} - P\frac{dw_1}{dx} = 0 \text{ at } x = y = L$$

and if

$$\frac{d^2\xi_1}{dx^2} = 0$$

$$EI\frac{d^4w_1}{dx^4} + P\frac{d^2w_1}{dx^2} = 0$$

$$\frac{d^2\eta_1}{dy^2} = 0 \qquad\qquad (7.33)$$

$$\frac{d^4v_1}{dy^4} = 0$$

Equations (7.32) are the natural boundary conditions for the frame, and Eqs. (7.33) are the stability equations.

Integration of the stability equations and use of the four forced and two natural boundary conditions at $x = y = 0$ gives, for the displacement components, the expressions

$$\beta_A \xi_1 = C_1 \frac{x}{EA}$$

$$\beta_A \eta_1 = C_2 \frac{y}{EA}$$

$$\beta_A w_1 = C_3 \sin kx + C_4 x$$

$$\beta_A v_1 = C_5 y + C_6 y^3$$

where $k^2 \equiv P/EI$ and $C_1, C_2, C_3, C_4, C_5, C_6$ are integration constants. Introduction of these expressions into the four forced and three natural boundary conditions at $x = y = L$ gives

$$C_3 \sin kL + C_4 L = \frac{L}{EA} C_2$$

$$C_5 + C_6 L^2 = -\frac{1}{EA} C_1$$

$$kC_3 \cos kL + C_4 = -\beta_A$$

$$C_5 + 3C_6 L^2 = -\beta_A$$

$$-k^2 C_3 \sin kL + 6C_6 L = 0$$

$$C_1 + 6EIC_6 = 0$$

$$C_2 - PC_4 = 0$$

These seven homogeneous equations in $\beta_A, C_1, \ldots, C_6$ determine the critical load and buckling mode. The quantities P/EA and $6EI/EAL^2$ are negligibly small in comparison with unity. For a nontrivial solution the determinant of the coefficients of the homogeneous equations must equal zero. In this manner we again are led to the characteristic equation [compare Eq. (2.27)]

$$\tan kL = \frac{kL}{1 + \frac{1}{3}(kL)^2} \qquad (7.34)$$

As noted in Sec. 2.2, the smallest eigenvalue is found to be

$$k_{CL} L = 3.72 \qquad (7.35)$$

where $k_{CL}^2 \equiv P_{CL}/EI$. The normalized classical buckling mode given by these equations is found to be

$$\xi_1 = \frac{3P_{CL}/EA}{(k_{CL}L)^2}\, x$$

$$\eta_1 = \frac{3P_{CL}/EA}{(k_{CL}L)^2}\, y$$

$$w_1 = \frac{3}{(k_{CL}L)^2} \left(x - L\,\frac{\sin k_{CL}x}{\sin k_{CL}L}\right)$$

$$v_1 = \tfrac{1}{2}y\left(1 - \frac{y^2}{L^2}\right)$$

(7.36)

The approximate expression for the potential energy increment ΔV now is obtained by substituting Eqs. (7.31) and (7.36) into Eqs. (7.27) to (7.30). In deriving the new expression for the second variation, we note that the applied load P may be written in the form $P = P_{CL} + (\lambda - 1)P_{CL}$, where $\lambda \equiv P/P_{CL}$. Introduction of this expression and Eqs. (7.31) into Eq. (7.28) and rearrangement gives

$$\tfrac{1}{2}\delta^2 V = \tfrac{1}{2}(1 - \lambda)P_{CL}\left[\int_0^L \left(\frac{dw_1}{dx}\right)^2 dx\right]\beta_A^2 \qquad (7.37)$$

Introduction of the expression for w_1 from Eqs. (7.36) and integration gives

$$\tfrac{1}{2}\delta^2 V = \tfrac{1}{2}(1 - \lambda)P_{CL}\frac{9L}{(k_{CL}L)^2}\,[1 + \tfrac{1}{18}(k_{CL}L)^2]\,\beta_A^2$$

Now let

$$A_2 \equiv P_{CL}L\frac{9}{2(k_{CL}L)^2}\,[1 + \tfrac{1}{18}(k_{CL}L)^2] \qquad (7.38)$$

Then the approximate second-variation expression may be written

$$\tfrac{1}{2}\delta^2 V = (1 - \lambda)A_2\beta_A^2 \qquad (7.39)$$

For $k_{CL}L = 3.72$, from Eq. (7.35),

$$A_2 = 0.574 P_{CL}L \qquad (7.40)$$

A corresponding expression for the third variation may be obtained by introducing Eqs. (7.31) and (7.36) into Eq. (7.29). Integration gives

$$\frac{1}{3!}\delta^3 V = \frac{3P_{CL}L}{(k_{CL}L)^2}\left\{\frac{9}{2(k_{CL}L)^2}\,[1 + \tfrac{1}{18}(k_{CL}L)^2] + \tfrac{1}{10}\right\}\beta_A^3$$

Now let

$$A_3 \equiv \frac{3P_{CL}L}{(k_{CL}L)^2}\left\{\frac{9}{2(k_{CL}L)^2}[1 + \tfrac{1}{18}(k_{CL}L)^2] + \tfrac{1}{10}\right\} \qquad (7.41)$$

Then the approximate third-variation expression becomes

$$\frac{1}{3!}\delta^3 V = A_3 \beta_A{}^3 \qquad (7.42)$$

For $k_{CL}L = 3.72$,

$$A_3 = 0.1456 P_{CL} L \qquad (7.43)$$

The corresponding expression for the fourth variation is quartic in β_A and may be neglected for sufficiently small values of β_A.

The final, approximate expression for the potential energy increment ΔV is the sum of the expressions for the second and third variations in Eqs. (7.39) and (7.42):

$$\Delta V = (1 - \lambda)A_2 \beta_A{}^2 + A_3 \beta_A{}^3 \qquad (7.44)$$

where A_2 and A_3 are constants given by Eqs. (7.40) and (7.43), respectively.

For equilibrium, $d(\Delta V)/d\beta_A = 0$. For $\beta_A \neq 0$ we obtain

$$\lambda = 1 + 0.380\beta_A \qquad (7.45)$$

A plot of the equilibrium path given by this equation is shown in Fig. 7.11. As emphasized in Sec. 7.3a, this approximate relationship represents the tangent to the actual secondary path at the bifurcation point.

Analysis of slightly imperfect frames For an eccentric load applied at a distance $f = \phi L$ to the right of point A, as illustrated in Fig. 7.7, the potential energy expression in Eq. (7.25) must be modified by the addition of a term

$$\Omega_\phi = - PL\phi\beta_A \qquad (7.46)$$

Equation (7.46) represents the change in potential energy due to the couple Pf during the rotation β_A. Since $\beta_A = 0$ prior to buckling, the expression in Eq. (7.44) for the potential energy increment is merely replaced by the equation

$$\Delta V = (1 - \lambda)A_2 \beta_A{}^2 + A_3 \beta_A{}^3 - PL\phi\beta_A \qquad (7.47)$$

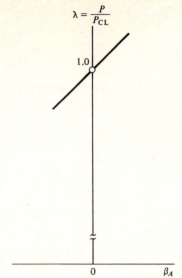

FIGURE 7.11
Theoretical secondary paths for initially perfect frame.

Although the last term in this expression is linear in β_A, it is of the order of magnitude of the other terms in the equation for sufficiently small values of the eccentricity parameter ϕ.

For convenience let

$$B_1 \equiv P_{\mathrm{CL}} L \qquad (7.48)$$

[see Eqs. (7.40) and (7.43)]. Introduction of this notation and the definition $P \equiv \lambda P_{\mathrm{CL}}$ into the last term gives, for the potential energy increment, the approximate expression

$$\Delta V = (1 - \lambda) A_2 \beta_A{}^2 + A_3 \beta_A{}^3 - \lambda B_1 \phi \beta_A \qquad (7.49)$$

Application of the equilibrium criterion $d(\Delta V)/d\beta_A = 0$ and rearrangement now gives

$$\lambda = 1 + \frac{3A_3}{2A_2} \beta_A - \frac{B_1}{2A_2} \frac{\phi}{\beta_A} \lambda \qquad (7.50)$$

For values of the applied load P sufficiently close to P_{CL}, λ may be replaced by unity on the right side of Eq. (7.50). Then introduction of the values for A_2, A_3, and B_1 from Eqs. (7.40), (7.43), and (7.48) gives

$$\lambda = 1 + 0.380\beta_A - 0.871 \frac{\phi}{\beta_A} \qquad (7.51)$$

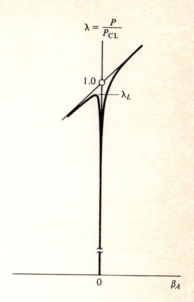

$$\lambda = \frac{P}{P_{CL}}$$

FIGURE 7.12
Typical theoretical equilibrium paths for
imperfect frame.

This is the expression for the equilibrium path of a slightly imperfect frame. For $\phi = 0$, it reduces to the corresponding expression for an initially perfect frame in Eq. (7.45). A sketch of typical equilibrium paths given by this equation is shown in Fig. 7.12.

The limit-point load factor λ_L for small negative values of ϕ is obtained from Eq. (7.50) by application of the relation $d\lambda/d\beta_A = 0$. The resulting expression is found to be

$$\lambda_L = 1 - \frac{(3A_3 B_1 \lambda_L)^{1/2}}{A_2}(-\phi)^{1/2} \qquad (7.52)$$

Replacement of λ_L by unity on the right side of the equation and introduction of the values for A_2, A_3, and B_1 gives, approximately,

$$\lambda_L = 1 - 1.15(-\phi)^{1/2} \qquad (7.53)$$

A sketch of limit-point load versus magnitude of imperfection parameter is shown in Fig. 7.13. The curve has the characteristic vertical tangent at $\phi = 0$.

Comparison of theory and experiment   Theoretical results for the initial portion of the secondary equilibrium path of a perfect structure are compared

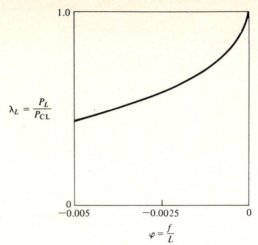

FIGURE 7.13
Theoretical results for the influence of
imperfection amplitude.

$$\lambda_L = \frac{P_L}{P_{CL}}$$

$$\varphi = \frac{f}{L}$$

in Fig. 7.14 with test data from Fig. 7.8. Agreement is seen to be extremely close for small values of rotation. Theoretical values given by Eq. (7.53) for the limit-point load factor λ_L are compared in Fig. 7.15 with corresponding test data from Fig. 7.9. Again the agreement is remarkably close.

7.3c Other Examples

Many other investigations of initial-postbuckling behavior have been reported in the literature in recent years. An excellent summary of such analyses is included in a review article by Hutchinson and Koiter (Ref. 7.15). Results of some of the investigations are briefly summarized in the following paragraphs.

Cylindrical panel   An initial-postbuckling analysis of a narrow cylindrical panel subjected to axial compression was reported by Koiter in Ref. 7.16. Results for the primary and secondary equilibrium paths of initially perfect panels are shown in Fig. 7.16 (Ref. 7.16, fig. 3), where $\lambda \equiv P/P_{CL}$, and δ is the *axial* displacement of one end of the panel relative to the other (the *shortening*) as a fraction of the corresponding value at the bifurcation point (compare Fig. 7.1). The parameter $\bar{\theta}$ in the figure is a measure of the initial curvature of the panel, with $\bar{\theta} \leq 1$ for a narrow panel. The limiting value $\bar{\theta} = 0$ represents a flat plate. The slope of the secondary equilibrium path for $\bar{\theta} = 0$ in Fig. 7.16 is in agreement

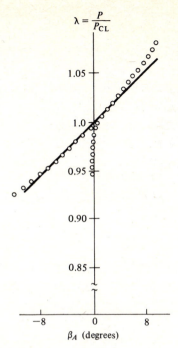

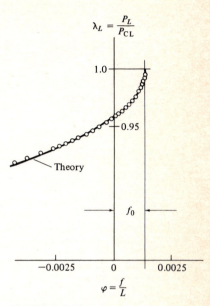

FIGURE 7.14
Comparison of theoretical and experimental results for the equilibrium paths.

FIGURE 7.15
Comparison of theoretical and experimental results for limit-point load.

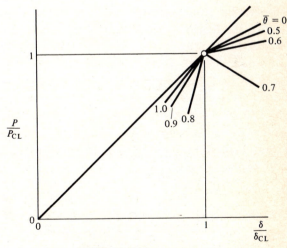

FIGURE 7.16
Equilibrium paths for initially perfect panels.

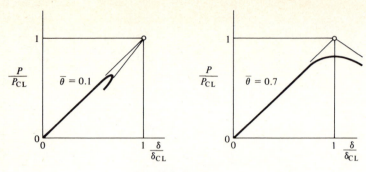

FIGURE 7.17
Equilibrium paths for slightly imperfect panels.

with the initial slope of the secondary path for the flat plate in Fig. 3.6a. As the initial curvature of the panels increases, the slopes of the secondary paths decrease, as shown in the figure. For values of $\bar{\theta}$ greater than approximately 0.64, the slopes are negative. Thus the very narrow panels for which $\bar{\theta} < 0.64$ are not imperfection-sensitive, but the somewhat wider panels for which $\bar{\theta} > 0.64$ are imperfection-sensitive. Calculated equilibrium paths for slightly imperfect panels are shown in Fig. 7.17 (Ref. 7.16, figs. 4 and 5) for two values of panel curvature. These curves are based on the simplifying assumption that the initial imperfection is of the form of the classical buckling mode. The magnitude of the imperfection parameter μ in both cases is the same, namely, $\mu = 0.1$. For $\bar{\theta} = 0.7$, the limit-point load λ_L is seen to be approximately 85 percent of the bifurcation-point load, whereas, for $\bar{\theta} = 0.1$, it is only 65 percent of that load. Limit-point loads for other values of the imperfection parameter are shown in Fig. 7.18 (Ref. 7.16, fig. 6), where λ_L is plotted versus μ for $\bar{\theta} = 0.7$ and for $\bar{\theta} = 0.1$. The curves are seen to be similar in general form to those in Figs. 7.3 and 7.13, indicating that the panels are sensitive to small imperfections. Unlike the initial-postbuckling analysis of the two-bar frame, this investigation provides a continuous spectrum of examples from models that are imperfection-sensitive to those that are not.

Cylindrical shell subjected to external pressure Many structural shells of practical interest have secondary equilibrium paths whose initial shape is characterized by the curves of Fig. 7.5b. For that case, called Case II in Sec. 7.3a, the coefficient a_1 equals zero in the expression for secondary equilibrium

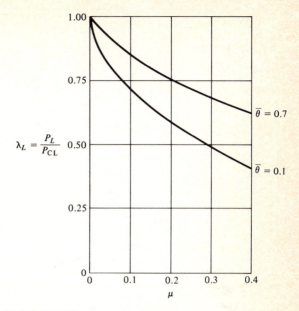

FIGURE 7.18
Influence of imperfection magnitude for cylindrical panels.

paths in Eq. (7.22). Then, for sufficiently small values of the lateral displacement parameter δ, the equation becomes, approximately,

$$\lambda = 1 + a_2 \delta^2 \qquad (7.54)$$

where $a_2 < 0$. The coefficient a_2 is seen to be a measure of the sharpness of curvature of the secondary equilibrium path and therefore to be, to some extent, a measure of the degree of imperfection sensitivity of the structure. It is sometimes called the imperfection-sensitivity parameter.

An example of the treatment of a_2 as an imperfection-sensitivity parameter is given by Budiansky and Amazigo in Ref. 7.17. Initial-postbuckling analyses of several models are reported, including cylindrical shells subjected to external hydrostatic pressure. In that analysis the displacement parameter δ of Eq. (7.54) is the ratio of amplitude of lateral displacement to shell-wall thickness, and λ is the ratio of applied pressure to classical critical pressure. Theoretical and experimental buckling loads for this method of loading are compared in the present work in Fig. 5.17. Theoretical and experimental results are seen to be in much closer agreement for long cylinders (values of the Batdorf parameter Z equal to, say, 1,000) than for short ones (Z equal to, say, 10). Cal-

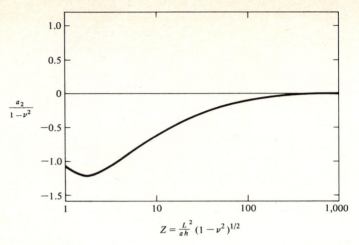

FIGURE 7.19
Imperfection sensitivity of cylinders subjected to lateral pressure.

culated values of the imperfection-sensitivity parameter a_2 are plotted versus the Batdorf parameter Z in Fig. 7.19 (Ref. 7.17, fig. 3, for cylinders with simply supported ends). These theoretical results for the degree of imperfection sensitivity are seen to be in qualitative agreement with the experimental evidence in Fig. 5.17.

Even though the values given in Ref. 7.17 for the parameter a_2 are exact, quantitative comparisons of theoretical and experimental results such as those for the two-bar frame in Figs. 7.14 and 7.15 are not possible. Quantitative information about the initial imperfections of the test cylinders is not available and there is no general agreement on how accidental imperfections in structural shells should be characterized for analytical purposes. In Ref. 7.17 initial imperfections are assumed to be of the form of the classical buckling mode. In terms of that assumption, the equation given for the relationship among limit-point load, imperfection magnitude, and the imperfection-sensitivity parameter a_2 is

$$(1 - \lambda_L)^{3/2} = \frac{(3)(3)^{1/2}}{2}(-a_2)^{1/2}\lambda_L|\mu| \qquad (7.55)$$

where $a_2 < 0$

λ_L = ratio of applied pressure at the limit point to corresponding classical critical pressure

μ = ratio of imperfection amplitude to shell-wall thickness

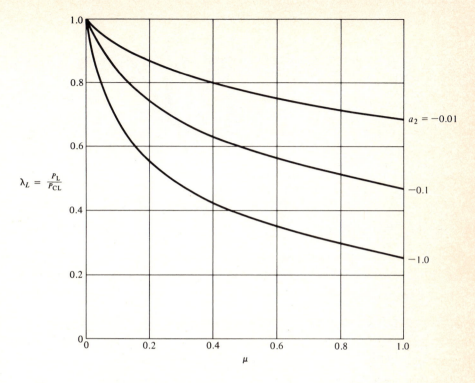

FIGURE 7.20
Influence of imperfection magnitude for cylinders subjected to lateral pressure.

A plot of λ_L versus μ for selected values of the imperfection-sensitivity parameter is shown in Fig. 7.20. The curves have the characteristic vertical tangent at $\mu = 0$ that is exhibited in the earlier graphs of limit-point load versus imperfection magnitude.

Oval cylindrical shells   Tests of oval (noncircular) cylindrical shells subjected to axial compression indicate that initial buckling loads of such shells may be much smaller than corresponding values based on classical stability theory. Tests also show, however, that some oval cylinders support loads *after* buckling that are significantly larger than the classical critical load. Flat plates and other structures of the Case III type also support postbuckling loads that are larger than the classical critical load, but for such structures initial buckling does not occur at loads lower than the classical critical value. Thus oval cylinders are Case II structures in their initial-buckling behavior, but conditions on the

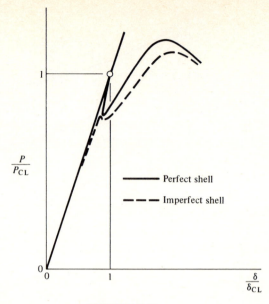

FIGURE 7.21
Equilibrium paths for oval cylinders.

secondary path evidently change rapidly as we move away from the bifurcation point. As a result, behavior of oval cylinders in the far postbuckling range resembles that of Case III structures. An initial-postbuckling analysis of oval cylinders reported by Hutchinson in Ref. 7.18 confirms the observed reduction in initial-buckling load due to imperfection sensitivity. A related large-deflection analysis in Ref. 7.19 shows that the large postbuckling loads are a consequence of the fact that the secondary equilibrium path initially drops downward and then curves upward, as shown in Fig. 7.21. In such a case, of course, knowledge of only initial-postbuckling behavior is not a suitable basis for determination of the ultimate failure load of the structure. The postbuckling behavior of oval cylinders is examined in greater detail in Chap. 9.

7.4   CONCLUSIONS

The investigations reported in this chapter provide rigorous confirmation of the conclusion that buckling loads can be reduced substantially by initial imperfections. The Koiter initial-postbuckling theory, in particular, is one of the most important contributions in recent years to our general understanding of the buckling behavior of structures. The theory, however, does not constitute a quantitative basis for the design of imperfection-sensitive structures. For the oval cylinder, for example, the theory can determine the first maximum point in the load-displacement relationship in Fig. 7.21 (if the imperfections are quite small), but it cannot determine the more important second maximum point. Application of the Koiter theory also may not be practical for cases in which the respective eigenvalues corresponding to a large number of buckling modes are clustered in a narrow range slightly above the classical critical load. The lowest eigenvalue may correspond to a stable secondary path, while slightly higher bifurcation points correspond to imperfection-sensitive buckling. This occurs, for example, in the case of an axially compressed cylindrical shell with moderate internal pressure.

Recent summaries of initial postbuckling analysis have been given by Seide (Ref. 7.20) and Budiansky (Ref. 7.21). It should be emphasized that much of the current theoretical and experimental work on the influence of initial imperfections is unrelated to initial-postbuckling theory (for example, Ref. 7.22).

Because of the failure to find inexpensive approximate methods for calculation of the critical loads of slightly imperfect structures, and the difficulty in predicting the shape and size of probable imperfections, purely analytical methods for determination of the critical loads of imperfection-sensitive structures are not available. Semiempirical methods using statistically determined knockdown factors, such as those discussed in Chap. 5, are used in the practical design of imperfection-sensitive structures to account for the effects of imperfections. Only for circular cylindrical and shallow spherical shells, however, are the test data numerous enough for adequate analysis. Conclusions are extended to other shell configurations on a more or less intuitive basis. Nevertheless, with the increasing availability of high-speed computers and more efficient methods of numerical analysis, it is becoming more common to support the design criterion by a theoretical nonlinear analysis, including the effect of particular imperfections. Such theoretical analyses are discussed in Chap. 9.

REFERENCES

7.1 FLÜGGE, W.: Die Stabilität der Kreiszylinderschale, *Ing. Arch.*, vol. 3, pp. 463–506, 1932.

7.2 LUNDQUIST, E. E.: Strength Tests of Thin-walled Duralumin Cylinders in Compression, *NACA Rep.* 473, 1933.

7.3 DONNELL, L. H.: A New Theory for the Buckling of Thin Cylinders under Axial Compression and Bending, *ASME Trans.*, vol. 56, pp. 795–806, 1934.

7.4   VON KÁRMÁN, TH., and H.-S. TSIEN: The Buckling of Thin Cylindrical Shells under Axial Compression, *J. Aeronautical Sci.*, vol. 8, pp. 303–312, August 1941.

7.5   DONNELL, L. H., and C. C. WAN: Effects of Imperfections on Buckling of Thin Cylinders and Columns under Axial Compression, *J. Appl. Mech.*, vol. 17, pp. 73–83, 1950.

7.6   KOITER, W. T.: On the Stability of Elastic Equilibrium (in Dutch with English summary), thesis, Delft, H. J. Paris, Amsterdam, 1945. English translation, *Air Force Flight Dyn. Lab. Tech. Rep.* AFFDL-TR-70-25, February 1970.

7.7   KOITER, W. T.: Elastic Stability and Postbuckling Behavior, *Proc. Symp. Non-linear Probl.*, University of Wisconsin Press, Madison, 1963, pp. 257–275.

7.8   KOITER, W. T.: The Effect of Axisymmetric Imperfections on the Buckling of Cylindrical Shells under Axial Compression, *Proc. K. Ned. Akad. Wet.*, Amsterdam, ser. B, vol. 6, 1963; also, *Lockheed Missiles and Space Co. Rep.* 6-90-63-86, Palo Alto, Calif., August 1963.

7.9   FLÜGGE, W.: "Stresses in Shells," 2d ed., Springer, Berlin, 1973.

7.10   BUDIANSKY, B., and J. W. HUTCHINSON: Buckling of Circular Cylindrical Shells under Axial Compression, *Contrib. Theory Aircraft Structures*, Rotterdam University Press, 1970, pp. 239–259.

7.11   KOITER, W. T.: General Equations of Elastic Stability for Thin Shells, *Proc. Symp. Theory of Shells to Honor Lloyd Hamilton Donnell*, University of Houston, Houston, Tex., 1967, pp. 187–223.

7.12   ROORDA, J., and A. H. CHILVER: Frame Buckling: An Illustration of the Perturbation Technique, *Intl. J. Non-Linear Mech.*, vol. 5, pp. 235–246, 1970.

7.13   KOITER, W. T.: Postbuckling Analysis of a Simple Two-Bar Frame, *Recent Prog. Appl. Mech.*, Almquist and Wiksell, Stockholm, 1966.

7.14   ROORDA, J.: Stability of Structures with Small Imperfections, *J. Eng. Mech. Div.*, ASCE, vol. 91, no. EM1, pp. 87–106, 1965.

7.15   HUTCHINSON, J. W., and W. T. KOITER: Postbuckling Theory, *Appl. Mech. Rev.*, vol. 23, pp. 1353–1366, 1970.

7.16   KOITER, W. T.: Buckling and Post-buckling Behavior of a Cylindrical Panel under Axial Compression, *Natl. Aeronaut. Res. Inst.*, Amsterdam, Rep. S. 476, May 1956.

7.17   BUDIANSKY, B., and J. C. AMAZIGO: Initial Postbuckling Behavior of Cylindrical Shells under External Pressure, *J. Math. Phys.*, vol. 47, pp. 223–235, 1968.

7.18   HUTCHINSON, J. W.: Buckling and Initial Postbuckling Behavior of Oval Cylindrical Shells under Axial Compression, *J. Appl. Mech.*, vol. 35, pp. 66–72, 1968.

7.19   ALMROTH, B. O., F. A. BROGAN, and M. B. MARLOWE: Collapse Analysis for Elliptic Cones, *AIAA J.*, vol. 9, pp. 32–37, January 1971.

7.20   SEIDE, P.: A Reexamination of Koiter's Theory of Initial Postbuckling Behaviour and Imperfection Sensitivity of Structures, in *Thin-Shell Structures*, edited by Y. C. Fung and E. E. Sechler, Prentice-Hall, Englewood Cliffs, N. J., 1974.

7.21 BUDIANSKY, B.: Theory of Buckling and Post-Buckling Behavior of Elastic Structures, to appear in *Advances in Applied Mechanics*, vol. 14, edited by C. S. Yih, Academic Press, New York, 1974.

7.22 SINGER, J., J. ARBOCZ, and C. D. BABCOCK: Buckling of Imperfect Stiffened Cylindrical Shells under Axial Compression, *AIAA J.*, vol. 9, pp. 68–75, January 1971.

PROBLEMS

7.1 By use of Eqs. (7.2) to (7.5), derive the compatibility condition in Eq. (7.6) for an imperfect cylindrical shell.

7.2 By use of Eq. (7.22), plot secondary equilibrium paths for three examples of initially perfect structures: (*a*) $a_1 = 0.2$, $a_2 = 0.4$, (*b*) $a_1 = 0$, $a_2 = 0.4$, and (*c*) $a_1 = 0$, $a_2 = -0.4$. Let $-0.1 \le \delta \le 0.1$.

7.3 By use of Eqs. (7.24), plot limit-point load λ_L versus imperfection magnitude μ for the imperfection-sensitive structures represented by Cases I and II. Let $\alpha = 1.32$, and $-0.01 \le \mu \le 0$.

7.4 An expression for the total potential energy of a rectangular flat plate subjected to in-plane compression and lateral pressure is given in Eqs. (3.24) and (3.25). An expression for the second variation of the total potential energy is given in Eqs. (3.41) and (3.42). Derive corresponding expressions for the third and fourth variations of the potential energy for the plate.

7.5 Show that the addition of small incremental displacement components $\bar{\xi}$, $\bar{\eta}$, $\bar{v}$, $\bar{w}$ in the potential energy expression for the two-bar frame in Eq. (7.25) leads to the expressions in Eqs. (7.28) to (7.30) for the second, third, and fourth variations of the potential energy.

7.6 For the two-bar frame, derive the stability equations and natural boundary conditions in Eqs. (7.33) and (7.32) from the second-variation expression in Eq. (7.28).

7.7 For the two-bar frame, show that introduction of the expressions for the buckling mode components in Eqs. (7.36) into the third-variation expression in Eq. (7.29) leads to the approximate expression for the third variation in Eqs. (7.41) and (7.42). Show that, for $k_{CL} L = 3.72$, the expression simplifies to the equation

$$\frac{1}{3!} \delta^3 V = 0.1456 P_{CL} L \beta_A{}^3$$

7.8 For a slightly imperfect two-bar frame, an approximate expression for the potential energy increment ΔV is given in Eq. (7.49), where A_2, A_3, and B_1 are given by Eqs. (7.40), (7.43), and (7.48), respectively. Specialize the equations for $k_{CL} L = 3.72$. Then, by application of the stationary potential energy criterion, $d(\Delta V)/d\beta_A = 0$, derive the equation for the secondary equilibrium path, $\lambda(\beta_A, \phi)$. Plot λ versus β_A for the imperfection parameter $\phi = -0.001$ and for $\phi = -0.0001$. Let $-0.01 \leq \beta_A \leq 0$ for both plots.

7.9 By use of Eq. (7.54) and Fig. 7.19, plot the secondary equilibrium path λ versus δ for a simply supported cylindrical shell that is subjected to external hydrostatic pressure and has a Batdorf parameter Z value of 10. Let $\nu = 0.3$, and let $-0.5 \leq \delta \leq 0.5$. Repeat for $Z = 1,000$.

8

NUMERICAL SOLUTION METHODS

8.1 INTRODUCTION

Most plate or shell stability problems, as they are posed in practical engineering, cannot be solved analytically. Consequently, numerical methods must be applied. The distinction between the terms "analytical" and "numerical" sometimes is not entirely clear. As used here, analytical methods refer to methods by which an exact closed-form solution of the governing differential equations is obtained. Hence solutions such as those based on trigonometric series with undetermined coefficients are considered here to be numerical solutions. Numerical methods have in common the characteristic that they approximate the continuum by a system with a finite number of degrees of freedom.

In earlier chapters the majority of the examples were solved by use of analytical methods. The reason for this was that simple examples were chosen or that simplifying assumptions were introduced that made the application of analytical methods possible. Solutions to such problems have long served the designer as guidelines. He never really encounters models such as an infinitely long cylinder, but the critical axial load for such a cylinder guides his choice of thickness for any relatively long cylinder.

With the advent of high-speed computing equipment and the subsequent rapid development of numerical methods, the complexity of the models subjected to analysis has changed rather drastically. The designer has become able to address to an ever increasing degree a precise conceptual representation of the actual physical problem he is facing. Practical structures usually are extremely complicated, however, and modeling of a structure so that it will become amenable for analysis will always be an important part of the design procedure.

If the stress resultants, material properties, and shell thickness are independent of the surface coordinates, and the prebuckling equilibrium configuration is free from rotations, the stability equations are linear differential equations with constant coefficients. If, in addition, certain boundary conditions prevail (classical simple-support conditions, for example), it is possible to obtain a closed-form analytical solution for the critical load. In almost all other cases, numerical methods must be used. Emphasis in this chapter is on methods for the solution of problems in which the stability equations are linear but have variable coefficients. The special problems encountered if the primary equilibrium path is nonlinear are discussed in Chap. 9.

Four commonly used numerical solution methods are discussed in the following sections: series expansion, including the Rayleigh-Ritz and the Galerkin methods; finite differences; numerical integration; and finite element analysis. Special emphasis is placed on finite difference and finite element methods, because these appear to be particularly well suited for computer solution of problems based on partial differential equations. For such equations, of course, the dependent variables are functions of two space variables. Mixed methods, such as the use of a Fourier series expansion with respect to one space variable and direct numerical integration with respect to the other, have been used successfully for special geometries, but are not discussed here.

8.2 MATRIX NOTATION

In the preceding chapters we have avoided the use of vector or tensor notation. This has to some degree encumbered the presentation of the material, but, on the other hand, it has made the text available to those who are not familiar with such notations. The use of matrix notation in this chapter therefore may seem to be somewhat inconsistent. However, to present a discussion of finite element methods without use of this compact notation would be awkward. Furthermore, the material presented in this chapter is on an elementary level, and deeper knowledge of the theory of matrix algebra is not needed. The notations used are presented in this section, together with the few matrix operations involved in the derivations.

A matrix is an array of numbers organized in rows and columns. The notations used here are illustrated by the example

$$[A] \equiv [a_{ij}] \equiv \begin{bmatrix} a_{11} & a_{12} & a_{13} \\ a_{21} & a_{22} & a_{23} \\ a_{31} & a_{32} & a_{33} \\ a_{41} & a_{42} & a_{43} \end{bmatrix} \tag{8.1}$$

The matrix $[A]$ has four rows and three columns. We say that $[A]$ has the dimension four by three, or that it is a 4×3 matrix. Two matrices with identical dimensions can be added simply by addition of corresponding elements. For example,

$$[C] = [A] + [B] \tag{8.2}$$

means that
$$c_{ij} = a_{ij} + b_{ij}$$

for all combinations of i and j within the dimensions of the matrices. Addition of two matrices with different dimensions is meaningless.

A matrix with only one *column* is called a *vector*. The notation for a vector is shown in the following example:

$$\{x\} \equiv \{x_i\} \equiv \begin{Bmatrix} x_1 \\ x_2 \\ x_3 \\ x_4 \end{Bmatrix} \tag{8.3}$$

To transpose a matrix means to interchange its columns with its rows. Thus, if $[A]$ is of the dimension $n \times m$, then its transpose, denoted $[A]^T$, is of the dimension $m \times n$. For example,

$$\begin{bmatrix} 4 & 5 & 6 & 7 \\ 11 & 3 & 2 & 1 \\ 0 & 14 & 0 & 8 \end{bmatrix}^T = \begin{bmatrix} 4 & 11 & 0 \\ 5 & 3 & 14 \\ 6 & 2 & 0 \\ 7 & 1 & 8 \end{bmatrix} \tag{8.4}$$

The transpose of a vector is a matrix containing only one row:

$$\{a\}^T = [a_1, a_2, a_3, \ldots] \tag{8.5}$$

Two matrices can be multiplied by one another if and only if the number of rows in the *second* is equal to the number of columns in the *first* matrix. The elements in the product matrix are defined as follows. If

$$[C] = [A][B] \tag{8.6}$$

then
$$c_{ij} = \sum_{q=1}^{Q} a_{iq} b_{qj} \tag{8.7}$$

where Q is the number of columns in $[A]$ (and number of rows in $[B]$). The process of matrix multiplication is not commutative, i.e., in general,

$$[A][B] \neq [B][A] \qquad (8.8)$$

If $[A]$ is an $n \times q$ matrix and $[B]$ is a $q \times m$ matrix, then the dimensions of C are $n \times m$.

We notice that a matrix can always be premultiplied by its transpose; i.e.,

$$[D] = [A]^T[A] \qquad (8.9)$$

is always a meaningful operation.

Any quadratic form can be written as a product of matrices,

$$\sum_{i,\,j=1}^{I} c_{ij} x_i x_j = \{x\}^T[C]\{x\} \qquad (8.10)$$

In this case the matrix $[C]$ is a square matrix, that is, $n = m$. Another example of a square matrix is encountered if we write a system of linear algebraic equations in matrix notation. For instance, the system

$$a_{11} x_1 + a_{12} x_2 + a_{13} x_3 = b_1$$
$$a_{21} x_1 + a_{22} x_2 + a_{23} x_3 = b_2 \qquad (8.11)$$
$$a_{31} x_1 + a_{32} x_2 + a_{33} x_3 = b_3$$

may be written [compare Eq. (8.6)]

$$[A]\{x\} = \{b\} \qquad (8.12)$$

This example shows that the definition of matrix multiplication was a rational choice. It serves our purpose also to define the inverse of a matrix, so that if Eq. (8.12) holds, then

$$\{x\} = [A]^{-1}\{b\} \qquad (8.13)$$

where $[A]^{-1}$ is called the *inverse* of $[A]$. Equations (8.12) and (8.13) express the identical relations if we define the inverse of a matrix so that

$$[A]^{-1}[A] = [I]$$

where I is called the *unity matrix*, i.e., a square matrix in which all elements on the diagonal ($i = j$) are equal to 1 and all other elements vanish. The inverse is a meaningful definition only for square matrices.

In this chapter we occasionally use the form shown in Eq. (8.13). This does not mean that the elements of the inverse must be determined. The relation simply means that $\{x\}$ represents the solution to the equation system defined in Eq. (8.12). The solution can, of course, be obtained by means of defining the elements of $[A]^{-1}$, but this is generally not an economic procedure, and we do not define here the operations that lead to the elements of the inverse.

8.3 SERIES EXPANSIONS

Numerical methods for solution of differential equations are based on a reduction of the continuum with its infinitely many degrees of freedom to a system with a finite number of degrees of freedom. Such reduction can be achieved, for example, through a restriction of the deformation patterns that are considered possible.

As an example, we may assume that a column under axial compression can be deformed only in such a way that the lateral displacement is of the form

$$w = a_1 + a_2 \frac{x}{L} + a_3 \left(\frac{x}{L}\right)^2 \qquad (8.14)$$

where a_1, a_2, a_3 are constants. If we require that both ends of the column be restrained from lateral displacement ($w = 0$ for $x = 0, L$), then

$$w = k(\xi - \xi^2) \qquad (8.15)$$

where $\xi \equiv x/L$. Differentiation of Eq. (8.15) with respect to x gives

$$w' = \frac{k}{L}(1 - 2\xi) \qquad w'' = \frac{k}{L^2}(-2) \qquad (8.16)$$

For inextensional buckling, an expression for the second variation of the total potential energy of a column may be obtained from Eq. (A.42), as follows:

$$\delta^2 V = \int_0^L [EI(w'')^2 - P(w')^2] \, dx \qquad (8.17)$$

where w is the incremental displacement. Substitution of Eqs. (8.16) into Eq. (8.17) yields

$$\delta^2 V = L \int_0^1 \left[\frac{4k^2}{L^4} EI - \frac{k^2}{L^2} P(1 - 2\xi)^2\right] d\xi \qquad (8.18)$$

After the integration is carried out, we have

$$\delta^2 V = \frac{k^2}{L}\left(\frac{4EI}{L^2} - \tfrac{1}{3}P\right) \qquad (8.19)$$

Thus the second variation vanishes for

$$P_{cr} = \frac{12EI}{L^2} \qquad (8.20)$$

The critical load for the column as computed with this approximate method is about 20 percent larger than the classical value given in Eq. (1.18). By use of additional terms in Eq. (8.14), the assumed displacement can be made less restrictive, and more accurate results can be obtained.

8.3a   The Rayleigh-Ritz Method

The type of approach discussed in the preceding section was used by Lord Rayleigh in 1877 (Ref. 8.2) in the analysis of free vibrations. Ritz (Ref. 8.3) used the same approach for both equilibrium and eigenvalue problems. The procedure applied by Ritz, usually referred to as the Rayleigh-Ritz method, is discussed in this section. Equation (8.15) is an admissible displacement function because it is continuous and satisfies the forced boundary conditions. It was shown in the preceding section that the critical load of the column based on this function is too high. The exclusion of components of deformation other than those given by Eq. (8.15) results in a stiffening of the system. The chosen relation does not accurately represent the form of the buckled column, and therefore the computed critical load is not close to the exact solution. If we had chosen instead to approximate the displacement function by the relation

$$w = a \sin \frac{\pi x}{L} \qquad (8.21)$$

application of Rayleigh's method would have led to the exact solution [Eq. (1.18)]

$$P_{cr} = \frac{\pi^2 EI}{L^2} \qquad (8.22)$$

A suitable displacement function sometimes can be chosen by aid of observation of the buckling mode in tests. It should be realized, however, that the deformation pattern observed in tests corresponds to a state in which the incremental displacements are finite, and that this deformation pattern may differ considerably from the initial-buckling mode. Furthermore, the availability of high-speed computers has made it possible to obtain more accurate solutions, and has considerably reduced the interest in this type of procedure.

   If additional degrees of freedom are included in the displacement function, the computed critical load becomes smaller. A better result ($P_{cr} = 9.875 EI/L^2$) would have been obtained if, instead of the displacement function in Eq. (8.15), we had assumed

$$w = k_1(\xi - \xi^2) + k_2(\xi - \xi^2)^2 \qquad (8.23)$$

A displacement function can be assumed in the form of a series of functions with undetermined coefficients. It is shown in Ref. 8.4, sec. 112, for instance, that if such a series is complete, the computed critical load converges, from above, with an increasing number of terms to the exact solution.

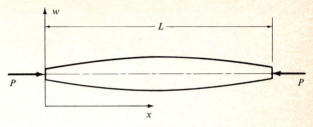

FIGURE 8.1
Column with variable cross section.

As an example of the use of the Rayleigh-Ritz method, we consider the problem of the buckling of a column (Fig. 8.1) with a moment of inertia that varies with the axial coordinate as follows:

$$I = I_0 \left(1 + \alpha \sin \frac{\pi x}{L} \right) \qquad (8.24)$$

where α is a constant. The second variation of the total potential energy of the column is [compare Eq. (8.17)]

$$\delta^2 V = EI_0 \int_0^L \left[\left(1 + \alpha \sin \frac{\pi x}{L} \right)(w'')^2 - \lambda \left(\frac{\pi}{L} \right)^2 (w')^2 \right] dx \qquad (8.25)$$

where

$$\lambda \equiv \frac{P}{EI_0} \left(\frac{L}{\pi} \right)^2 \qquad (8.26)$$

We let the lateral displacements along the column during buckling be represented by the function

$$w = \sum_{n=1}^{N} k_n \sin \frac{n\pi x}{L} \qquad (8.27)$$

Then

$$w' = \sum_{n=1}^{N} \frac{n\pi}{L} k_n \cos \frac{n\pi x}{L}$$

$$w'' = - \sum_{n=1}^{N} \left(\frac{n\pi}{L} \right)^2 k_n \sin \frac{n\pi x}{L} \qquad (8.28)$$

Writing

$$\delta^2 V = \delta^2 V_1 + \delta^2 V_2 + \delta^2 V_3$$

$$= EI_0 \int_0^L (w'')^2 dx - EI_0 \left(\frac{\pi}{L} \right)^2 \lambda \int_0^L (w')^2 dx$$

$$+ EI_0 \alpha \int_0^L \sin \frac{\pi x}{L} (w'')^2 dx \qquad (8.29)$$

we find that

$$\delta^2 V_1 = EI_0 \int_0^L \left[\sum_{n=1}^{N} - \left(\frac{n\pi}{L}\right)^2 k_n \sin \frac{n\pi x}{L} \right]^2 dx$$

$$= EI_0 \int_0^L \left[\sum_{n=1}^{N} \sum_{m=1}^{N} \left(\frac{n\pi}{L}\right)^2 \left(\frac{m\pi}{L}\right)^2 k_n k_m \sin \frac{n\pi x}{L} \sin \frac{m\pi x}{L} \right] dx \qquad (8.30)$$

Since

$$\int_0^1 \sin nx \sin mx \, dx = \begin{cases} 0 & \text{if } n \neq m \\ \tfrac{1}{2} & \text{if } n = m \end{cases} \qquad (8.31)$$

we have

$$\delta^2 V_1 = EI_0 \frac{L}{2} \left(\frac{\pi}{L}\right)^4 \sum_{n=1}^{N} n^4 k_n^2 \qquad (8.32)$$

Similarly, we find that

$$\delta^2 V_2 = -EI_0 \frac{L}{2} \left(\frac{\pi}{L}\right)^4 \lambda \sum_{n=1}^{N} n^2 k_n^2 \qquad (8.33)$$

For the limiting case of a column with a constant moment of inertia ($\alpha = 0$), $\delta^2 V_3 = 0$ and

$$\delta^2 V = \frac{L}{2} EI_0 \left(\frac{\pi}{L}\right)^4 \sum_{n=1}^{N} (n^4 - \lambda n^2) k_n^2 \qquad (8.34)$$

Setting the second variation equal to zero, we find that $\lambda = n^2$ and the lowest eigenvalue (for $n = 1$) is $\lambda_{cr} = 1$. Consequently, from Eq. (8.26) [compare Eq. (1.18)],

$$P_{cr} = \frac{\pi^2 EI_0}{L^2} \qquad (8.35)$$

For a nonzero value of α,

$$\delta^2 V_3 = EI_0 \left(\frac{\pi}{L}\right)^4 \alpha \int_0^L \sum_{n=1}^{N} \sum_{m=1}^{N} \left(n^2 m^2 k_n k_m \sin \frac{\pi x}{L} \sin \frac{n\pi x}{L} \sin \frac{m\pi x}{L}\right) dx \qquad (8.36)$$

We find that for n odd and m even, or for n even and m odd, this integral is zero. The result of this is that, in the stability matrix, there is no coupling between terms corresponding to odd and even values of n. The odd values correspond to a buckling mode that is symmetric about the midpoint of the column. Clearly, the antisymmetric buckling pattern, corresponding to even values of n, will lead to a higher eigenvalue, and we can restrict the analysis to include only odd values of n.

If both n and m are odd (or both even),

$$\int_0^L \sin\frac{\pi x}{L}\sin\frac{n\pi x}{L}\sin\frac{m\pi x}{L}\,dx$$

$$= \frac{L}{2\pi}\left(\frac{1}{1+n-m}+\frac{1}{1-n+m}-\frac{1}{1+n+m}-\frac{1}{1-n-m}\right) \qquad (8.37)$$

Equations (8.25) and (8.28) define the second variation of the total potential energy as a quadratic form in k_n. The critical load is represented by the value of P for which the coefficient matrix ceases to be positive definite, i.e. (Ref. 8.1, p. 27),

$$\det\,[A + \lambda B] = 0$$

where

$$A_{nm} = \delta_{nm}\left[n^4 + \frac{\alpha}{\pi}n^2 m^2\left(\frac{1}{1+n-m}+\frac{1}{1-n+m}-\frac{1}{1+n+m}-\frac{1}{1-n-m}\right)\right]$$

$$B_{nm} = \delta_{nm}\,n^2$$

$$n = 2p - 1$$

$$m = 2q - 1 \qquad\qquad (8.38)$$

$$\delta_{nm} = \begin{cases} 0 & \text{if } n \neq m \\ 1 & \text{if } n = m \end{cases}$$

$$p, q = 1, 2, 3, \ldots$$

If only one term is included in the series, $p = q = n = m = 1$, and

$$A_{1,1} = 1 + \frac{\alpha}{\pi}(1 + 1 - \tfrac{1}{3} + 1) = 1 + \frac{8\alpha}{3\pi}$$

$$B_{1,1} = 1 \qquad\qquad (8.39)$$

and

$$\lambda_{cr} = 1 + \frac{8\alpha}{3\pi}$$

This result is slightly less than $1 + \alpha$, that is, slightly less than the critical load for a column with a constant moment of inertia, $I = I_0(1 + \alpha)$. For the case with $\alpha = 1$, the variation of the critical load with the number of terms in the series is shown in Fig. 8.2. The result for the one-term solution is about 1.3 percent too high.

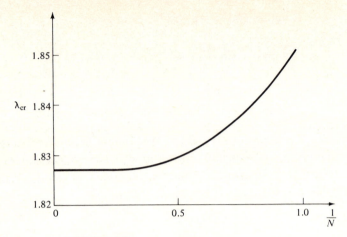

FIGURE 8.2
Critical load for column with variable moment of inertia.

8.3b The Galerkin Method

Another method based on series expansion was developed by Galerkin in 1915 (Ref. 8.5). It was originally proposed by Bubnov and is sometimes referred to as the Bubnov-Galerkin method. A brief description of the method follows here.

Let us write the differential equation governing our problem in the form

$$L(u) = 0 \qquad (8.40)$$

where L is a differential operator and u represents the displacement field. We shall attempt to find a solution of the form

$$u = \sum_{n=1}^{N} a_n \varphi_n \qquad (8.41)$$

where the φ_n represent a sequence of kinematically admissible functions. Galerkin's method is based on the fact (see, for instance, Ref. 8.4) that some measure of the error in $L(u) = 0$ is minimized for any fixed value of N if we satisfy simultaneously for $m = 1, N$ the conditions

$$\int_V L(u)\varphi_m \, dV = \int_V L\left(\sum_{n=1}^{N} a_n \varphi_n\right)\varphi_m \, dV = 0 \qquad m = 1, 2, 3, \ldots \qquad (8.42)$$

For the variational problem with a quadratic functional, it is shown in Ref. 8.4 that the Galerkin and the Rayleigh-Ritz methods are equivalent. Indeed, Galerkin refers to this equivalence to verify the validity of his method. If the set of functions used in the Galerkin procedure satisfies only the forced boundary conditions, an integral over the boundaries of the domain must also be included, so that deviations from the natural boundary conditions will be minimized as well.

Therefore, if a set of functions satisfying natural as well as forced boundary conditions cannot readily be chosen, it is preferable to use the Rayleigh-Ritz method. On the other hand, the Galerkin approach can be applied to a more general class of problems, since it is not necessary that the differential equation $L(u) = 0$ be derivable by use of a variational approach.

Applying the Galerkin method to the column with variable cross section (Fig. 8.1), we utilize the equilibrium equation [compare Eq. (1.11b)]

$$L(w) = \left[EI_0\left(1 + \alpha \sin \frac{\pi x}{L}\right)w''\right]'' + Pw'' = 0 \qquad (8.43)$$

A solution is sought in the form

$$w_N = \sum_{n=1}^{N} k_n \sin \frac{n\pi x}{L} \qquad (8.44)$$

Substituting this solution into Eq. (8.43), we obtain

$$
\begin{aligned}
L(w) = {} & EI_0\left(1 + \alpha \sin \frac{\pi x}{L}\right)\left(\frac{\pi}{L}\right)^4 \sum_{n=1}^{N} n^4 k_n \sin \frac{n\pi x}{L} \\
& - 2EI_0\,\alpha\left(\frac{\pi}{L}\right)^4 \sum_{n=1}^{N} n^3 k_n \cos \frac{\pi x}{L} \cos \frac{n\pi x}{L} \\
& + EI_0\,\alpha\left(\frac{\pi}{L}\right)^4 \sum_{n=1}^{N} n^2 k_n \sin \frac{\pi x}{L} \sin \frac{n\pi x}{L} - P\left(\frac{\pi}{2}\right)^2 \sum_{n=1}^{N} n^2 k_n \sin \frac{n\pi x}{L} \qquad (8.45)
\end{aligned}
$$

and the condition expressed by Eq. (8.42) takes the form

$$
\begin{aligned}
L(w_n)\varphi_m = {} & EI_0\left(\frac{\pi}{L}\right)^4 \sum_{n=1}^{N} n^4 k_n \int_0^L \left(1 + \alpha \sin \frac{\pi x}{L}\right) \sin \frac{n\pi x}{L} \sin \frac{m\pi x}{L}\,dx \\
& - 2EI_0\,\alpha\left(\frac{\pi}{L}\right)^4 \sum_{n=1}^{N} n^3 k_n \int_0^L \cos \frac{\pi x}{L} \cos \frac{n\pi x}{L} \sin \frac{m\pi x}{L}\,dx \\
& + EI_0\,\alpha\left(\frac{\pi}{L}\right)^4 \sum_{n=1}^{N} n^2 k_n \int_0^L \sin \frac{\pi x}{L} \sin \frac{n\pi x}{L} \sin \frac{m\pi x}{L}\,dx \\
& - P\left(\frac{\pi}{L}\right)^2 \sum_{n=1}^{N} n^2 k_n \int_0^L \sin \frac{n\pi x}{L} \sin \frac{m\pi x}{L}\,dx = 0 \qquad (8.46)
\end{aligned}
$$

For $m = 1$, N, we obtain a set of N homogeneous equations in the m_i's, and the lowest value of P for which the coefficient determinant vanishes defines the critical load. If the integrations are carried out, it will be found that the equations are identical with those derived by use of the Rayleigh-Ritz approach in Sec. 8.3a.

8.3c Buckling of a Cylinder under Bending

In Ref. 8.6 the Galerkin method was applied to the problem of buckling of a cylindrical shell subjected to bending combined with axial compression. It is assumed in that analysis that the prebuckling behavior is linear and that effects of prebuckling rotations may be neglected. The analysis is summarized in this section. An analysis of cylindrical shell bending in which prebuckling nonlinearity is taken into account is included in Chap. 9.

The analysis of Ref. 8.6 is based on a modification in Ref. 8.7 of the Donnell-Mushtari-Vlasov equations as follows [cf. Eq. (5.27)]:

$$L(w) = D\nabla^4 w + \frac{Eh}{a^2}\nabla^{-4}w_{,xxxx} - \left(N_x w_{,xx} + \frac{2}{a}N_{x\theta}w_{,x\theta} + \frac{1}{a^2}N_\theta w_{,\theta\theta}\right) = 0 \quad (8.47)$$

where
$$\nabla^4 f = f_{,xxxx} + \frac{2}{a^2}f_{,xx\theta\theta} + \frac{1}{a^4}f_{,\theta\theta\theta\theta}$$

as before, and $\nabla^{-4}f$ is defined by $\nabla^{-4}(\nabla^4 f) = f$.

The prebuckling stress field is approximated by the membrane solution. Hence $N_\theta = N_{x\theta} = 0$ and

$$N_x = -\frac{P}{2\pi a} - \frac{M}{\pi a^2}\cos\theta \equiv -N_c - N_b\cos\theta \quad (8.48)$$

where P is the applied axial compressive force and M is the applied bending moment. The linear differential equation governing the stability of the shell is one with variable coefficients since N_x [Eq. (8.47)] is a function of the circumferential coordinate θ. It is seen that the displacement function

$$w = \sin\frac{m\pi x}{L}\sum_{n=0}^{\infty}k_n\cos n\theta \quad (8.49)$$

satisfies the boundary conditions for simply supported edges. Application of the Galerkin method leads to

$$\int_0^{2\pi}\int_0^L L(w)\sin\frac{p\pi x}{L}\cos q\theta\, dx\, d\theta = 0 \quad \text{for } p = 1, 2, 3, \ldots \text{ and } q = 0, 1, 2, \ldots$$

$$(8.50)$$

Substituting Eqs. (8.48) and (8.49) into Eq. (8.47), we find that Eq. (8.50) can be written in the form

$$\int_0^{2\pi} \int_0^L \sin \frac{m\pi x}{L} \sin \frac{p\pi x}{L} f(n,m,\theta) \cos q\theta \, dx \, d\theta = 0 \qquad (8.51)$$

where, with $\lambda \equiv am\pi/L$,

$$f(n,m,\theta) = \sum \left[\frac{D}{a^4}(\lambda^2 + n^2)^2 + \frac{Eh\lambda^4/a^2}{(\lambda^2 + n^2)^2} - (N_c + N_b \cos \theta)\frac{\lambda^2}{a^2} \right] k_n \cos n\theta$$

The integration with respect to the axial coordinate is easily carried out. If $m \neq p$, the integral vanishes. As a consequence, we can solve Eq. (8.51) for different values of m and thus obtain one value of the critical load for any fixed value of the number of half waves in the axial direction.

For each value of q we have one equation in the unknown coefficients k_n. The homogeneous equation system obtained after integration can be written

$$-r_b(1 + \delta_{1n} - \delta_{0n})k_{n-1} + \left[\frac{(\lambda + n^2/\lambda)^2}{S} + \frac{S}{(\lambda + n^2/\lambda)^2} - 2r_c \right] k_n - r_b k_{n+1} = 0$$

$$\text{for } n = 0, 1, 2, \ldots$$

where

$$r_c = \frac{N_c}{N}$$

$$r_b = \frac{N_b}{N}$$

$$S = [12(1 - v^2)]^{1/2} \frac{a}{h} \qquad (8.52)$$

$$\lambda = \frac{m\pi a}{L}$$

$$\delta_{ij} = \begin{cases} 1 & \text{if } i = j \\ 0 & \text{if } i \neq j \end{cases}$$

After the equation system is truncated in a suitable way, critical values of r_b and r_c may be computed as eigenvalues.

Numerical results are presented in Ref. 8.6. The equation system was truncated so that inclusion of one more harmonic would not have affected the computed eigenvalue in the fifth figure. The critical bending stress generally was found to exceed only very slightly the critical stress in pure compression. Of course, the critical value of r_c (with $r_b = 0$) is equal to 1. Critical values of r_b (with $r_c = 0$) are listed in Table 8.1.

Reference 8.8 shows that, for pressurized cylinders, on the other hand, the critical bending stress can be considerably above the critical stress in pure compression. Even without internal pressure, the critical bending stress significantly exceeds the critical uniform axial stress for certain orthotropic cylinders (Ref. 8.9).

8.4 FINITE DIFFERENCE ANALYSIS

The finite difference method for solution of differential equations constitutes another method for the reduction of the continuum to a system with a finite number of degrees of freedom. It was formalized by Boole and others in the nineteenth century. The reduction is accomplished by substitution of algebraic expressions for the unknown functions and their derivatives. These expressions contain function values at discrete points. Finite difference methods have been discussed extensively in textbooks on numerical analysis. For instance, the books by Collatz (Ref. 8.10) and Forsythe and Wasow (Ref. 8.11) contain thorough discussions of the application of the finite difference methods to initial-value, as well as boundary-value, problems. A brief discussion is given here of the application of finite difference techniques to shell stability problems. The discussion is restricted to formulations of the problem in which the unknown functions are the displacement components. For a deeper insight into the subject, the reader is referred to Ref. 8.11 and to more recent studies such as those by Noor (Ref. 8.12).

For the derivation of finite difference expressions, we shall use a Taylor series expansion, and for simplicity we first consider the one-dimensional case. In the following discussion we refer to points at which finite difference approximations are desired as *reference points*. For distinction, points at which the

Table 8.1

(a/h)	$(r_b)_{cr}$
100	1.015
200	1.009
500	1.006
1,000	1.003

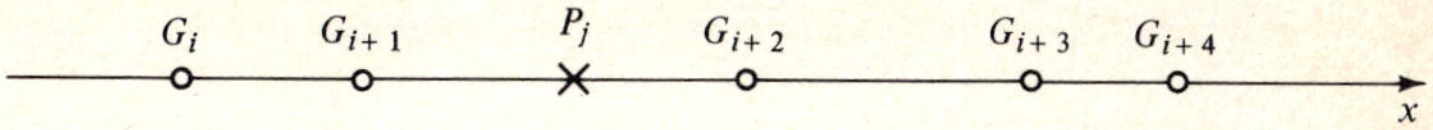

FIGURE 8.3
Finite difference grid.

functions are defined are called *grid points*. The discrete values of the unknown functions at the grid points are the degrees of freedom of the system. The set of grid points may be, but need not be, identical with the set of reference points.

Figure 8.3 shows one of the reference points P_j and a number of neighboring grid points, G_i to G_{i+4}. An expansion of a function f in a Taylor series about the reference point yields

$$f_{(\text{at } G_i)} = (f + \alpha_{ij}f' + \tfrac{1}{2}\alpha_{ij}^2 f'' + \tfrac{1}{6}\alpha_{ij}^3 f''' + \cdots)_{(\text{at } P_j)} \qquad (8.53)$$

where a prime denotes differentiation with respect to x, and where

$$\alpha_{ij} = x_{(\text{at } G_i)} - x_{(\text{at } P_j)}$$

If by use of Eq. (8.53) we express the function values at a number of grid points in terms of the function and its derivatives at P_j, an equation system is defined from which the finite difference expressions at P_j can be determined; i.e., the solution of the equation system expresses f, f', f'', etc., at P_j in terms of the discrete values of the function at G_i, G_{i+1}, G_{i+2}, ...

We consider as an example the special case in which the sets of grid points and reference points are identical and are equally spaced at distances Δ. If the Taylor series is truncated after the term including the fourth-order derivative, the equations are

$$f_{-2} = f - 2\Delta f' + \frac{(2\Delta)^2}{2!}f'' - \frac{(2\Delta)^3}{3!}f''' + \frac{(2\Delta)^4}{4!}f^{\text{iv}}$$

$$f_{-1} = f - \Delta f' + \frac{\Delta^2}{2!}f'' - \frac{\Delta^3}{3!}f''' + \frac{\Delta^4}{4!}f^{\text{iv}}$$

$$f_0 = f \qquad\qquad (8.54)$$

$$f_{+1} = f + \Delta f' + \frac{\Delta^2}{2!}f'' + \frac{\Delta^3}{3!}f''' + \frac{\Delta^4}{4!}f^{\text{iv}}$$

$$f_{+2} = f + 2\Delta f' + \frac{(2\Delta)^2}{2!}f'' + \frac{(2\Delta)^3}{3!}f''' + \frac{(2\Delta)^4}{4!}f^{\text{iv}}$$

The equation system (8.54) has the solution

$$f = f_0$$

$$f' = \frac{1}{\Delta}\left(-\tfrac{1}{12}f_{+2} + \tfrac{2}{3}f_{+1} - \tfrac{2}{3}f_{-1} + \tfrac{1}{12}f_{-2}\right)$$

$$f'' = \frac{1}{\Delta^2}\left(-\tfrac{1}{12}f_{+2} + \tfrac{4}{3}f_{+1} - \tfrac{5}{2}f_0 + \tfrac{4}{3}f_{-1} - \tfrac{1}{12}f_{-2}\right) \qquad (8.55)$$

$$f''' = \frac{1}{\Delta^3}\left(\tfrac{1}{2}f_{+2} - f_{+1} + f_{-1} - \tfrac{1}{2}f_{-2}\right)$$

$$f^{iv} = \frac{1}{\Delta^4}\left(f_{+2} - 4f_{+1} + 6f_0 - 4f_{-1} + f_{-2}\right)$$

More accurate finite difference expressions can be obtained for these derivatives if more terms are used in the Taylor series and more function values at grid points are included. However, use of higher-order approximations involves other disadvantages, and the lowest possible approximation is more commonly used. If derivatives are needed only through the second order, use of three grid points is sufficient. The corresponding finite difference expressions are

$$f' = \frac{1}{2\Delta}\left(f_{+1} - f_{-1}\right)$$

$$f'' = \frac{1}{\Delta^2}\left(f_{+1} - 2f_0 + f_{-1}\right)$$

(8.56)

It is possible to determine error bounds for the finite difference approximations. The lowest-order term (in Δ), which is not included in the expression, is easily obtained if a Taylor series representation of the function values is substituted back into the finite difference expression. For instance, if the expressions

$$f_{+1} = f_0 + \Delta f' + \tfrac{1}{2}\Delta^2 f'' + \tfrac{1}{6}\Delta^3 f''' + \cdots$$

$$f_{-1} = f_0 - \Delta f' + \tfrac{1}{2}\Delta^2 f'' - \tfrac{1}{6}\Delta^3 f''' + \cdots$$

(8.57)

are substituted into the expression for the first-order derivative in Eq. (8.56), we find

$$\frac{1}{2\Delta}\left(f_{+1} - f_{-1}\right) = \frac{1}{2\Delta}\left(2\Delta f' + \tfrac{1}{3}\Delta^3 f''' + \cdots\right)$$

or

$$f' = \frac{1}{2\Delta}\left(f_{+1} - f_{-1}\right) - \tfrac{1}{6}\Delta^2 f''' + \cdots$$

(8.58)

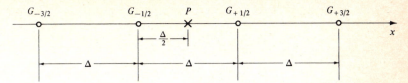

FIGURE 8.4
Finite difference grid with uniform spacing.

It can be shown (Ref. 8.10) that if the third-order derivative in Eqs. (8.58) is represented by its maximum value within the interval, the term $\frac{1}{6}\Delta^2 f'''$ represents a bound on the error. We say, therefore, that the first-order derivative in Eqs. (8.56) is of second-order accuracy; the error term E is $0(\Delta^2)$. In the same way it can be shown that, in the finite difference expressions given by Eqs. (8.55), $E = 0(\Delta^4)$ for the first- and second-order derivatives and $E = 0(\Delta^2)$ for the third- and fourth-order derivatives. Use of the third- and fourth-order derivatives as defined by Eqs. (8.55), together with the expressions in Eqs. (8.56) for the lower-order derivatives, results in a set of expressions that is homogeneous in accuracy.

It may be advantageous to position the reference points midway between two grid points rather than at the grid points. Such an arrangement is shown in Fig. 8.4. With a constant grid spacing, the following finite difference expressions are readily derived:

$$f' = \frac{1}{\Delta}(f_{+1/2} - f_{-1/2}) - \tfrac{1}{24}\Delta^2 f''' + \cdots$$

$$f'' = \frac{1}{2\Delta^2}(f_{+3/2} - f_{+1/2} - f_{-1/2} + f_{-3/2}) - \tfrac{5}{24}\Delta^2 f^{\text{iv}} + \cdots$$

(8.59)

For the case in which reference points and grid points coincide, we have

$$f' = \frac{1}{2\Delta}(f_{+1} - f_{-1}) - \tfrac{1}{6}\Delta^2 f''' + \cdots$$

$$f'' = \frac{1}{\Delta^2}(f_{+1} - 2f_0 + f_{-1}) - \tfrac{1}{12}\Delta^2 f^{\text{iv}} + \cdots$$

(8.60)

Clearly, first-order derivatives are best approximated midway between grid points, and second-order derivatives are more accurately determined at the grid points. It can be shown that this conclusion holds, not only for first- and second-order derivatives, but more generally for odd- and even-order derivatives.

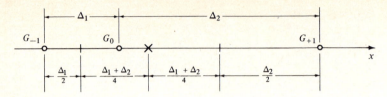

FIGURE 8.5
Finite difference grid with variable spacing.

A finite difference grid with variable spacing is shown in Fig. 8.5. With the grid point G_0 chosen as reference point, the first two derivatives are found by use of a Taylor expansion to be

$$w' = -\frac{\Delta_2}{\Delta_1(\Delta_1 + \Delta_2)} w_{-1} + \frac{\Delta_2 - \Delta_1}{\Delta_1 \Delta_2} w_0 + \frac{\Delta_1}{\Delta_2(\Delta_1 + \Delta_2)} w_{+1} - \frac{\Delta_1 \Delta_2}{6} w''' + \cdots$$

$$w'' = \frac{2}{\Delta_1(\Delta_1 + \Delta_2)} w_{-1} - \frac{2}{\Delta_1 \Delta_2} w_0 + \frac{2}{\Delta_2(\Delta_1 + \Delta_2)} w_{+1} - \frac{\Delta_2 - \Delta_1}{3} w''' + \cdots$$

$$(8.61)$$

If we consider the point P as a reference point, the finite difference expressions are

$$w' = -\frac{1}{2\Delta_1} w_{-1} + \frac{\Delta_2 - \Delta_1}{\Delta_1 \Delta_2} w_0 + \frac{1}{2\Delta_2} w_{+1}$$

$$+ \left[\frac{5}{24} \left(\frac{\Delta_1 + \Delta_2}{2} \right)^2 - \tfrac{1}{24} \Delta_1 \Delta_2 \right] w'' + \cdots \qquad (8.62)$$

$$w'' = \frac{2}{\Delta_1(\Delta_1 + \Delta_2)} w_{-1} - \frac{2}{\Delta_1 \Delta_2} w_0 + \frac{2}{\Delta_2(\Delta_1 + \Delta_2)} w_{+1} + \frac{\Delta_2 - \Delta_1}{12} w''' + \cdots$$

We notice that, for the first-order derivative, we still have a second-order approximation $[E = 0(\Delta^2)]$, but for the second-order derivatives the grid with variable spacing gives us only a first-order approximation. However, the lowest order of the omitted terms contains the difference between two adjacent spacings rather than the spacing itself. Experience shows that variable-spacing grids can often be used to expedite the analysis. We notice also that the error term in the second derivative is smaller by a factor of 4 if P is used as reference point rather than G_0.

Once the derivatives have been expressed in terms of the function values at discrete points, the finite difference approach can be used either in connection with the differential equations of equilibrium or with the energy functional.

The use of finite differences in connection with the energy approach is discussed in Ref. 8.11. The equilibrium equations must be enforced, in principle, at each of the reference points. If the energy approach is used, the energy density is determined at each of the reference points, and the total potential energy is obtained through numerical integration. Methods for such integration are available with various degrees of accuracy, and it is important to choose one that is consistent with the accuracy of the selected finite difference expressions. The final set of algebraic equations is obtained through minimization of the energy with respect to the displacements at the grid points.

In the finite difference schemes discussed above [Eqs. (8.55) and (8.56)] the reference point is located at the center of the range of grid points included in the expression. Such schemes are referred to as *central difference approximations*. If the reference point is located at an end point of the range, away from the center, we have either *backward* or *forward* difference approximations. If central difference approximations are used in connection with equilibrium equations, we can define displacements at fictitious grid points outside the structure in order that derivatives may be determined at or near the boundary. In this case the number of reference points is equal to the number of real grid points, and the boundary conditions correspond to the fictitious points, so that the number of equations equals the number of unknowns. By use of forward and backward differences at the boundaries, the need for fictitious points can be eliminated. Since some boundary conditions generally must be enforced, we can, in this case, balance the number of equations with the number of unknowns by not enforcing equilibrium at boundary points. If the energy method is used, the balance between the number of equations and unknowns is automatic, and the number of reference points is, in principle, independent of the number of grid points. This situation occurs because, in energy methods, we need not enforce natural boundary conditions.

8.4a Application to Column Buckling

Let us consider as an example the buckling of a column under axial load with both ends simply supported. The critical load is given by solution of the eigenvalue problem defined by Eq. (1.12):

$$EIw^{iv} + Pw'' = 0 \qquad (8.63)$$

After the finite difference expressions are substituted for the derivatives, the equation corresponding to reference point i is

$$EI\frac{1}{\Delta^4}(w_{i+2} - 4w_{i+1} + 6w_i - 4w_{i-1} + w_{i-2}) + P\frac{1}{\Delta^2}(w_{i+1} - 2w_i + w_{i-1}) \qquad (8.64)$$

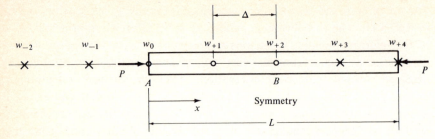

FIGURE 8.6
Finite difference grid for a column.

Each equation contains only five unknowns; i.e., the semibandwidth of the coefficient matrix is 3. The semibandwidth indicates a number of spaces from the diagonal in the matrix beyond which all elements are zero. To solve such an equation system on a digital computer requires very little time, and the number of degrees of freedom can always be chosen so that an accurate result is obtained. However, for demonstration, we choose a very coarse grid, as shown in Fig. 8.6.

Due to symmetry about the midplane, only half of the column length need be considered. Real points are marked by small circles in the figure and fictitious points by x's. The grid spacing equals one-fourth of the column length.

The boundary conditions at A are

$$w = w'' = 0 \qquad (8.65)$$

and at B,

$$w' = w''' = 0 \qquad (8.66)$$

With the first two equations represented by boundary conditions at A and the last two by boundary conditions at B, we obtain the following equation system:

$$w_0 = 0$$

$$w_{-1} - 2w_0 + w_1 = 0$$

$$\frac{EI}{\Delta^4}(w_{-2} - 4w_{-1} + 6w_0 - 4w_1 + w_2) + \frac{P}{\Delta^2}(w_{-1} - 2w_0 + w_1) = 0$$

$$\frac{EI}{\Delta^4}(w_{-1} - 4w_0 + 6w_1 - 4w_2 + w_3) + \frac{P}{\Delta^2}(w_0 - 2w_1 + w_2) = 0 \qquad (8.67)$$

$$\frac{EI}{\Delta^4}(w_0 - 4w_1 + 6w_2 - 4w_3 + w_4) + \frac{P}{\Delta^2}(w_1 - 2w_2 + w_3) = 0$$

$$w_3 - w_1 = 0$$

$$-\tfrac{1}{2}w_0 + w_1 - w_3 + \tfrac{1}{2}w_4 = 0$$

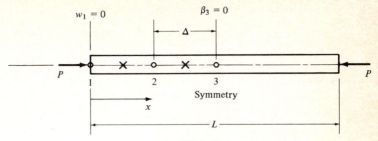

FIGURE 8.7
Modified finite difference grid for a column.

Equations (8.67) are homogeneous, and the critical load is represented by the lowest value of P for which the coefficient determinant equals zero. Thus

$$P_{cr} = (2 - \sqrt{2})16\frac{EI}{L^2} \qquad (8.68)$$

This value is 5 percent lower than that indicated by the exact solution in Eq. (1.18).

There are advantages to using the finite difference approximation in association with a so-called *direct energy method*, i.e., a method in which the second variation is minimized numerically without resort to the Euler equations of the calculus of variations. Most important of these advantages may be that it is not necessary to enforce the natural boundary conditions. Here we shall illustrate the method by application to the column problem that was just solved by numerical solution of the differential equation. The column is shown in Fig. 8.7. Lateral displacements are denoted by w, and rotations by β. Subscripts refer to the gridpoint number defined in the figure.

The second variation of the total potential energy is given by [compare Eq. (8.17)]

$$\delta^2 V = \int_0^L [EI(w''^2) - P(w')^2]\, dx \qquad (8.69)$$

To obtain better accuracy in the first-order derivatives [see Eqs. (8.59) and (8.60)], the second term will be evaluated at *half-stations* (marked by + in the figure). The finite difference expression for the rotation then is

$$w'_{(i+\frac{1}{2})} = \frac{w_{i+1} - w_i}{\Delta} \qquad (8.70)$$

and the contribution of this term to the total potential energy can be written

$$\delta^2 V_1 = \int_0^L P(w')^2 \, dx = P \Delta \sum_{i=1}^{L/2n+1} \left[\frac{1}{\Delta}(w_{i+1} - w_i) \right]^2 \tag{8.71}$$

The second-order derivatives are most accurately approximated by the difference expressions given in Eq. (8.59). Therefore we shall evaluate the first term in Eq. (8.69) at the regular stations, while the half-stations will mark the divisions between integration intervals. To avoid use of fictitious points, we define as unknowns of the system, in addition to the displacements at all the grid points, the rotations at the two end points (β_1 and β_N). The second-order derivatives at the internal points are given by

$$(w'')_i = \frac{1}{\Delta^2}(w_{i+1} - 2w_i + w_{i-1}) \tag{8.72}$$

At the two end points we have

$$w_2 = w_1 + \Delta\beta_1 + \frac{\Delta^2}{2}(w'')_1$$

and

$$w_{N-1} = w_N - \Delta\beta_N + \frac{\Delta^2}{2}(w'')_N \tag{8.73}$$

where the Nth grid point is located at midlength of the column.

Using trapezoidal integration, we find

$$\delta^2 V_2 = \int_0^L [EI(w'')^2] \, dx = EI \left\{ \frac{\Delta}{2} \left[\frac{2}{\Delta^2}(w_2 - w_1 - \Delta\beta_1) \right]^2 \right.$$

$$+ \Delta \sum_{i=2}^{N-1} \left[\frac{1}{\Delta^2}(w_{i+1} - 2w_i + w_{i-1}) \right]^2$$

$$+ \left. \frac{\Delta}{2} \left[\frac{2}{\Delta^2}(w_{N-1} - w_N + \Delta\beta_N) \right]^2 \right\} \tag{8.74}$$

Since the displacement pattern is symmetric about the plane through the column at midlength, we have $\beta_N = 0$, and with simply supported ends, $w_1 = 0$. The rotation β_1 occurs only in the first term of Eq. (8.74), and a minimum of the energy will be obtained if this term vanishes [i.e., if $\beta_1 = (w_2 - w_1)/\Delta$]. Therefore the first term may be dropped, corresponding to enforcement of the natural boundary condition of zero bending moment at the end of the column. The remaining natural boundary condition implies that the transverse shear force is zero at the symmetry plane. This is a more complicated condition, and therefore we take advantage of the fact that with the energy method it need not be enforced.

If the same coarse net is chosen as was used in the application of the equilibrium approach [Eqs. (8.67)], $N = 3$, then with

$$x_1 = \frac{w_2}{\Delta}$$

$$x_2 = \frac{w_3}{\Delta}$$

(8.75)

we have

$$\delta^2 V = \delta^2 V_1 + \delta^2 V_2 = EI[(x_2 - 2x_1)^2 + 2(x_1 - x_2)^2] - P\,\Delta^2[x_1{}^2 + (x_2 - x_1)^2]$$

(8.76)

Noting that the quadratic form for the second variation of the total potential energy ceases to be positive definite at the lowest value of P for which the determinant vanishes, we can define an eigenvalue problem of the form (see Sec. 8.2 for notations)

$$[A]\{x\} + P[B]\{x\} = 0$$

or

$$\det\,([A] + P[B]) = 0 \qquad (8.77)$$

The critical value of the axial load is again found to be

$$P_{cr} = 16\bigl(2 - \sqrt{2}\bigr)\frac{EI}{L^2} \qquad (8.78)$$

More accurate values are obtained for the critical load if a closer grid spacing is used. With $\bar{P}_{cr} = P_{cr}/[EI(\pi/L)^2]$, we obtain the results shown in Table 8.2.

Table 8.2 CRITICAL LOAD AS FUNCTION OF NUMBER OF GRID POINTS

Number of grid points (on one-half of column)	Number of degrees of freedom	Critical load, $\bar{P}_{cr}$
3	2	0.9495
4	3	0.9774
5	4	0.9872
7	6	0.9943
9	8	0.9968
11	10	0.9979
∞	∞	1.0000

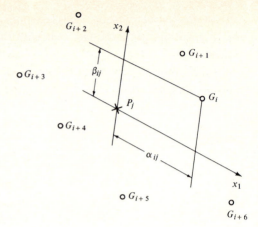

FIGURE 8.8
Finite difference grid in two dimensions.

The energy method can be applied in many different ways, with or without fictitious points, and with different types of finite difference approximations. The possibility of using higher-order finite difference expressions, including the so-called Hermitian method in which function derivatives at grid points are added as unknowns (Ref. 8.12), has received little attention in the literature on numerical methods for structural analysis.

8.4b Two-dimensional Finite Difference Expressions

The finite difference expressions derived above were restricted to the case in which the unknown function is dependent on only one spatial coordinate. Although finite difference expressions can be derived corresponding to a space of any dimensions, the two-dimensional expressions are sufficient for the purpose of thin plate or shell analysis. The derivation of such expressions closely follows that in the one-dimensional case. Figure 8.8 shows a reference point P_j at which appropriate finite difference expressions are to be defined in terms of the discrete values of the unknown function at a number of grid points G_i, G_{i+1}, etc. The distances of grid points G_i from the reference point P_j in the two coordinate directions (x_1, x_2) are α_{ij} and β_{ij}. A Taylor series approach leads to the expression

$$f_{(\text{at } G_i)} = (f + \alpha_{ij} f' + \beta_{ij} \dot{f} + \tfrac{1}{2} \alpha_{ij}^2 f'' + \tfrac{1}{2} \beta_{ij}^2 \ddot{f} + \alpha_{ij} \beta_{ij} \dot{f}' + \cdots)_{(\text{at } P_j)} \quad (8.79)$$

where a prime indicates differentiation with respect to the x_1 coordinate, and a dot with respect to the x_2 coordinate.

Let us assume that only derivatives up to the second order and of the lowest possible approximations are desired. Then it is sufficient to specify for each reference point a suitable set of six neighboring grid points (G_i to G_{i+5}). By applying Eq. (8.79) at each of these six grid points, we obtain the following set of equations:

$$\{f\} = [A]\{g\} \qquad i = 1,6 \qquad (8.80)$$

where

$$[A] = \begin{bmatrix} 1 & \alpha_{1j} & \beta_{1j} & \frac{1}{2}\alpha_{1j}^2 & \alpha_{1j}\beta_{1j} & \frac{1}{2}\beta_{1j}^2 \\ 1 & \alpha_{2j} & \beta_{2j} & \frac{1}{2}\alpha_{2j}^2 & \alpha_{2j}\beta_{2j} & \frac{1}{2}\beta_{2j}^2 \\ 1 & \alpha_{3j} & \beta_{3j} & \frac{1}{2}\alpha_{3j}^2 & \alpha_{3j}\beta_{3j} & \frac{1}{2}\beta_{3j}^2 \\ 1 & \alpha_{4j} & \beta_{4j} & \frac{1}{2}\alpha_{4j}^2 & \alpha_{4j}\beta_{4j} & \frac{1}{2}\beta_{4j}^2 \\ 1 & \alpha_{5j} & \beta_{5j} & \frac{1}{2}\alpha_{5j}^2 & \alpha_{5j}\beta_{5j} & \frac{1}{2}\beta_{5j}^2 \\ 1 & \alpha_{6j} & \beta_{6j} & \frac{1}{2}\alpha_{6j}^2 & \alpha_{6j}\beta_{6j} & \frac{1}{2}\beta_{6j}^2 \end{bmatrix} \qquad (8.81)$$

and

$$\{g\} = [f, f', f^{\,\cdot}, f'', f^{\,\cdot\cdot}]^T \qquad (8.82)$$

Taking the inverse of Eqs. (8.81), we find the derivatives at the reference point j in terms of the function values at the chosen neighboring grid points, i.e.,

$$\{g\} = [C]\{f\} \qquad (8.83)$$

where the coefficient matrix is determined from

$$[C] = [A]^{-1} \qquad (8.84)$$

If the reference point P_j coincides with a grid point, this grid point should be included as one of the neighbors.

As a special case we now let the sets of reference points and grid points be identical. In addition, these points are located at the intersections of uniformly spaced surface coordinate lines (see Fig. 8.9). The points numbered 1 to 6 represent one way of choosing the set of neighboring grid points for the reference point P (at point 4). We have

$$\alpha_{14} = \alpha_{24} = \Delta_1$$
$$\alpha_{34} = \alpha_{44} = \alpha_{54} = 0$$
$$\alpha_{64} = -\Delta_1$$
$$\beta_{14} = \beta_{34} = \Delta_2 \qquad (8.85)$$
$$\beta_{24} = \beta_{44} = \beta_{64} = 0$$
$$\beta_{54} = -\Delta_2$$

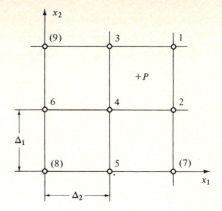

FIGURE 8.9
Uniform finite difference grid in two
dimensions.

If these values are substituted into Eq. (8.81) and the inversion indicated by
Eq. (8.84) is carried out, we find that

$$f' = \frac{1}{2\Delta_1} (f_2 - f_6)$$

$$f'' = \frac{1}{\Delta_1{}^2} (f_2 - 2f_4 + f_6)$$

$$f^{\cdot} = \frac{1}{2\Delta_2} (f_3 - f_5) \tag{8.86}$$

$$f^{\cdot\cdot} = \frac{1}{\Delta_2{}^2} (f_3 - 2f_4 + f_5)$$

$$f'^{\cdot} = \frac{1}{\Delta_1 \Delta_2} (f_1 - f_2 - f_3 + f_4)$$

There is, of course, no special reason to include point 1 here rather than
any of the other corner points. A different choice would lead to a change only
in the definition of $f'^{\cdot}$. This derivative is often approximated by

$$f'^{\cdot} = \frac{1}{4\Delta_1 \Delta_2} (f_1 - f_7 - f_9 + f_8) \tag{8.87}$$

The second-order mixed derivative would be more accurately determined at
points such as P in the figure.

Because the grid points are located at intersections between lines with
constant values of the surface coordinates, many of the α_{ij} and β_{ij} [Eq. (8.85)]
vanish. This leads to an uncoupling of the equation system, so that the deriva-

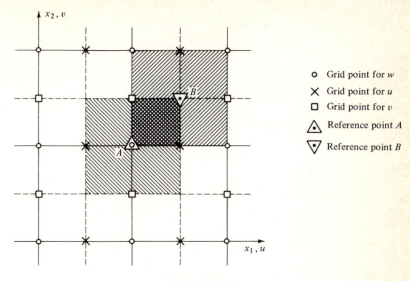

FIGURE 8.10
Modified finite difference grid in two dimensions.

tives with respect to the two coordinates are obtained from two separate equation systems. The finite difference expressions, except of course for the mixed derivative, are therefore identical with the expressions given above for the one-dimensional case. The first-order derivatives are represented by the less favorable expressions [see Eqs. (8.59) and (8.60)].

In order that the more accurate expressions may be used, other schemes for two-dimensional finite differences have been devised. For instance, Noor (Ref. 8.12) suggests the scheme illustrated in Fig. 8.10. Here a different set of grid points is used for each of the three displacement components. The shaded areas surrounding these points are used in the numerical integration procedure.

A different scheme was suggested in Ref. 8.13 and is shown here in Fig. 8.11. In this scheme we have only one set of grid points, but the bending strain energy is integrated over the reference points A, and the twisting and membrane strain energies over reference points B.

The two-dimensional finite difference schemes can be defined in many different ways; the presence of coupling terms between bending and membrane strains as well as anisotropic material properties will further complicate the choice of a suitable scheme. However, it is important for computer economy that adequate finite difference expressions be chosen and that the method of numerical integration be consistent with the chosen expressions.

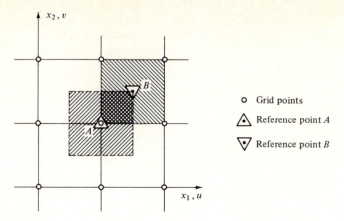

FIGURE 8.11
Modified finite difference grid in two dimensions (with all displacement components defined at same point).

In the cases discussed here, the grid points are located at the intersections of lines with constant values of the shell coordinates. This means that the shell can be mapped onto a rectangular region with a rectangular mesh. For more complicated shell structures it is desirable to remove such restrictions. A more general approach, still based on a Taylor series expansion for determination of finite difference coefficients, was used in Ref. 8.14. An approach based on Green's theorem is presented in Ref. 8.15, and was applied successfully in a linear shell analysis.

In the application of two-dimensional finite difference analysis, the most critical problem is connected with the solution of large equation systems. The bandwidth (see Sec. 8.4a) of the equation system is necessarily large, and excessive computer run time is the major factor limiting the state of the art. Discussion of methods for solution of large linear equation systems is beyond the scope of this book. The problem is discussed in many textbooks, such as Ref. 8.1.

8.5 NUMERICAL INTEGRATION

In the preceding sections, the problem of buckling of columns, plates, and shells has been directly attacked as a boundary-value problem. A different approach, that sometimes has been used in shell stability analysis, is based on a conversion of the boundary-value problem into an initial-value problem. The method has been applied only to ordinary differential equations, and it does not seem profitable to extend it to partial differential equations. The technique, sometimes referred to as the *shooting method*, is based on application of the numerical integration methods that have been developed for initial-value problems. Assumed values for some of the unknowns at one end of the domain are used to get the procedure started. These unknowns are ultimately determined by use of the boundary conditions at the other end.

Methods for numerical integration are discussed in many textbooks. For an introduction to the subject, Ref. 8.16 is recommended. The simplest of the methods, Euler's method, is seldom used because its convergence with decreasing step size is poor, but its simplicity makes it suitable for demonstration of the technique of numerical integration. Other more sophisticated methods may be considered to be extensions of Euler's method.

Euler's method is used for solution of a differential equation of the form

$$y' = f(x,y) \qquad (8.88)$$

with the initial condition

$$y_0 = \alpha \qquad (8.89)$$

The solution is based on the recurrence formula

$$y_{n+1} = y_n + y'_n \, \Delta = y_n + f(x_n,y_n) \, \Delta \qquad (8.90)$$

Clearly, this solution is based on the assumption that the intervals or steps are chosen so small that the function can be assumed to vary linearly within the interval. The extension to the solution of a set of simultaneous first-order differential equations is obvious. Higher-order differential equations can always be broken down into sets of first-order equations.

Euler's method is applied here to the case of column buckling. The governing differential equation

$$EIw^{iv} + Pw'' = 0 \qquad (8.91)$$

is first written as a system of first-order equations. With $r \equiv w'$, $s \equiv w''$, $t \equiv w'''$, we have

$$EIt' + Ps = 0$$
$$t = s' \qquad (8.92)$$
$$s = r'$$

The boundary conditions at one end of the simply supported column give us

$$s_0 = (w'')_0 = 0 \qquad (8.93)$$

In addition, we assume that

$$r_0 = \alpha_1$$
$$t_0 = \alpha_2 \tag{8.94}$$

The function values at discrete points along the column with a spacing Δ can now be computed in terms of α_1 and α_2. We use the recurrence formulas

$$\left.\begin{aligned} r_i &= r_j + (r')_j\Delta = r_j + s_j\Delta \\ s_i &= s_j + (s')_j\Delta = s_j + t_j\Delta \\ t_i &= t_j + (t')_j\Delta = t_j - \lambda s_j\Delta \end{aligned}\right\} \quad i = j+1 \tag{8.95}$$

where $\lambda = P/EI$ and Δ is the increment in the x coordinate. Application of Eqs. (8.93) and (8.94), together with the recurrence formulas, gives the result shown in Table 8.3.

If one-half of the length of the column is divided into two intervals, we have $\Delta = L/4$, and the boundary conditions at midlength are

$$r_2 = w_{,x}(x = 2\Delta) = 0$$
$$t_2 = w_{,xxx}(x = 2\Delta) = 0 \tag{8.96}$$

These yield the homogeneous equation system (see Table 8.3)

$$\alpha_1 + \Delta^2\alpha_2 = 0$$
$$(1 - \Delta^2\lambda)\alpha_2 = 0 \tag{8.97}$$

and the stability matrix is

$$\begin{bmatrix} 1 & \Delta^2 \\ 0 & 1 - \Delta^2\lambda \end{bmatrix} \tag{8.98}$$

Setting the determinant of this matrix equal to zero, we find

$$\lambda_{cr} = \frac{4}{L} \tag{8.99}$$

Table 8.3

Step i	x	r_i	s_i	t_i
0	0	α_1	0	α_2
1	Δ	α_1	$\Delta\alpha_2$	α_2
2	2Δ	$\alpha_1 + \Delta^2\alpha_2$	$2\Delta\alpha_2$	$\alpha_2(1 - \Delta^2\lambda)$
3	3Δ	$\alpha_1 + 3\Delta^2\alpha_2$	$\Delta\alpha_2(3 - \Delta^2\lambda)$	$\alpha_2(1 - 3\Delta^2\lambda)$
4	4Δ	$\alpha_1 + 6\Delta^2\alpha_2 - \Delta^4\lambda\alpha_2$	$\Delta\alpha_2(4 - 4\Delta^2\lambda)$	$\alpha_2(1 - 6\Delta^2\lambda + \Delta^4\lambda^2)$
5	5Δ	$\alpha_1 + 10\Delta^2\alpha_2 - 5\Delta^4\lambda\alpha_2$	$\Delta\alpha_2(5 - 10\Delta^2\lambda + \Delta^4\lambda^2)$	$\alpha_2(1 - 10\Delta^2\lambda + 5\Delta^4\lambda^2)$
6	6Δ	$\alpha_1 + 15\Delta^2\alpha_2 - 15\Delta^4\lambda\alpha_2 + \Delta^6\lambda^2\alpha_2$	$\Delta\alpha_2(6 - 20\Delta^2\lambda + 6\Delta^4\lambda^2)$	$\alpha_2(1 - 15\Delta^2\lambda + 15\Delta^4\lambda^2 - \Delta^6\lambda^3)$

and consequently,

$$P_{cr} = 16\frac{EI}{L^2} \qquad (8.100)$$

This result is seen to be approximately 60 percent too high. Results shown in Table 8.4 for up to six intervals indicate a rather slow convergence.

The Euler method is classified as a one-step method. This means that the function value at $x_{n+1} = x_n + \Delta$ can be computed by use of only the function value and its derivative at $x = x_n$. Once it is assumed that y' is constant over the interval, the numerical integration is trivial. More accurate one-step methods can be obtained if the function value y is estimated at a number of points in the interval so that corresponding values of the derivative can be determined and numerically integrated over the interval. Such methods are referred to as being of the Runge-Kutta type. The classical fourth-order Runge-Kutta method is of the form

$$y_{n+1} = y_n + \frac{\Delta}{6}(k_1 + 2k_2 + 2k_3 + k_4) \qquad (8.101)$$

where
$$k_1 = f(x_n, y_n)$$
$$k_2 = f(x_n + \tfrac{1}{2}\Delta, \; y_n + \tfrac{1}{2}k_1\Delta)$$
$$k_3 = f(x_n + \tfrac{1}{2}\Delta, \; y_n + \tfrac{1}{2}k_2\Delta)$$
$$k_4 = f(x_n + \Delta, \; y + k_3\Delta)$$

The Runge-Kutta methods do not take advantage of all available information and are therefore less efficient than the so-called multistep methods. In the latter methods the function $f[x, y(x)]$, is replaced by a polynomial $p(x)$ that fits to a number of previously obtained solutions. The polynomial is then integrated over the interval, and we find the new function value from the expression

$$y(x + \Delta) = y(x) + \int_x^{x+\Delta} p(t)\, dt \qquad (8.102)$$

Table 8.4

i	$P_{cr}L^2/EI$
2	16
3	12
4	11.0
5	10.54
6	10.33

If the power series includes only the previously computed function values, $y(x + \Delta)$ is obtained explicitly. In this case the method is called an explicit scheme, or a *predictor*. If the power series includes an estimated value of $f(x_{n+1}, y_{n+1})$, this estimated value can be gradually refined through iteration. The method is then called an implicit scheme, or a *corrector*. All multistep methods can be written in the form

$$\sum_{i=1}^{m} \alpha_i y_{n+i} = \Delta \sum_{i=1}^{m} \beta_i y'_{n+i} \qquad n = -1, 0, 1, \ldots \qquad (8.103)$$

If $\beta_m = 0$, we have a predictor method, and with $\beta_m \neq 0$, a corrector method. A multistep method is of a different order, depending on how many previously computed function values are included. Values of α_i and β_i for different orders of integration are available in textbooks (Ref. 8.16). A commonly used corrector method is the so-called *Adams-Moulton method*.

A multistep method cannot be used to start the procedure; hence the first solutions are often obtained by use of a Runge-Kutta procedure. The most efficient way to solve a two-point boundary problem by use of numerical integration appears to be a procedure in which a Runge-Kutta method is used for the first solutions, and for later steps a multistep predictor method gives a first estimate of new function values that are then iteratively improved by use of a corrector method.

The use of numerical integration procedures has the advantage that error control is more readily applied than in the finite difference solution. The error can be evaluated at any time, and the step size adjusted if found necessary. On the other hand, there are some disadvantages with the procedure. The most important may be that it cannot be adapted to two-dimensional analysis (partial differential equations). In shell stability analysis numerical integration methods have been used in Refs. 8.17 and 8.18.

8.6   FINITE ELEMENT ANALYSIS

8.6a   Introduction

The finite element method has much in common with the energy-based finite difference method discussed in Sec. 8.4. In fact, the finite element method can be defined in a sufficiently general way to include the finite difference energy method as one of its forms.

In contrast to the finite difference method, which was developed in the realm of applied mathematics, the finite element method was derived by means of physical considerations in the field of structural mechanics. Later, a mathematical interpretation of the finite element method was added, which led to considerable refinement of the method and to extension of its use beyond the field of structural mechanics. The relative merits of the physical and the mathematical interpretations of the method are discussed in Ref. 8.19.

The finite element method is an extension of the methods for analysis of statically indeterminate structures that were developed within the aircraft industry in the early 1950s. For the swept-wing structures introduced at that time, the simple beam theory previously used for wing analysis was clearly inadequate. Stress distributions in such structures could be accurately determined only after high-speed computers became available. The first efforts to improve the situation were based on analog computers, but these did not operate with sufficient accuracy, and the results were not entirely satisfactory.

The formulation of the matrix theory for structural analysis presented by Langefors in 1952 (Ref. 8.20) provided a very efficient means for organization of the computations and made it possible to utilize the digital computers being introduced at that time. The aircraft structures considered consisted of an assemblage of individual structural components (bars, beams, shear panels) that were connected at a number of nodal points. That is, the structure was naturally divided into individual elements with simple structural behavior. We might say that the finite element method was born when a continuum was first analyzed, after the structure had been artificially subdivided into individual elements. The first published successful effort along those lines appears to be a paper by Turner et al., in 1956 (Ref. 8.21), but the name of the method was first used by Clough, in 1960 (Ref. 8.22).

We intend to give the reader a basic understanding of the method, and in particular to show how it is applied in structural stability analysis. For a more thorough treatment of the subject, the reader is referred to the textbooks on the topic, such as Refs. 8.23 and 8.24.

The systems for structural analysis first used by Langefors and others were based on the so-called *force method*. In this application, internal forces (and moments) at the nodes are the basic unknown variables. In the *displacement method*, first presented by Levy (Ref. 8.25) and Argyris (Ref. 8.26), displacements (and rotations) at the node points are the basic unknowns. Primarily due to its simplicity, the displacement method was subsequently preferred in structural analysis. The force method is usually associated with application of the principle of minimum complementary energy, and the displacement method with the principle of minimum potential energy. Mixed formulations can be

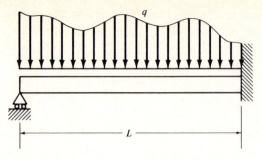

FIGURE 8.12
Beam with lateral load.

obtained by use of the Reissner potential (Ref. 8.27). For simplicity, the present discussion is limited to the displacement method. A discussion of the relative merits of the different methods is included in Ref. 8.24.

Here the mathematical interpretation of the method is presented; i.e., the finite element method is considered as a special form of the Rayleigh-Ritz method. In application of the Rayleigh-Ritz method, the displacements are generally expressed in terms of a complete set of linearly independent functions defined within the boundaries of the structure. The coefficients for those functions are the unknowns of the system. Using the finite element method, we subdivide the structure into elements by use of a set of grid lines, and the node-point displacement (and rotation) components are the basic unknowns. The displacements in the interior of each of the elements are expressed in terms of functions in which the node-point variables appear as coefficients. If we define these functions so that complete compatibility is enforced at element interfaces, the method is a form of the Rayleigh-Ritz method, and we can conclude that, with a gradually refined grid, the strain energy (and the critical load in stability analysis) will converge from above (Ref. 8.4, sec. 112). More rapid convergence can sometimes be achieved by use of nonconforming elements. If nonconforming elements are used, the finite element is not a form of the Rayleigh-Ritz procedure, and convergence from above is not assured. A discussion of nonconforming elements is beyond the scope of this book.

8.6b   Bending of a Beam

Let us consider first as an example a linear stress problem. A prismatic beam that is simply supported at one end and clamped at the other and is subjected to a distributed lateral load is shown in Fig. 8.12. In this example we must evaluate

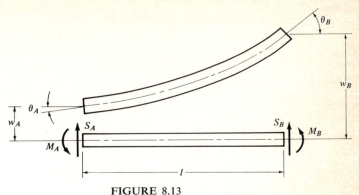

FIGURE 8.13
Beam element in deformed and undeformed state.

the stiffness properties of the beam element shown in Fig. 8.13. The loads acting on the element (notations in Fig 8.13) are represented by the force vector

$$\{P\} = \left[S_A, \frac{M_A}{l}, S_B, \frac{M_B}{l}\right]^T \qquad (8.104)$$

and the displacements (Fig. 8.13) by the vector

$$\{u\} = [w_A, l\theta_A, w_B, l\theta_B]^T \qquad (8.105)$$

We shall attempt to determine the elements of a matrix $[k^e]$ that defines the force-displacement relationship for the element in accordance with the equation

$$\{P\} = [k^e]\{u\} \qquad (8.106)$$

For this purpose we can use the equilibrium equations for the element:

$$S_A + S_B = 0$$
$$M_A - lS_A + M_B = 0 \qquad (8.107)$$

and the stiffness properties

$$\theta_B = \theta_A + S_B \frac{l^2}{2EI} + M_B \frac{l}{EI}$$

$$w_B = w_A + l\theta_A + S_B \frac{l^3}{3EI} + M_B \frac{l^2}{2EI} \qquad (8.108)$$

or in terms of the vector components defined in Eqs. (8.104) and (8.105),

$$P_1 + P_3 = 0$$

$$-P_1 + P_2 + P_4 = 0$$

$$\frac{l^3}{EI}(\tfrac{1}{2}P_3 + P_4) = -u_2 + u_4 \tag{8.109}$$

$$\frac{l^3}{EI}(\tfrac{1}{3}P_3 + \tfrac{1}{2}P_4) = -u_1 - u_2 + u_3$$

Solving these equations, we find the relation

$$\{P\} = [k^e]\{u\} \qquad [k^e] = \frac{EI}{l^3}\begin{bmatrix} 12 & 6 & -12 & 6 \\ 6 & 4 & -6 & 2 \\ -12 & -6 & 12 & -6 \\ 6 & 2 & -6 & 4 \end{bmatrix} \tag{8.110}$$

To emphasize the relation to the Rayleigh-Ritz method, we now evaluate the stiffness matrix $[k^e]$ of a beam element by use of energy considerations. The contribution from one element to the total potential energy of the system can be written

$$V^e = \frac{1}{2}\left[\int_0^1 EI(w'')^2 \, d\xi - \sum_{i=1}^{2} P_i w_i\right] \tag{8.111}$$

where $\xi \equiv x/l$, the subscript in the summation indicates number of the node, and P_i is a set of point forces applied at the nodes and equivalent to the distributed forces on the element.

We assume first a displacement function for the beam element of the form

$$\frac{w}{l} = a_1 + a_2\,\xi + a_3\,\xi^2 + a_4\,\xi^3 \tag{8.112}$$

This is equivalent to the assumption of a linearly varying bending moment within the element boundaries. The rotation w' is represented by the function

$$w' = \frac{w_{,\xi}}{l} = a_2 + 2a_3\,\xi + 3a_4\,\xi^2 \tag{8.113}$$

The displacement vector u defined in Eq. (8.105) can be expressed in terms of the coefficients in Eq. (8.112):

$$u_1 = w_A = w(\xi = 0) = la_1$$
$$u_2 = l\theta_A = lw'(\xi = 0) = la_2 \qquad (8.114)$$
$$u_3 = w_B = w(\xi = 1) = l(a_1 + a_2 + a_3 + a_4)$$
$$u_4 = l\theta_B = lw'(\xi = 1) = l(a_2 + 2a_3 + 3a_4)$$

or

$$\{u\} = [A]\{a\} \qquad [A] = l \begin{bmatrix} 1 & 0 & 0 & 0 \\ 0 & 1 & 0 & 0 \\ 1 & 1 & 1 & 1 \\ 0 & 1 & 2 & 3 \end{bmatrix} \qquad (8.115)$$

Taking the inverse of this relation, we find [the form of the inverse is evident after solution of Eqs. (8.115)]

$$\{a\} = [B]\{u\} \qquad [B] = \frac{1}{l} \begin{bmatrix} 1 & 0 & 0 & 0 \\ 0 & 1 & 0 & 0 \\ -3 & -2 & 3 & -1 \\ 2 & 1 & -2 & 1 \end{bmatrix} \qquad (8.116)$$

Since $w'' = (2a_3 + 6a_4 \xi)/l$ and $\{a\} = [B]\{u\}$, we have

$$w'' = [F]\{a\} = [F][B]\{u\} \qquad (8.117)$$

where

$$[F] = \frac{1}{l} [0,0,2,6\xi] \qquad (8.118)$$

The bending strain energy in the element thus can be written

$$V^e = \tfrac{1}{2}EIl \int_0^1 (w'')^2 \, d\xi = \frac{EI}{2l} \int_0^1 ([F]^T[B]\{u\})^2 \, d\xi$$

$$= \{u\}^T[k^e]\{u\} \qquad (8.119)$$

where

$$[k^e] = EIl \int_0^1 [B]^T[F]^T[F][B] \, d\xi \qquad (8.120)$$

After the matrix multiplications and integration are carried out, we find the same stiffness matrix as was given by Eq. (8.110).

The contribution from one element to the first variation of the total potential energy is a vector with four components. Each of these represents the variation with respect to one of the degrees of freedom of the element:

$$\delta V^e = \tfrac{1}{2}[k_{ij}^{\,e}]\{u\} - \tfrac{1}{2}[P_1,0,P_2,0]^T \qquad (8.121)$$

where P_1 and P_2 are point forces, applied at the nodes, equivalent to the distributed forces applied to the element.

A summation over all the elements gives the first variation of the total potential energy of the system. If all the vector components in this expression are set equal to zero, an equation system results, the solution of which gives the displacements and rotations at each of the node points. In the assembly we notice that

$$\begin{aligned} u_1^{i+1} &= u_3^{i} \\ u_2^{i+1} &= u_4^{i} \end{aligned} \qquad (8.122)$$

where superscripts indicate element number. A sequence of the assembled stiffness matrix thus is given by

...

$$\left|\begin{array}{c|c|c|c|c|c|c|c}
k_{31}^{i-1} & k_{32}^{i-1} & k_{33}^{i-1}+k_{11}^{i} & k_{34}^{i-1}+k_{12}^{i} & k_{13}^{i} & k_{14}^{i} & & \\
k_{41}^{i-1} & k_{42}^{i-1} & k_{43}^{i-1}+k_{21}^{i} & k_{44}^{i-1}+k_{22}^{i} & k_{23}^{i} & k_{24}^{i} & & \\
 & & k_{31}^{i} & k_{32}^{i} & k_{33}^{i}+k_{11}^{i+1} & k_{34}^{i}+k_{12}^{i+1} & k_{13}^{i+1} & k_{14}^{i+1} \\
 & & k_{41}^{i} & k_{42}^{i} & k_{43}^{i}+k_{21}^{i+1} & k_{44}^{i}+k_{22}^{i+1} & k_{23}^{i+1} & k_{24}^{i+1}
\end{array}\right|$$

$$... (8.123)$$

For this particular example the procedure of assembling the stiffness matrix for the system is very simple and straightforward. For the more complicated cases, in which several elements meet at one node, it is desirable to have a systematic assembly procedure. Such methods are discussed in textbooks on the topic (see Ref. 8.24, for example). Enforcement of the geometric boundary conditions is also easy in this case; some of the degrees of freedom of the system are suppressed. These constraints are enforced by elimination of a row and the corresponding column in the k matrix, and of the corresponding element in the load vector. For more complicated situations it is easier to enforce boundary constraints by use of lagrangian multipliers (Ref. 8.28).

8.6c   Column Buckling

Once we have established that the finite element method (with conforming elements) may be regarded as a form of the Rayleigh-Ritz method, its extension to stability analysis or to nonlinear equilibrium analysis is straightforward. For illustration we now consider the buckling problem for the simply supported column under axial compression. The second variation of the total potential energy for the column is [Eq. (8.17)]

$$\delta^2 V = \int_0^L [EI(w'')^2 - P(w')^2]\, dx \qquad (8.124)$$

where P is the applied axial load.

We find that the first term in the integral is the same as the integral evaluated in Sec. 8.6b for the bending of the beam element. Equation (8.113) may be written

$$w' = [G]\{a\} \qquad (8.125)$$

where
$$[G] = [0,1,2\xi,3\xi^2] \qquad (8.126)$$

and the contribution to the second variation of the potential energy from the second term in Eq. (8.24) is

$$\delta^2 V_2{}^e = P\{u\}^T [n^e]\{u\} \qquad (8.127)$$

where
$$[n^e] = l \int_0^1 [B]^T [G]^T [G][B]\, d\xi \qquad (8.128)$$

After the matrix multiplication and integration, we find

$$[n^e] = \frac{1}{l}\begin{bmatrix} \frac{6}{5} & \frac{1}{10} & -\frac{6}{5} & \frac{1}{10} \\ \frac{1}{10} & \frac{2}{15} & -\frac{1}{10} & -\frac{1}{30} \\ -\frac{6}{5} & -\frac{1}{10} & \frac{6}{5} & -\frac{1}{10} \\ \frac{1}{10} & -\frac{1}{30} & -\frac{1}{10} & \frac{2}{15} \end{bmatrix} \qquad (8.129)$$

Assembling the matrices for the total structure, we notice that

$$u_3{}^i = u_1^{i+1} \qquad u_4{}^i = u_2^{i+1} \qquad (8.130)$$

and observe the boundary condition

$$w(\text{at } x = 0) = u_1{}^1 = 0$$

$$w_{,x}\left(\text{at } x = \frac{L}{2}\right) = u_4{}^J = 0 \qquad (8.131)$$

where J is the number of elements on half the column.

With all elements of equal length ($l = L/J$), we obtain the following matrices with $2(J+1) - 2 = 2J$ rows:

$$
k=\frac{EI}{l^{3}}
\begin{bmatrix}
4 & -6 & 2 & & & \\
-6 & 12+12 & -6+6 & -12 & 6 & \\
2 & -6+6 & 4+4 & -6 & 2 & \\
 & -12 & -6 & 12+12 & -6+6 & -12 \\
 & 6 & 2 & -6+6 & 4+4 & -6 \\
 & & & -12 & -6 & 12 \\
 & & & & \cdots &
\end{bmatrix}
\tag{8.132}
$$

$$
n=\frac{1}{l}
\begin{bmatrix}
\dfrac{2}{15} & -\dfrac{1}{10} & -\dfrac{1}{30} & & & \\[4pt]
-\dfrac{1}{10} & \dfrac{6}{5}+\dfrac{6}{5} & -\dfrac{1}{10}+\dfrac{1}{10} & -\dfrac{6}{5} & \dfrac{1}{10} & \\[4pt]
-\dfrac{1}{30} & -\dfrac{1}{10}+\dfrac{1}{10} & \dfrac{2}{15}+\dfrac{2}{15} & -\dfrac{1}{10} & -\dfrac{1}{30} & \\[4pt]
 & -\dfrac{6}{5} & -\dfrac{1}{10} & \dfrac{6}{5}+\dfrac{6}{5} & -\dfrac{1}{10}+\dfrac{1}{10} & -\dfrac{6}{5} \\[4pt]
 & \dfrac{1}{10} & -\dfrac{1}{30} & -\dfrac{1}{10}+\dfrac{1}{10} & \dfrac{2}{15}+\dfrac{2}{15} & -\dfrac{1}{10} \\[4pt]
 & & & -\dfrac{6}{5} & -\dfrac{1}{10} & \dfrac{6}{5} \\[4pt]
 & & & & \cdots &
\end{bmatrix}
\tag{8.133}
$$

With one element only (on the half column), we obtain an eigenvalue problem with two degrees of freedom. The corresponding critical load is found to be 0.75 percent too high. With two elements (four degrees of freedom), the error is only 0.049 percent.

8.6d Shells of Revolution

Finite element analysis of shells of revolution was first performed through approximation of the shell by a series of truncated cones (see, for instance, Ref. 8.29). The stiffness matrix for the cone element is determined from analytically computed influence coefficients. The element behavior is then exactly determined, and the approximation in the analysis lies entirely in the modeling of the shell as a series of truncated cones. This approach was later abandoned because the computation of the influence coefficients, in terms of Kelvin functions, becomes rather tedious. Later analyses, such as the buckling analysis in Ref. 8.30, used simple polynomial representations for the displacement components within the element. This approach leads to fewer computations, and it is more typical of finite element analysis. Inclusion of a curved-element meridian does not complicate the problem considerably. Geometrical quantities occurring in the energy expression such as the curvatures may be expressed in terms of simple polynomials. If the same-order polynomials are used to represent the geometrical properties of the element and displacement components within the element, we refer to the element as an *isoparametric element*.

In this section we discuss the derivation of such elements for bifurcation-buckling analysis. For simplicity, it is assumed that prebuckling stresses vary linearly with the applied load and that the values of the stress resultants are known. They could have been obtained, for instance, from a linear stress analysis based on finite elements. The implications of the assumption of linearity is discussed in Chap. 9. We also assume that prebuckling rotations have a negligible effect on the critical load (see Sec. 6.3).

The stiffness matrix is derived by use of the expression for the elastic strain energy. For this purpose we first define some vectors and matrices. The incremental strains and curvature changes and the incremental rotations during buckling are components of the vectors

$$\{\varepsilon\} = [\varepsilon_\varphi, \varepsilon_\theta, \gamma_{\varphi\theta}, \kappa_{\varphi\varphi}, \kappa_{\theta\theta}, 2\kappa_{\varphi\theta}]^T \qquad (8.134)$$

and

$$\{\beta\} = [\beta_\varphi, \beta_\theta, \beta_z]^T \qquad (8.135)$$

The results presented in this section were obtained by Bushnell (Ref. 8.31). The kinematic relations used in the analysis were based on those presented by Novo-zhilov (Ref. 6.13). These differ somewhat from the relations derived in Chap. 6 [Eq. (6.26)]. For example, the rotation component β_z in Eq. (8.135) was as-sumed in that derivation to be of negligible influence, and consequently it was disregarded. Other differences are due to the use of the arc length as one of the shell coordinates in Ref. 8.31. Defining the components of these vectors in terms of the displacements, we recognize that, for shells of revolution with axisym-metric loading, the incremental displacements vary harmonically with the cir-cumferential shell coordinate. Under those circumstances the kinematic relations in Ref. 8.31 are

$$\varepsilon_\varphi = u_{,s} + \frac{w}{R_1}$$

$$\varepsilon_\theta = -n\frac{v}{r} + \frac{u}{r}r_{,s} + \frac{w}{R_2}$$

$$\gamma_{\phi\theta} = v_{,s} - \frac{v}{r}r_{,s} + n\frac{u}{r}$$

$$\kappa_{\varphi\varphi} = \beta_{\varphi,s} \tag{8.136}$$

$$\kappa_{\theta\theta} = -\frac{n}{r}\beta_\theta + \frac{\beta_\varphi}{r}r_{,s}$$

$$\kappa_{\varphi\theta} = -\frac{n}{r}\beta_\varphi + \frac{\beta_\theta}{r}r_{,s} + \frac{v_{,s}}{R_2}$$

and

$$\beta_\varphi = w_{,s} - \frac{u}{R_1}$$

$$\beta_\theta = n\frac{w}{r} - \frac{v}{R_2} \tag{8.137}$$

$$\beta_z = \frac{1}{2}\left(n\frac{u}{r} - v_{,s} - \frac{v}{r}r_{,s}\right)$$

where n is the number of circumferential waves in the buckling pattern, and s is arc length along the meridian.

The coefficients in the constitutive equations for the shell wall are represented in a 6×6 symmetric matrix $[C]$. For an isotropic shell wall the nonzero coefficients in this matrix are

$$C_{11} = C_{22} = \frac{Eh}{1 - v^2}$$

$$C_{12} = C_{21} = vC_{11}$$

$$C_{33} = \frac{1 - v}{2} Eh \qquad (8.138)$$

$$C_{44} = C_{55} = D$$

$$C_{45} = C_{54} = vD$$

$$C_{66} = 2(1 - v)D$$

The prebuckling membrane stress resultants corresponding to a unit load are represented by the components of a 3×3 diagonal matrix:

$$[N_0] = \begin{bmatrix} N_\varphi & 0 & 0 \\ 0 & N_\theta & 0 \\ 0 & 0 & N_\varphi + N_\theta \end{bmatrix} \qquad (8.139)$$

The stiffness matrix of a finite element for the buckling analysis of a shell of revolution with axisymmetric loading can now be defined from the expression for the second variation of the total potential energy:

$$\tfrac{1}{2}\delta^2 V = \tfrac{1}{2}\int_A^B (\{\varepsilon\}^T[C]\{\varepsilon\} + \lambda\{\beta\}^T[N_0]\{\beta\})r \, ds \qquad (8.140)$$

Three different elements were considered in Ref. 8.31. These elements will be discussed here for illustration of some of the principles involved in finite element analysis. For the simplest possible element the degrees of freedom of the system are the node-point displacement components and the derivatives of the lateral displacement component with respect to the arc length. These degrees of freedom for each element are represented by the components of the displacement vector

$$\{u\} = [u_A, u_B, v_A, v_B, w_A, w_B, (w_{,s})_A, (w_{,s})_B]^T \qquad (8.141)$$

The displacements within the element are given in terms of a vector $\{a\}$ by the relations

$$u = a_1 + a_2 s$$

$$v = a_3 + a_4 s \qquad (8.142)$$

$$w = a_5 + a_6 s + a_7 s^2 + a_8 s^3$$

If the length of the meridian of the element is l, we have $s = 0$ at A and $s = l$ at B. Then, following the same procedure as was used in the derivation of Eq. (8.115), we find that

$$\{u\} = [A]\{a\} \qquad (8.143)$$

where

$$[A] = \begin{bmatrix} 1 & 0 & 0 & 0 & 0 & 0 & 0 & 0 \\ 1 & l & 0 & 0 & 0 & 0 & 0 & 0 \\ 0 & 0 & 1 & 0 & 0 & 0 & 0 & 0 \\ 0 & 0 & 1 & l & 0 & 0 & 0 & 0 \\ 0 & 0 & 0 & 0 & 1 & 0 & 0 & 0 \\ 0 & 0 & 0 & 0 & 1 & l & l^2 & l^3 \\ 0 & 0 & 0 & 0 & 0 & 1 & 0 & 0 \\ 0 & 0 & 0 & 0 & 0 & 1 & 2l & 3l^2 \end{bmatrix} \qquad (8.144)$$

For the inverse of $[A]$ (see Sec. 8.2), we introduce the notation

$$[B] = [A]^{-1} \qquad (8.145)$$

By use of the strain displacement relations [Eqs. (8.136) and (8.137)] and the displacement functions [Eqs. (8.142)] we can now express the components of the strain and rotation vectors in terms of the components of the vector $\{a\}$. These relations are readily written in matrix form as follows:

$$\{\varepsilon\} = [F]\{a\}$$

and

$$\{\beta\} = [G]\{a\} \qquad (8.146)$$

The elements of the matrices $[F]$ and $[G]$ are functions of the arc length s as follows:

$$[F] = \begin{bmatrix}
0 & 1 & 0 & 0 & \dfrac{1}{R_1} & \dfrac{s}{R_1} & \dfrac{s^2}{R_1} & \dfrac{s^3}{R_1} \\[2ex]
\dfrac{r_{,s}}{r} & \dfrac{sr_{,s}}{r} & -\dfrac{r}{n} & -\dfrac{ns}{r} & \dfrac{1}{R_2} & \dfrac{s}{R_2} & \dfrac{s^2}{R_2} & \dfrac{s^3}{R_2} \\[2ex]
\dfrac{n}{r} & \dfrac{ns}{r} & -\dfrac{r_{,s}}{r} & 1-\dfrac{sr_{,s}}{r} & 0 & 0 & 0 & 0 \\[2ex]
-\left(\dfrac{1}{R_1}\right)_{,s} & s\left[\dfrac{1}{R_1}-\left(\dfrac{1}{R_1}\right)_{,s}\right] & 0 & 0 & 0 & 0 & 2 & 6s \\[2ex]
\dfrac{-r_{,s}}{rR_1} & \dfrac{-sr_{,s}}{rR_1} & \dfrac{n}{rR_2} & \dfrac{ns}{rR_2} & -\left(\dfrac{n}{r}\right)^2 & \dfrac{r_{,s}}{r}-s\left(\dfrac{n}{r}\right)^2 & s\left[\dfrac{2r_{,s}}{r}-s\left(\dfrac{n}{r}\right)^2\right] & s^2\left[\dfrac{3r_{,s}}{r}-s\left(\dfrac{n}{r}\right)^2\right] \\[2ex]
\dfrac{n}{rR_1} & \dfrac{sn}{rR_1} & \dfrac{-r_{,s}}{rR_2} & \dfrac{-sr_{,s}}{rR_2} & \dfrac{nr_{,s}}{r^2} & \dfrac{n}{r}\left(\dfrac{sr_{,s}}{r}-1\right) & s\dfrac{n}{r}\left(\dfrac{sr_{,s}}{r}-2\right) & s^2\dfrac{n}{r}\left(\dfrac{sr_{,s}}{r}-3\right)
\end{bmatrix}$$

(8.147)

$$[G] = \begin{bmatrix} \dfrac{-1}{R_1} & \dfrac{-s}{R_1} & 0 & 0 & 0 & 1 & 2s & 3s^2 \\[3mm] 0 & 0 & \dfrac{-1}{R_2} & \dfrac{-s}{R_2} & \dfrac{n}{r} & \dfrac{ns}{r} & \dfrac{ns^2}{r} & \dfrac{ns^3}{r} \end{bmatrix} \qquad (8.148)$$

In view of Eqs. (8.143) and (8.145), we can express the strain and rotation vectors in terms of the degrees of freedom of the system:

$$\{\varepsilon\} = [F][B]\{u\}$$
$$[\beta] = [G][B]\{u\} \qquad (8.149)$$

We notice that, in the column buckling analysis, the vectors $\{\varepsilon\}$ and $\{\beta\}$ each contain only one element, w'' and w', respectively. Therefore the matrices corresponding to $[F]$ and $[G]$ degenerate to the vector forms $\{f\}$ and $\{g\}$ [Eqs. (8.118) and (8.126)].

The integration of the energy over the shell element will be carried out numerically by use of Gauss' method (Ref. 8.23). It is shown in Ref. 8.31 that, for the element we are concerned with here, we should use two integration points. More accurate integration would be inconsistent with the accuracy of the displacement function and would not lead to improvement of the final results. Gaussian integration over two points corresponds to setting

$$\int_0^l f(s)\, ds = \frac{l}{2}[f(s_1) + f(s_2)] \qquad (8.150)$$

where
$$s_1 = 0.211325l$$
$$s_2 = 0.788675l$$

Consequently, the matrices $[F]$ and $[G]$ must be evaluated at $s = s_1$ and at $s = s_2$. We define

$$[F_1] = [F(s_1)], \ [F_2] = [F(s_2)]$$
$$[G_1] = [G(s_1)], \ [G_2] = [G(s_2)] \qquad (8.151)$$

The functional relationship defining the shell meridian does not, in general, readily supply explicit expressions for the curvatures as functions of the arc length. The easiest way to determine the geometrical quantities appearing in $[F]$ and $[G]$ at the integration points may be by use of polynomial expressions. At the two end points and a number of internal points we can determine the arc length s, as well as the quantities r, $1/R_1$, $1/R_2$, appearing in Eqs. (8.147) and

(8.148). Based on these values, polynomial approximations for the needed quantities are readily derived.

The expression for the second variation of the potential energy of the element obtained by use of Eqs. (8.140), (8.147), (8.148), and (8.151) is

$$\delta^2 V^e = \{u\}^T([k^e] + \lambda[n^e])\{u\} \qquad (8.152)$$

where

$$[k^e] = l([B]^T[F_1]^T[C][F_1][B] + [B]^T[F_2]^T[C][F_2][B]) \qquad (8.153)$$

and

$$[n^e] = l([B]^T[G_1]^T[N_0][G_1][B] + [B]^T[G_2]^T[N_0][G_2][B]) \qquad (8.154)$$

The stiffness matrix $[K]$ and the load-geometric matrix $[N]$ for the total structure can now be assembled from $[k^e]$ and $[n^e]$. In this procedure we must observe that the assembled displacement matrix is contracted by use of the relations

$$u_{2i-1}^{n+1} = u_{2i}^n \qquad (8.155)$$

where the subscript indicates the component number in the element displacement vector, and the superscript indicates the element number.

After appropriate boundary conditions have been enforced, we obtain the critical value of the load parameter λ as the lowest eigenvalue of the equation

$$[K]\{u\} + \lambda[N]\{u\} = 0 \qquad (8.156)$$

Two higher-order elements are also discussed in Ref. 8.31. The displacement function for both these elements is

$$u = a_1 + a_2 s + a_3 s^2 + a_4 s^3$$
$$v = a_5 + a_6 s + a_7 s^2 + a_8 s^3 \qquad (8.157)$$
$$w = a_9 + a_{10} s + a_{11} s^2 + a_{12} s^3$$

Consequently, each of the two elements must have 12 degrees of freedom. In one of them the derivatives of the in-plane displacement (u,v) with respect to the arc length at the two end points constitute the four additional degrees of freedom. In the other element the in-plane displacements are defined as degrees of freedom at two points internal to the element (extra nodes). Reference 8.31 shows that, with a curved meridian of the element, one of these higher-order elements must be used in order that a rigid-body motion of the element be

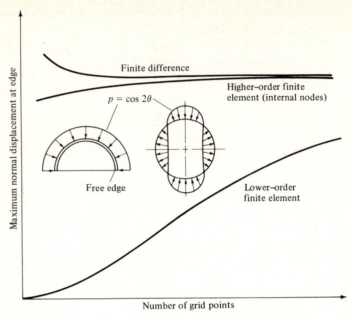

FIGURE 8.14

Normal displacement of free edge of hemisphere under uniform pressure.

free from strain energy. Also, for these elements the numerical integration should utilize three gaussian points.

Some results from Ref. 8.31 are shown in Figs. 8.14 and 8.15. The equations presented above and the computer program used to obtain the numerical results of Ref. 8.31 are valid for any shell of revolution. The results presented here, however, are for the special case of spherical shells. Figure 8.14 shows results for a hemisphere with a nonuniform external pressure and a free edge. The displacements at the edge are shown as functions of the number of grid points. Results are obtained by use of a (lowest-order) finite difference scheme as well as from finite element analyses with the lowest-order element and with the element with extra nodes. The poor performance of the lowest-order element is presumably a result of the rigid body motion problem.

Figure 8.15 shows that the low-order element is more useful for analysis of buckling of a hemisphere with clamped edge. In comparison with the finite difference solution, the convergence with grid size is somewhat faster. On the other hand, the solution for the same grid size requires more computer time.

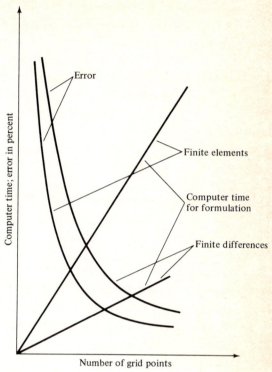

FIGURE 8.15
Computer times and rates of convergence for clamped hemisphere with uniform pressure.

REFERENCES

8.1 WILKINSON, J. H.: "The Algebraic Eigenvalue Problem," Clarendon, Oxford, 1965.

8.2 RAYLEIGH, T. W. S., LORD: "The Theory of Sound," 2d ed., vol. 2, App. A, Dover, New York, 1965.

8.3 RITZ, W.: Über eine neue Methode zur Lösung gewisser Variationsprobleme der matematischen Physik, *J. reine angew. Math.*, vol. 135, pp. 1–61, 1909.

8.4 SOKOLNIKOFF, I. S.: "Mathematical Theory of Elasticity," 2d ed., McGraw-Hill, New York, 1956.

8.5 GALERKIN, B. G.: Series Solutions of Some Problems of Equilibrium of Rods and Plates (in Russian), *Vestn. Inzh.*, vol. 1, pp. 879–908, 1915.

8.6 SEIDE, P., and V. I. WEINGARTEN: On the Buckling of Circular Cylindrical Shells under Pure Bending, *J. Appl. Mech.*, vol. 28, pp. 112–116, 1961.

8.7 BATDORF, S. B.: A Simplified Method of Elastic-Stability Analysis for Thin Cylindrical Shells, II, Modified Equilibrium Equation, *NACA* TN 1342, 1947.

8.8 WEINGARTEN, V. I.: Effects of Internal Pressure on the Buckling of Circularcylindrical Shells under Bending, *J. Aerospace Sci.*, vol. 29, pp. 804–807, July 1962.

8.9 BLOCK, D.: Buckling of Eccentrically Stiffened Orthotropic Cylinders under Pure Bending, *NASA* TN D-3351, 1966.

8.10 COLLATZ, L.: "Functional Analysis and Numerical Mathematics," Academic, New York, 1966.

8.11 FORSYTHE, G. E., and W. R. WASOW: "Finite-Difference Methods for Partial Differential Equations," Wiley, New York, 1960.

8.12 NOOR, A. K.: Improved Multilocal Finite-Difference Variant for the Bending Analysis of Arbitrary Cylindrical Shells, University of South Wales, *UNICIV Rep.* R-63, 1971.

8.13 BUSHNELL, D., and B. O. ALMROTH: Finite Difference Energy Method for Nonlinear Shell Analysis, *J. Comput. Structures*, vol. 1, pp. 361–387, 1971.

8.14 ALMROTH, B. O., F. A. BROGAN, and M. B. MARLOWE: Stability Analysis of Cylinders with Circular Cutouts, *AIAA J.*, vol. 11, pp. 1582–1584, November 1973.

8.15 JOHNSON, D. E.: A Difference-based Variational Method for Shells, *Intl. J. Solids* Structures, vol. 6, pp. 699–723, June 1970.

8.16 HENRICI, P.: "Discrete Variable Methods in Ordinary Differential Equations," 2d ed., Wiley, New York, 1964.

8.17 ARBOCZ, J., and C. D. BABCOCK: The Effect of General Imperfections on the Buckling of Cylindrical Shells, *J. Appl. Mech.*, vol. 36, pp. 28–38, 1969.

8.18 COHEN, G. A.: Computer Analysis of Asymmetric Buckling of Ring-stiffened Orthotropic Shells of Revolution, *AIAA J.*, vol. 6, pp. 141–149, January 1968.

8.19 FELIPPA, C. A., and R. W. CLOUGH: The Finite Element Method in Solid Mechanics, *ASM Symp. Numer. Solution Field Probl. Continuum Mech.*, Durham, N.C., 1968.

8.20 LANGEFORS, B.: Analysis of Elastic Structures by Matrix Transformation with Special Regard to Semi-monocoque Structures, *J. Aeronaut. Sci.*, vol. 19, pp. 451–458, July 1952.

8.21 TURNER, M. J., et al.: Stiffness and Deflection Analysis of Complex Structures, *J. Aerospace Sci.*, vol. 23, pp. 805–823, September 1956.

8.22 CLOUGH, R. W.: The Finite Element in Plane Stress Analysis, *Proc. 2d ASCE Conf. Electronic Comput.*, Pittsburgh, Pa., 1960.

8.23 ZIENKIEWICZ, O. C.: "The Finite Element Method in Engineering Science," McGraw-Hill, London, 1971.

8.24 GALLAGHER, R. H.: "Finite Element Analysis: Fundamentals," Prentice Hall, Englewood Cliffs, N.J., 1975.

8.25 LEVY, S.: Structural Analysis and Influence Coefficients for Delta Wings, *J. Aeronaut. Sci.*, vol. 20, pp. 449–454, 1953.

8.26 ARGYRIS, J. H.: "Energy Theorems and Structural Analysis," Butterworth, London, 1960 (reprinted from *Aircraft Eng.*, 1954–1955).

8.27 REISSNER, E.: On a Variational Theorem in Elasticity, *J. Math. Phys.*, vol. 29, pp. 90–95, 1950.

8.28 MELOSH, R.: Basis for Derivation of Matrices for the Direct Stiffness Method, *AIAA J.*, vol. 1, pp. 1631–1637, July 1963.

8.29 MEYER, R., and M. HARMON: Conical Segment Method for Analyzing Open-Crown Shells of Revolution for Edge Loading, *AIAA J.*, vol. 1, pp. 886–891, April 1963.

8.30 NAVARATNA, D. R., T. H. H. PIAN, and E. A. WITMER: Stability Analysis of Shells of Revolution by the Finite-Element Method, *AIAA J.*, vol. 6, pp. 355–361, February 1968.

8.31 BUSHNELL, D.: Finite Difference Energy Methods versus Finite Element Models: Two Variational Approaches in One Computer Program, *Proc. ONR Symp. Numer. Methods Struct. Mech.*, University of Illinois, Academic, New York, 1973.

PROBLEMS

8.1 Use a fourth-order power series [Eq. (8.23)] to determine the critical load of a column under axial compression.

8.2 Carry out the integrations in Eq. (8.46) to show that the equations derived are identical with those obtained by use of the Rayleigh-Ritz approach [Eq. (8.38)].

8.3 Given a set of three grid points on equal spacing, determine finite difference expressions (including error estimates) for the first- and second-order derivatives at a reference point located at one end of the interval (forward differences).

8.4 Determine the critical load for the column based on numerical integration. Use the Euler method with seven intervals (i.e., add one more row in Table 8.2).

8.5 Show that the stiffness matrix for a beam element [Eq. (8.110)] can be obtained by use of the variational approach, i.e., from the formulation given in Eq. (8.120).

8.6 Determine the critical load of a column, clamped at one end and free at the other (Prob. 2.3), by use of finite elements.

9

NONLINEAR ANALYSIS

9.1 INTRODUCTION

Throughout the preceding chapters we have emphasized that buckling is a non-linear phenomenon and that linearized analyses are used only for reasons of analytical convenience. For many applications a linearized analysis gives results that are suitable for design use. There are three situations, however, in which a nonlinear analysis is needed:

1 The linearized theory of elastic stability in Chaps. 1 to 6 is limited to the determination of a bifurcation point at which the primary equilibrium path in the load-displacement relationship is intersected by a secondary equilibrium path. As noted, equilibrium on the primary path becomes unstable at such a point, and structural behavior beyond the bifurcation point is governed by conditions on the secondary path. A linearized analysis indicates the load level at which a new deformation pattern begins to develop and determines the initial mode of this pattern, but it gives no information about the shape of the secondary path. Sometimes the behavior of a structure can be understood only if the shape of the secondary path is known. Such knowledge explains, for example, why a flat plate develops postbuckling strength but a cylindrical shell buckles abruptly and even explosively. Furthermore, the shape of the secondary path governs the sensitivity of the structure to the influence of initial imperfections. Information about the shape of the secondary path in the immediate neighborhood of the bifurcation point is given by Koiter's linearized theory for initial postbuckling behavior, but that information is limited. Consequently, a nonlinear analysis sometimes is necessary in order to determine the shape of the secondary path.

2 The prebuckling deformation of straight columns, flat plates, and circular rings is rotation-free for the loading cases treated in earlier chapters, and the primary equilibrium paths can be obtained from equations in which the nonlinear terms are deleted. The prebuckling deformation of cylindrical and general shells entails rotation of structural elements, however, and the primary equilibrium paths are nonlinear from the outset. In the applications sections of Chaps. 5 and 6, we assumed that the critical load nevertheless could be determined with sufficient accuracy when prebuckling nonlinearity was neglected. That assumption often is satisfactory. In other cases, however, prebuckling nonlinearity must be taken into account if the critical load is to be determined with accuracy. In such cases the stability equations for the determination of the bifurcation point are linear, but the variable coefficients in the stability equations are governed by the nonlinear equations of equilibrium. Then nonlinear analysis is necessary for the determination of the coefficients in the stability equations.

3 Finally, in the most general case, loss of stability occurs at a limit point rather than at a bifurcation point. In such cases the critical load must be determined through solution of the nonlinear equations of equilibrium.

Examples of all these kinds of behavior are included in an excellent survey by Hutchinson and Koiter (Ref. 9.1) and are also examined in this chapter. As a first example we consider a structure that is very simple yet displays the complexities of shell behavior.

9.2 TWO-COLUMN STRUCTURE

A structure consisting of two flexible bars is shown in Fig. 9.1.* In addition to the notations shown in the figure, let $B =$ arcsin (b/L). Also, N is the force

---

* Figures 9.1 to 9.5 also appeared in B. Almroth's paper, Nonlinear Behavior of Shells, in "Numerical Solution of Nonlinear Structural Problems," AMD, Vol. 6, American Society of Mechanical Engineers, November, 1973, edited by R. F. Hartung.

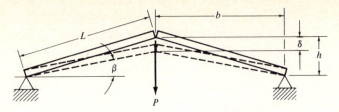

FIGURE 9.1
Two-column structure.

in each column, and L' is the length of each column under load. Then

$$\frac{N}{EA} = \frac{1}{L}(L - L') = 1 - \frac{\cos B}{\cos \beta} \qquad (9.1)$$

and
$$P = 2N \sin \beta \qquad (9.2)$$

We introduce the notation

$$\lambda = \frac{\pi^2 EI}{L^2} \qquad \rho = \frac{I}{A(\pi/L)^2} \qquad (9.3)$$

and
$$P^* = \frac{P}{\lambda} \qquad N^* = \frac{N}{\lambda} \qquad (9.4)$$

The relation between deformation δ/L and load P^* is readily obtained in the parametric representation

$$\frac{\delta}{L} = \sin B - \cos B \tan \beta$$
$$P^* = \frac{2}{\rho} \sin \beta \left(1 - \frac{\cos B}{\cos \beta}\right) \qquad (9.5)$$

By substitution of different values of β into Eqs. (9.5), we can determine the displacement parameter δ/L as a function of the load parameter P^*. The results are shown in Fig. 9.2 for two different structures, both with $h/L = 0.1$, one with $\rho = 2 \times 10^{-4}$ and one with $\rho = 4 \times 10^{-4}$. With increasing load the stiffness $\partial P^*/\partial \delta$ decreases, and at a deformation corresponding to $\delta/L = 0.04$ ($\delta/h = 0.4$), a maximum occurs in each of the two load-displacement curves. We consider the case in which load rather than the displacement is the controlled variable. Then, as the maximum is reached, the structure snaps into a position such that the load P subjects the two columns to tension.

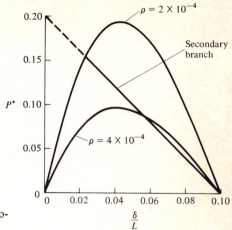

FIGURE 9.2
Load-displacement diagrams for two-column structure with $h/L = 0.1$.

We notice, however, that if $N^* > 1$, that is, $N > \pi^2 EI/L^2$, the columns will buckle. For relatively shallow structures ($h \ll L$), the shortening of the columns is moderate. In that case we can assume (see Chap. 1) that the column deforms under a constant load $N^* = 1$. A secondary equilibrium form with slightly bent bars is then represented by

$$P^* = \frac{\tan B - \delta/b}{[1 - (\tan B - \delta/b)^2]^{1/2}} \qquad (9.6)$$

This equilibrium form corresponds to $N^* = 1$ and exists only for values of δ larger than that for which buckling occurs. At the point of intersection between the primary solution [Eqs. (9.5)] and this secondary solution, the columns begin to buckle. From the graph we find that the more slender columns ($\rho = 2 \times 10^{-4}$) buckle at $P^* = 0.155$. The load cannot be increased beyond this value (the straight column represents an unstable equilibrium form). The structure snaps through, and the columns are bent in the process. For the structure with $\rho = 4 \times 10^{-4}$, the point of intersection occurs beyond the maximum in the primary load-displacement curve, indicating that the columns are straight at the inception of snap-through. The critical load of the two-column structure in that case is independent of the buckling load for the column.

In Chaps. 5 and 6 it was sometimes assumed that the effects of nonlinearities in the prebuckling range might be omitted. Such a procedure sometimes gives a close approximation even when not rigorously applicable. For the two-column structure we expect the effects of prebuckling nonlinearity to be small unless the structure is shallow ($h/L \ll 1$). If $h/L = 1$, the linear prebuckling analysis is

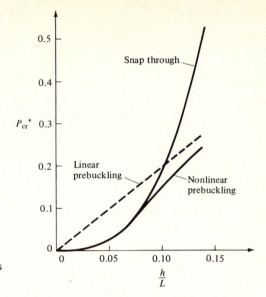

FIGURE 9.3
Critical load of two-column structure as function of h/L.

a rigorous solution (no rotations). In the linear analysis the equilibrium equations are defined by use of the geometry of the undeformed structure. Hence, with $N_{cr}^* = 1$, we find that

$$P_{cr}^* = 2 \sin B = \frac{2h}{L} \qquad (9.7)$$

Figure 9.3 shows how the critical load of the two-column structure varies with the parameter h/L. With h/L less than about 0.075, the structure will collapse at the maximum in the load-displacement curve—the columns are straight as snap-through begins. With higher values of h/L the critical load is represented by the curve marked "nonlinear prebuckling." For comparison, the critical load is also shown corresponding to the simplified analysis in which prebuckling deformations are omitted. For larger values of h/L this represents a close approximation.

We consider next a structure that has been slightly modified, as shown in Fig. 9.4, by addition of a linear spring that carries a part of the load. Figure 9.5 shows the load-displacement curves for two structures with $h/L = 0.1$ and $\rho = 2 \times 10^{-4}$. One is without a spring ($k_s = c\lambda = 0$), and the other includes a spring with the spring constant $k_s = 2.5$. The primary and secondary paths for the structure with a spring are obtained by addition of $c(\delta/L)$ to the value of P^* corresponding to $k_s = 0$.

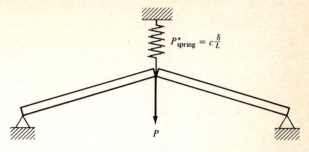

FIGURE 9.4
Two-column structure with spring.

With a spring, buckling occurs, of course, at the same value of δ/L. However, if the spring constant is sufficiently large, the slope of the curve for the secondary solution becomes positive. The increase of the load in the spring is more than sufficient to compensate for the decrease in the load carried by the columns.

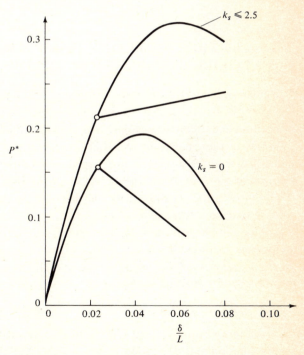

FIGURE 9.5
Load-displacement diagram for two-column structure with spring $h/L = 0.1$.

As noted in Chap. 7, the significance of this is that, for the structure without a spring (if load is the independent parameter), the bifurcation point indicates a load level at which the structure is set in motion. On the other hand, if the secondary path corresponds to stable equilibrium, as in the case with $k_s = 2.5$, the structure can take additional load beyond the bifurcation point. However, a new deformation pattern begins to develop, and the stiffness of the structure may be considerably reduced.

The two-column structure has only one degree of freedom, and numerical results are therefore readily obtained. In the nonlinear analysis of plates and shells, on the other hand, the solution will generally be based on one of the numerical procedures discussed in Chap. 8. With application of these procedures, the problem is reduced to the solution of a (generally large) nonlinear algebraic equation system. Methods of solutions of such systems that have been used for structural analysis problems are discussed in Sec. 9.3. Following that discussion, a number of examples of plate and shell stability problems are presented in which it is necessary to solve the nonlinear equations. All the different situations in which such analysis is needed (see Sec. 9.1) are illustrated.

9.3 SOLUTION METHODS FOR NONLINEAR ALGEBRAIC EQUATION SYSTEMS

The collapse load of a structure is defined by a maximum in the load-displacement curve. To find such a maximum, we usually solve the governing nonlinear differential equations for a stepwise increasing load or displacement. Analytical solutions of these equations generally cannot be obtained. All the methods of numerical analysis discussed in Chap. 8 reduce the nonlinear problem to the solution of a system of nonlinear algebraic equations at each load step. We shall briefly discuss the solution methods that most frequently have been used for such systems in structural analysis.

One simple way to find a solution is the so-called *incremental method*. Using this method, we substitute $u_0 + \Delta u$ for the unknown displacement u, where u_0 is a known solution and Δu is a small increment. A term u^n in the equation system then takes the form

$$u_0{}^n + c_1 u_0{}^{n-1} \Delta u + c_2 u_0{}^{n-2}(\Delta u)^2 + \cdots \qquad (9.8)$$

All terms containing u_0 only can be subtracted out because $u = u_0$ represents a solution to the equation system. If, in addition, it is assumed that higher-order terms in the increment Δu may be discarded, the displacement increments corresponding to a small increment in load can be obtained from a linear equation system.

The incremental method has the disadvantage that it accumulates errors, and the analyst has little control over them. The first known solution is the unloaded configuration with $u_0 = 0$. The subsequent u_0 configurations are subject to approximation. In general, the analyst must find a suitable size of the load step by trial and error. The analysis may be repeated, the load step being gradually decreased until stability in the results indicates that they are sufficiently accurate. The method is therefore usually not appealing from an economic viewpoint.

An improved, self-correcting version of the incremental method (Ref. 9.2) is more efficient. At each load step the unbalance in the nonlinear equations for the previous load step is added to the right-hand side of the equation system corresponding to the load increment. In this way accumulation of error still takes place, but it is minimized, and a bigger load step can be allowed.

Another simple method that has been applied frequently for solution of the equation systems occurring in connection with nonlinear structural analysis is the method of successive substitutions. The nonlinear terms are considered as pseudo loads that are evaluated by substitution of the values of the displacements obtained at the previous iteration. Considerable convergence difficulties are connected with the application of this method to collapse problems (the method is discussed in Ref. 9.3).

A powerful iterative method, the Newton-Raphson method, for the solution of the problem $g(x) = 0$ is defined by the recurrence formula

$$x_{n+1} = x_n - \frac{g_n}{(\partial g/\partial x)_n} \qquad (9.9)$$

where the subscript indicates iteration number. The iteration converges rapidly, provided that the initial estimate x_0 is sufficiently close. The extension to the case with several simultaneous equations is obtained by writing Eq. (9.9) in matrix form:

$$\{x_{n+1}{}^i\} = \{x_n{}^i\} - \left[\frac{\partial g^j}{\partial x^i}\right]_n^{-1} \{g^j\}_n \qquad (9.10)$$

A detailed discussion of the Newton-Raphson method and other similar approaches is presented in Ref. 9.4. Each iterative step requires the inversion or factoring of a matrix. In shell problems solved by use of finite difference or finite element methods, the number of unknowns is usually quite large and the factoring requires a substantial amount of computer time. Much of this time

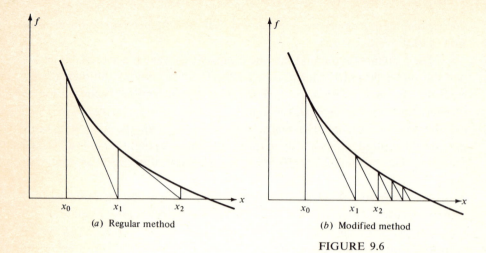

(a) Regular method   (b) Modified method

FIGURE 9.6
The Newton-Raphson method.

is avoided if a modified Newton-Raphson method is used, corresponding to the recurrence formula

$$\{x_{n+1}{}^i\} = \{x_n{}^i\} - \left[\frac{\partial g^j}{\partial x^i}\right]_0^{-1} \{g^j\}_n \qquad (9.11)$$

There is a simple geometrical interpretation of the Newton-Raphson method. In the case of only one unknown, the regular method and the modified method are illustrated in Fig. 9.6. From a point on the curve $g(x)$ that corresponds to the initial estimate, the tangent to the curve is extended to the x axis. The intersection between the x axis and the tangent gives the next approximation. In the regular Newton-Raphson method the tangent is determined again at this point. In the modified Newton-Raphson method the lines subsequently extended to the x axis all have the same slope as the tangent at the initial estimate. While the regular method converges faster, the modified method is frequently more economical, because repeated factoring of the coefficient matrix is avoided. Indeed, when a load-displacement curve is computed in a nonlinear structural analysis, it is often advantageous to retain the same factored matrix over a series of load steps. It is interesting to note that the improved incremental method discussed above is identical with a Newton-Raphson method with only one iteration per load step.

    Among other methods that have been used for solution of nonlinear shell behavior may be mentioned methods for direct minimization of the energy. These are called *minimization*, or *gradient*, methods. The most popular variants are the steepest-descent and conjugate-gradient schemes. A detailed theoretical treatment of these methods is given in Ref. 9.5.

A method called *dynamic relaxation* sometimes is used. In this method the dynamic problem is formulated. Damping is included, so that after the load is applied, the static equilibrium configuration is asymptotically approached. Interesting variations of this method and of the self-correcting method are presented in Ref. 9.6. Some comparisons between different solution methods are presented in Refs. 9.6 and 9.7.

Rapid developments in computer technology and in numerical analysis methods tend to make recommendations on the choice of method somewhat transitory. Perhaps it can be said with some confidence that, in general, one of the Newton-Raphson methods is to be preferred, but sometimes it may be worthwhile to utilize the option to save on computer time at the expense of the accuracy, which is possible with the self-correcting incremental method. For large problems with considerable nonlinearity, some form of dynamic relaxation may be advantageous.

9.4 POSTBUCKLING BEHAVIOR OF PLATES

An analysis of the postbuckling behavior of an infinitely long plate with supported edges that is subjected to a uniform shortening is presented in this section. Some results for this case were discussed in Sec. 3.8 in association with the failure loads of plates. After buckling, the central part of the plate bulges out, and an increasingly larger portion of the load is carried by the material close to the supported edges of the plate. It was suggested by von Kármán (Ref. 9.8) that, in the analysis of stresses and deformations of the semimonocoque designs used in airplanes, an effective width of the buckled skin might be included as load-carrying material at each stiffener. On a purely intuitive basis von Kármán suggested that the effective width should be the same as the width of a plate of the same thickness that would buckle at the given design load. In addition, he suggested that, for an efficient design, the stress at the design load should be relatively close to the yield strength of the material. With a Young's modulus of 10^7 psi and a yield strength of 40,000 psi, values that were typical of high-quality aluminum alloys of that time, it was found that the effective width should be around 30 times the plate thickness. As noted in Sec. 3.8, the appeal of the simplicity of the von Kármán equation is so great that it is still widely used in the engineering analysis of aerospace structures, although more accurate solutions have been made available.

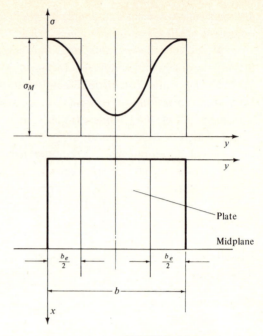

FIGURE 9.7
Stress distribution in plate.

The behavior of the plate in the postbuckling range can readily be determined through numerical solution of the governing nonlinear differential equations. From Eqs. (3.28) and (3.29) the governing equations are

$$\nabla^4 f = Eh(w_{,xy}{}^2 - w_{,xx}\, w_{,yy})$$
$$D\nabla^4 w = p + f_{,yy}\, w_{,xx} - 2f_{,xy}\, w_{,xy} + f_{,xx}\, w_{,yy} \tag{9.12}$$

Solutions of these equations were presented by several investigators around 1940; the best known are due to Marguerre (Ref. 9.9), Koiter (Ref. 9.10), and Cox (Ref. 9.11). Marguerre uses the associated potential energy expression [Eqs. (3.24) and (3.25)] and a Rayleigh-Ritz procedure in the numerical analysis. The results are given in terms of the effective width defined by von Kármán as a function of the maximum stress (see Fig. 9.7). Marguerre gives first a one-term solution in terms of the displacement function

$$w = c_1 \cos \frac{\pi x}{l} \cos \frac{\pi y}{b} \tag{9.13}$$

Here l represents the axial wavelength and b the plate width. Numerical solutions were obtained only for the case in which $l = b$. For an infinitely long plate, this would correspond to the assumption that the wavelength is the same in both directions. The results apply also to a square plate with simply supported edges. Through elimination of the wave amplitude c_1 between the two equations and after substitution of the stress [compare Eq. (3.50)]

$$\sigma_{cr} = \frac{\pi^2 E}{3(1 - v^2)} \left(\frac{h}{b}\right)^2 \qquad (9.14)$$

a solution is obtained in the form

$$\frac{b_e}{b} = \frac{1}{2}\left(\frac{1 + \sigma_{cr}}{\sigma_M}\right) \qquad (9.15)$$

where σ_M is the maximum stress (the stress in the stiffener). Although this equation gives quite close estimates in the early postbuckling range, it becomes less accurate as the buckling progresses. Marguerre therefore also presented a refined solution corresponding to the displacement function

$$w = c_1 \cos\frac{\pi x}{l} \cos\frac{\pi y}{b} - c_2 \cos\frac{3\pi x}{l}\left(\cos\frac{\pi y}{b} - \eta \cos\frac{3\pi y}{b}\right) \qquad (9.16)$$

Substitution of this expression into Eqs. (9.12) results in a nonlinear equation system with the three unknowns c_1, c_2, and η. For these equations approximate solutions are obtained.

Cox (Ref. 9.11) obtained a similar solution and found that, for plates with clamped edges, the effective width can be represented by the approximate formula

$$\frac{b_e}{b} = 0.14 + 0.86\left(\frac{\sigma_{cr}}{\sigma_M}\right)^{1/2} \qquad (9.17)$$

Koiter (Ref. 9.10) gives, for all edge conditions, the equation

$$\frac{b_e}{b} = \left[1.2 - 0.65\left(\frac{\sigma_{cr}}{\sigma_M}\right)^{2/5} + 0.45\left(\frac{\sigma_{cr}}{\sigma_M}\right)^{4/5}\right]\left(\frac{\sigma_{cr}}{\sigma_M}\right)^{2/5} \qquad (9.18)$$

Later treatments of the problem (such as Refs. 9.12 and 9.13) have included analyses with more degrees of freedom. The one-term solution in Ref. 9.9 is compared with the solution of Ref. 9.12 in Fig. 9.8.

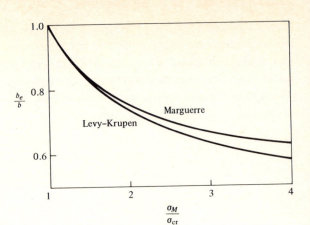

FIGURE 9.8
Effective width for buckled plates.

9.5   NONLINEAR ANALYSIS OF CYLINDRICAL SHELLS

Four nonlinear analyses of circular cylindrical shells are presented in this section: (1) axisymmetric deformation of finite-length cylinders under axial compression; (2) asymmetric bifurcation from the axisymmetric form in (1); (3) postbuckling behavior of infinitely long cylinders under axial compression, and (4) finite-length cylinders subjected to bending.

9.5a   Axisymmetric Behavior

The stability analysis of axially compressed cylindrical shells in Sec. 5.5 was carried out in terms of the simplifying assumption that a linear membrane analysis could be used for the prebuckling deformation. That solution is a rigorous one for infinitely long cylinders or for Poisson's ratio equal to zero. For these cases the prebuckling deformation is rotation-free. Actually, however, the presence of edge restraint, together with the Poisson ratio effect, causes rotations at the shell edge that result in nonlinearity in the primary path. The equation governing axisymmetric behavior of the cylindrical shell under axial compression is obtained by specialization for axial symmetry of the nonlinear equilibrium equations in Eqs. (5.8). The result is [compare Eq. (7.12)]

$$Dw^{iv} + Nw'' + \frac{Ehw}{a^2} - \frac{vN}{a} = 0 \qquad (9.19)$$

where $N = P/2\pi a$ and P is the applied compressive load. For axisymmetric deformation the lateral displacement w is a function of the axial coordinate x only. The shell behavior is nonlinear because of the presence of the term Nw'', but for any fixed value of the axial load N, the corresponding displacement configuration can be found through solution of a linear differential equation with constant coefficients. The solution to this equation is of the form

$$w = 2v\gamma R\overline{N}\left(1 + A_1 \sin\frac{a_1 x}{a} \sinh\frac{a_2 x}{a} + A_2 \cos\frac{a_1 x}{a} \cosh\frac{a_2 x}{a}\right) \qquad (9.20)$$

where
$$\gamma = \frac{h/a}{[12(1 - v^2)]^{1/2}} \qquad a_1 = k1(+ \overline{N})^{1/2}$$

$$\overline{N} = \frac{N}{2\gamma Eh} \qquad a_2 = k(1 - \overline{N})^{1/2} \qquad (9.21)$$

$$k = (2\gamma)^{-1/2}$$

The integration constants A_1 and A_2 are obtained through substitution into the boundary conditions.

The solution of this essentially nonlinear problem was first obtained by Föppl in 1926 (Ref. 9.14). For a cylinder with simply supported edges the results are discussed in detail in Ref. 9.15, sec. 8.2.5.1. The disturbance due to the restraint at the cylinder edge spreads over a larger part of the cylinder as the axial load increases. When $N = 2\gamma Eh$, the displacement pattern becomes purely sinusoidal ($a_2 = 0$) and the lateral displacements grow without bounds.

9.5*b* Bifurcation from a Nonlinear Prebuckling State

The value of the axial load at which the axisymmetric displacements for a cylinder under axial load become unlimited (Sec. 9.5*a*) is identical with the critical load obtained from a bifurcation analysis with linear prebuckling behavior [Eq. (5.52)]. However, nonlinear terms alter the prebuckling solution, and bifurcation buckling into asymmetric modes may occur before the axisymmetric collapse load is reached. The possible buckling modes are deformation patterns in which the lateral displacement varies harmonically in the circumferential direction; i.e., they are of the form

$$w = w(x) \cos n\theta \qquad n = 1, 2, 3, \ldots \qquad (9.22)$$

The value of $n = 1$ corresponds to buckling of the cylinder as a column with undistorted cross section and represents the critical mode for very long cylinders. The Donnell equations do not yield accurate solutions for that case, as noted in Sec. 5.5*b*.

The linear stability equations are readily obtained from the nonlinear equilibrium equations by use of the perturbation technique used in earlier chapters. After substitution of Eq. (9.22) for the incremental displacements, the stability equations are separated with respect to the space variables, and the stability problem can be solved for any value of the wavenumber n. Then in terms of a stress function f [see Eqs. (5.13)], the stability equations can be shown to be

$$D\left[w^{iv} - 2\left(\frac{n}{a}\right)^2 w'' + \left(\frac{n}{a}\right)^4 w\right] + \frac{f''}{a}$$

$$+ Nw'' + \left(\frac{n}{a}\right)^2 (w_0)''f - \left(\frac{n}{a}\right)^2 \left(\frac{DN - Ehw_0}{a}\right) w = 0 \qquad (9.23)$$

$$\frac{1}{Eh}\left[f^{iv} - 2\left(\frac{n}{a}\right)^2 f'' + \left(\frac{n}{a}\right)^4 f\right] - \frac{w''}{a} - \left(\frac{n}{a}\right)^2 (w_0)''w = 0$$

where w = incremental lateral displacement
f = increment in stress function
w_0 = prebuckling displacements computed from nonlinear analysis in Sec. 9.5a

This problem was first solved by Stein (Ref. 9.16) by use of an energy-based finite difference approach. Fischer (Ref. 9.17) presented a similar analysis but with different edge conditions for the in-plane displacements. In an extension of Stein's and Fischer's work, an analysis in Ref. 9.18 uses finite difference approximations to the stability equations [Eqs. (9.23)] in a study of the influence of different boundary conditions. In all cases considered, the shell edge was supported in the radial direction; that is, $w = 0$. Eight different sets of boundary conditions were considered, as shown in Table 9.1.

Table 9.1   BOUNDARY CONDITIONS

Case			
S1	$w'' = 0$	$u = 0$	$v = 0$
S2	$w'' = 0$	$N_x = 0$	$v = 0$
S3	$w'' = 0$	$u = 0$	$N_{xy} = 0$
S4	$w'' = 0$	$N_x = 0$	$N_{xy} = 0$
C1	$w' = 0$	$u = 0$	$v = 0$
C2	$w' = 0$	$N_x = 0$	$v = 0$
C3	$w' = 0$	$u = 0$	$N_{xy} = 0$
C4	$w' = 0$	$N_x = 0$	$N_{xy} = 0$

For any given load level, the prebuckling displacements w_0 can be computed at each of the grid points. The derivatives with respect to the axial coordinate in Eqs. (9.23) are replaced by their finite difference equivalents. The conditions of stability and compatibility at each of the grid points and the specified boundary conditions form a homogeneous system of linear equations in which the unknowns are the discrete values of the lateral displacements and the stress function. The discrete values of the prebuckling displacement w_0 appear as coefficients in these equations. The critical load is represented by the load level at which the determinant of the coefficient matrix corresponding to Eqs. (9.23) vanishes.

All the applications in Chaps. 1 to 6 were based on the assumption that the bifurcation load could be determined with sufficient accuracy by use of a linear prebuckling analysis. In such cases the analysis of stability is relatively simple. All prebuckling displacement components can be expressed as a product of a load factor λ and their values for a unit load. There is no need in that case to evaluate the determinant of the coefficient matrix for a series of increasing values of the load parameter. The critical value of λ can be directly determined as the eigenvalue λ_{cr} defined by an equation of the form

$$\det([A] + \lambda[B]) = 0 \qquad (9.24)$$

If the prebuckling displacements w_0 are not linearly dependent on the load parameter λ, a different procedure is needed. We may solve the prebuckling equations for stepwise increasing load, also computing for each step the stability determinant. The critical load is found when the first zero value of this determinant is bracketed with sufficient accuracy. This procedure is not entirely satisfactory in computer analysis, and more sophisticated methods have been developed (Ref. 9.19). These are based on linearization of the prebuckling behavior in the neighborhood of an estimated critical value. The critical load is obtained through consideration of a series of eigenvalue problems, with the solution converging toward the critical value of the load. It is noticed, however, that squares of prebuckling rotations appear in the stability equations [see Eqs. (6.18) and (6.19)]. If the stability analysis is based on these equations, the eigenvalue problems are of the form

$$\det([A] + \lambda[B] + \lambda^2[C]) = 0 \qquad (9.25)$$

The solution of this problem is complicated by the fact that complex eigenvalues exist. We may notice that this form is obtained also for the case in which a linear bending analysis is assumed for the prebuckling behavior, if the prebuckling rotations are retained in the formulation.

Table 9.2   CONVERGENCE WITH
GRID SIZE

No. of grid points	$\overline{N}_{cr}$
100	0.5153
200	0.5118
300	0.5109
400	0.5106
500	0.5104
600	0.5103
700	0.5102
800	0.5102

The numerical results obtained in Ref. 9.18 were based on computation of a sequence of values of the determinant. The critical load is obtained from a plot of the determinant as function of the load. The bandwidth of the matrix (see Chap. 8) is small, and for such systems economic methods of solution are available. A large number of grid points can be included at a moderate cost in computer time. Table 9.2 shows how the value of the critical load converges with decreasing mesh size for a cylinder with $a/h = 100$, $L/a = 3.2$, and boundary conditions corresponding to case S3.

The critical loads are shown in Table 9.3 for the same cylinder with the eight different sets of boundary conditions listed in Table 9.1.

It may be noticed that very low buckling loads are obtained in cases S3 and S4. This is caused not by the use of a nonlinear prebuckling analysis, but rather by the weak boundary support ($N_{xy} = 0$). For these cases similar results are obtained even if a linear membrane solution is used for prebuckling displacements (Ref. 9.20). Such boundary conditions are not likely to be encoun-

Table 9.3   CRITICAL LOADS FOR CYLINDERS WITH DIFFERENT BOUNDARY CONDITIONS

Case	$\overline{N}_{cr}$	Critical wavenumber
S1	0.868	9
S2	0.844	8
S3	0.510	2
S4	0.510	2
C1	0.928	9
C2	0.911	8
C3	0.928	9
C4	0.909	8

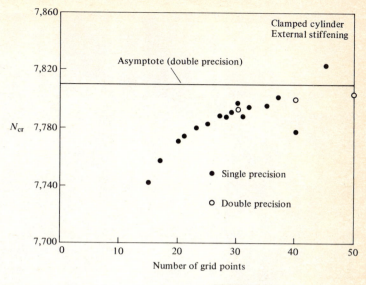

FIGURE 9.9
Buckling of stiffened cylinder.

tered in practical applications. For the other cases the analysis presented here, in comparison with the critical load for infinitely long cylinders, results in a reduction of the critical load of as much as 20 percent for simply supported ($w'' = 0$) shells and 10 percent for clamped ($w' = 0$) shells.

It is noticed from Table 9.2 that more accurate results are obtained with diminishing mesh size. However, the roundoff errors become more important with decreasing mesh size, primarily because the equation system becomes more ill-conditioned. The loss of accuracy due to the occurrence of roundoff errors is demonstrated by the stability analysis for an axially compressed cylinder with axial stiffening that was considered in Ref. 9.21. It is seen from Fig. 9.9 (Ref. 9.21, fig. 2) that, with an increasing number of points, the corresponding load appears to approach a certain limit; further increases result in somewhat erratic behavior. A much smaller mesh size can be used if the roundoff errors are minimized through introduction of double-precision arithmetic in the computer program (or by use of a computer with a larger word length).

9.5c   Postbuckling Behavior of Cylindrical Shells

Early attempts by Flügge (Ref. 9.22), Lundquist (Ref. 9.23), and Donnell (Ref. 9.24) failed to establish a correlation between theoretical and experimental

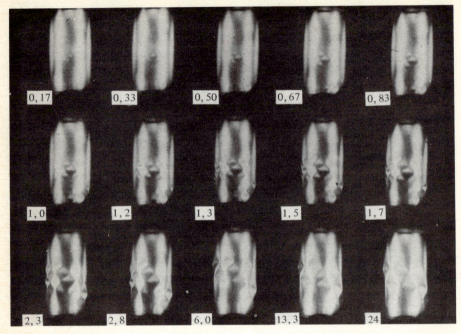

FIGURE 9.10
Buckling of an axially loaded cylinder.

buckling loads for axially compressed cylindrical shells. As discussed in Sec. 7.1, experimental results were scattered well below the analytical results. Also, the analysis indicates that the buckles are either axisymmetric or of the so-called checkerboard type $\{w = w_0 \, [\sin \, (mx\pi)/l] \, \sin \, n\theta\}$. In contrast, buckles observed in tests are diamond-shaped with primarily inward displacements. This discrepancy was noticed by the early investigators, but only later was it possible, by use of high-speed photography, to establish what actually happened in the transition to the buckled state. Figure 9.10 shows a few frames from a high-speed movie that was made in connection with a series of experiments reported in Ref. 9.25. These pictures show clearly that the final shape of the buckles does not resemble the buckle mode at incipient buckling that is the object of study in bifurcation buckling analysis.

An extensive summary of the efforts to explain these difficulties was presented by Hoff in Ref. 9.26. As a part of these efforts, nonlinear analyses in the postcritical range were first performed by Donnell (Ref. 9.24) and by von Kármán and Tsien (Ref. 9.27). The well-known postbuckling analysis by von Kármán and Tsien was based on the Rayleigh-Ritz approach and the nonlinear equilibrium equations [Eqs. (5.8)], or, rather, their potential energy counterpart in Eqs. (5.9) to (5.11). The same equations were used by Kempner (Ref. 9.28) in an analysis that removed some of the inadequacies in the earlier attempts. Kempner's analysis was based on too few degrees of freedom in the displacement function, however, and the results were still inaccurate.

A recent analysis by Hoff et al. (Ref. 9.29) included as many as 16 degrees of freedom, and it was concluded that even more terms would be needed if the secondary branch were to be accurately determined.

The analysis of the postbuckling behavior of axially loaded cylinders has contributed to a better understanding of the reasons for the scatter in test results and of the discrepancy between experimental and theoretical results (see discussion in Sec. 7.1). However, the conclusions are of a qualitative rather than quantitative character. The early postbuckling analyses indicated that a minimum occurs in the secondary path after the first sharp drop at the bifurcation point. This minimum was found to be somewhat less than one-third of the bifurcation buckling load, and it agreed reasonably well with the lower bound of the test results available at the time. It seemed reasonable then to assume that this minimum was a lower bound to the critical load, and that whenever the applied axial load exceeded this value, the shell might pass from the unbuckled state to a configuration on the secondary branch. The minimum postbuckling load unfortunately was termed the lower buckling load, and this expression is still sometimes used. However, it is misleading, because it has been shown that the lowest load in the postbuckling range for some shells, including the axially loaded cylinder, is far too small to serve as a guide in design. In Ref. 9.29 it is conjectured that, based on the Donnell equations and with sufficiently many degrees of freedom in the analysis, the secondary path may asymptotically approach zero. It has been shown also that, with very carefully manufactured cylinders, test results can be obtained that are quite close to the bifurcation buckling load (Ref. 9.30).

Without knowledge of the geometrical imperfections, we cannot determine a reasonably accurate value of the critical load, and empirical methods are generally used in design of cylindrical shells, as noted in Chaps. 5 and 7.

9.5d   Bending of Cylinders

In the discussion of stability of cylinders in bending in Chap. 8 it was tacitly assumed that a bifurcation analysis with linear prebuckling behavior is applicable. Such an analysis is a close approximation if the cylinder is relatively short

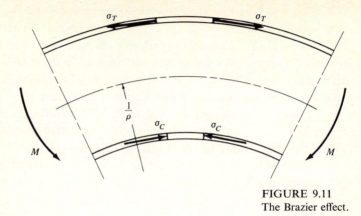

FIGURE 9.11
The Brazier effect.

and the shell edges are held circular. However, the prebuckling configuration is obviously not rotation-free and a bifurcation analysis based on linear prebuckling behavior is not a rigorous solution.

The most important effect of the nonlinear terms is an ovalization of the cross section. This problem was first studied by Brazier in 1927 (Ref. 9.31), and the ovalization of the shell is generally referred to as the *Brazier effect*. If a long cylindrical shell is subjected to a bending moment, it will assume a curvature as shown in Fig. 9.11. On account of this curvature, the longitudinal tension and compression stresses will have components directed toward the midplane of the tube. The effect of these components is to flatten the tube, and with this flattening, the resistance to the bending moment decreases. A diagram showing applied bending moment vs. radius of curvature must have a decreasing slope, as illustrated in Fig. 9.12. A maximum occurs in this diagram, and collapse of the shell results. Brazier's analysis is restricted to infinitely long shells, with the curvature ρ, applied to the cylinder axis, independent of the axial coordinate. With the displacement of the centerline of the cylinder equal to zero, the linear bending analysis for the shell results in the displacement pattern

$$w_0 = -v \frac{\rho}{2} a^2 \cos \theta$$

$$v_0 = -v \frac{\rho}{2} a^2 \sin \theta$$

(9.26)

where ρ is the curvature of the cylinder axis caused by a constant bending moment and a is the cylinder radius.

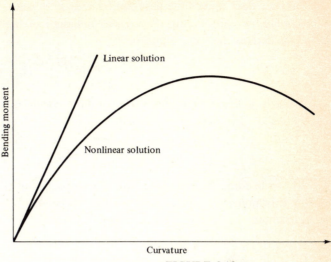

FIGURE 9.12
Equilibrium path for bent cylinder.

Brazier assumes that an incremental displacement field v_1, w_1 due to the nonlinear terms is inextensional, i.e., that (compare Sec. 4.8)

$$v_{1,\theta} + w_1 = 0 \qquad (9.27)$$

For a thin-walled shell the work due to shell-wall bending in the axial direction may be omitted in comparison with the membrane strain energy. The axial strain is

$$\varepsilon_x = \rho[(a + w) \cos \theta - v \sin \theta] \qquad (9.28)$$

and for minimum energy the circumferential stress is zero, i.e.,

$$\varepsilon_y = v\varepsilon_x \qquad (9.29)$$

For nonshallow cylinders the change of curvature in the circumferential direction is [compare Eqs. (5.28)]

$$\kappa_{\theta\theta} = \frac{1}{a} v_{,\theta} - \frac{1}{a^2} w_{,\theta\theta} \qquad (9.30)$$

and hence the total strain energy is

$$U = \frac{E}{2} \int_0^{2\pi} \left\{ \rho^2 ah[(a + w) \cos \theta - v \sin \theta]^2 + \frac{D}{a^3} (-w_{,\theta\theta} + av_{,\theta})^2 \right\} d\theta \qquad (9.31)$$

Brazier next makes the assumption that v, $w \ll a$. After use of the in-extensionality condition [Eq. (9.27)] and omission of higher-order terms in v and w, he derives the Euler equation

$$v^{vi} + 2v^{iv} + v'' = \frac{18\rho^2 a^5}{h^2}(1 - v^2)\sin 2\theta \qquad (9.32)$$

This is a linear equation with constant coefficients, and a solution is readily found. The maximum moment is

$$M_{cr} = \frac{2\sqrt{2}}{9}\frac{E\pi ah^2}{(1 - v^2)^{1/2}} \qquad (9.33)$$

If the bending stress is based on the properties of the undeformed cross section, this moment would correspond to a maximum axial stress (for $v = 0.3$) equal to about

$$\sigma_{cr} = 0.33E\frac{h}{a} \qquad (9.34)$$

This value is well below the critical bending stress computed from bifurcation theory. According to Table 8.1, this stress is insignificantly above the critical stress in pure compression [Eq. (5.53)]:

$$\sigma_{cr} = 0.605E\frac{h}{a} \qquad (9.35)$$

However, it appears from experiments on bending of thin shells that the shells fail through buckling in a wavy pattern rather than through collapse caused by the flattening of the cross section. There are two reasons for this. First, in tests, the edges are restrained from deformation, and the tested shells are generally too short for the Brazier effect to be important. The other reason is that, for buckling in the wavy mode, the shell is sensitive to imperfections, and the high stresses indicated by Eq. (9.35) are never reached in practical tests.

As yet we have not considered the possibility that bifurcation into a wavy pattern occurs at a reduced load after considerable flattening has taken place. The flattening of the cross section will result in relatively larger stresses for a fixed moment, and the stress at which bifurcation takes place is lower, due to an increase in the local radius at the point of maximum stress. Thus bifurcation from a nonlinear prebuckling state should be considered. This problem was first considered by Aksel'rad (Ref. 9.32). His somewhat approximate analysis indicates that buckling would occur just before the moment peaks, i.e., at

$$\sigma_{cr} = 0.295E\frac{h}{a} \qquad (9.36)$$

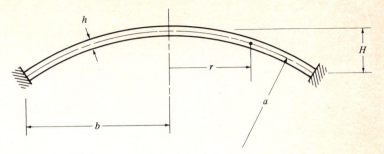

FIGURE 9.13
Spherical-cap geometry.

In view of imperfection sensitivity, we may expect a critical load for very long shells that is somewhat less than the value indicated by Eq. (9.36).

For cylindrical shells with an initial curvature, the maximum moment, in the direction of increasing curvature, occurs at a much lower axial stress. In such a case, bifurcation is not likely to occur before the maximum moment is reached. The bending properties of pipe bends have been studied by many investigators (for example, Ref. 9.33). An extensive study of critical bending loads for long cylinders was presented in Ref. 9.34. This study was based on two-dimensional finite difference approximations. The effects of internal or external pressure were included.

9.6 SPHERICAL CAPS

For a complete spherical shell under uniform external pressure, the primary path is linear, and bifurcation analysis yields the solution given in Eq. (6.52):

$$p_{cr} = \frac{2E}{[3(1 - v^2)]^{1/2}} \left(\frac{h}{R}\right)^2 \qquad (9.37)$$

For a spherical cap (Fig. 9.13), on the other hand, the rotation-free solution with a uniform radial displacement field generally does not satisfy the edge conditions. If the shell is clamped at the edge, the displacement pattern under uniform external pressure resembles the pattern shown in Fig. 9.14. As in the case of the two-column structure (Sec. 9.2), failure can occur through axisymmetric collapse or through bifurcation buckling from a nonlinear primary path. Figure 9.15a and b demonstrates the different possibilities in shell behavior.

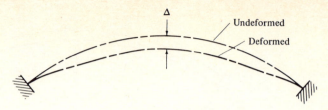

FIGURE 9.14
Spherical-cap deformation.

For very shallow shells there is no maximum in the load-displacement curve (Fig. 9.15a). The shell passes from the undeformed shape to a configuration with inverted curvature at the center without any snap-through. Despite the nonlinear behavior, there is no load level that can be characterized as a buckling load. For slightly deeper caps the load-displacement curve exhibits a maximum (Fig. 9.15b), and axisymmetric snap-through is the mode of failure. If the cap is even deeper, bifurcation from the nonlinear branch into an asymmetrical mode takes place before the maximum is reached (Fig. 9.15c).

An accurate solution of the nonlinear equation for axisymmetric behavior of spherical caps was first obtained by Budiansky (Ref. 9.35). This analysis was based on the equations

$$(x\theta_{,x})_{,x} - \frac{\theta}{x} + x\Phi = -2px^2 + \theta\Phi$$

$$(x\Phi_{,x})_{,x} - \frac{\Phi}{x} - x\theta = -\tfrac{1}{2}\theta^2$$

(9.38)

where p is the applied pressure normalized with respect to the critical pressure for the complete sphere [Eq. (9.37)], and

$$\Phi = \frac{12(1-v^2)b}{\lambda Eh^3}\psi$$

$$\lambda^4 = 48(1-v^2)\left(\frac{H}{h}\right)^2$$

$$\theta = \frac{\lambda b}{2H}\beta$$

$$x = \frac{\lambda r}{b}$$

(9.39)

Here β is rotation of the shell element, ψ is a stress function, and other quantities appearing in Eqs. (9.38) and (9.39) are as defined in Fig. 9.13.

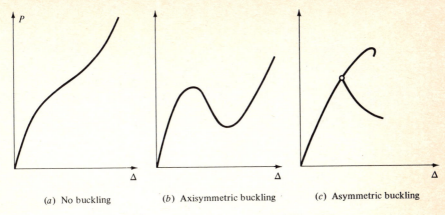

(a) No buckling   (b) Axisymmetric buckling   (c) Asymmetric buckling

FIGURE 9.15
Equilibrium paths for spherical caps.

Budiansky defined two integral equations equivalent to the equilibrium and compatibility equations, together with appropriate boundary conditions. He obtained solutions by use of an iterative procedure and a digital computer. Somewhat later Weinitschke (Ref. 9.36), using a power-series approach, solved the same differential equations, and obtained results that were in close agreement with those of Ref. 9.35.

The possibility of bifurcation from the nonlinear primary branch was first explored by Weinitschke in Ref. 9.37. A more accurate solution to this problem was obtained by Huang (Ref. 9.38). The approach is very similar to that used for cylindrical shells in Sec. 9.5b and will not be discussed in detail. Huang's results are shown in Fig. 9.16. Due to the presence of geometric imperfections, most test results are well below the theoretical solution. However, some test data obtained by use of very accurately manufactured shells (Ref. 9.39) are in reasonable agreement with the theoretical results. These test data are shown in Fig. 9.16.

Fitch and Budiansky (Ref. 9.40) obtained solutions for the case in which a clamped spherical cap is uniformly loaded over a circular region centered at the apex, but not extending over the entire cap. They found that, as the area of the loaded region increases, the deformation mode corresponding to loss of stability changes from asymmetric bifurcation to axisymmetric snap-through, and then back to asymmetric bifurcation. It is interesting to notice that, with a small loading area, bifurcation is connected with stable equilibrium on the secondary path, but for a relatively large loading area, this is not the case.

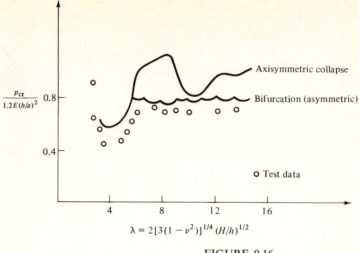

FIGURE 9.16
Critical loads for spherical caps.

9.7   SHELLS WITH ELLIPTIC CROSS SECTION

If a shell of revolution is subjected to an axisymmetric load, it can fail either axisymmetrically through the passing of a limit point or by bifurcation into a buckling mode in which the displacements vary sinusoidally in the circumferential direction. For shells of a more general shape, collapse at a limit point generally can be expected. However, if there are planes of symmetry in the displacement pattern, we must consider the possibility of buckling into patterns that are antisymmetric with respect to such planes. In the absence of axial symmetry the displacements are functions of two spatial coordinates, and the nonlinear analysis becomes quite expensive in terms of required computer time. For this reason a bifurcation analysis based on a linearized primary branch may sometimes be used instead of the nonlinear analysis as a reasonable approximation. It is often possible to judge a priori whether a bifurcation analysis will give satisfactory results. The examples discussed in the following paragraphs may shed some light on this problem. A more detailed discussion is presented in Ref. 9.41.

A bifurcation analysis of oval cylinders under axial compression was briefly discussed in Sec. 7.3c. A paper by Kempner and Chen (Ref. 9.42) also contains a nonlinear analysis of the postbuckling behavior based on the Rayleigh-Ritz method. The analysis is quite similar to the postbuckling analysis of circular cylinders (Sec. 9.5c). The shells considered were not elliptic, but were charac-

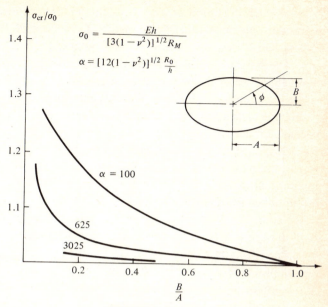

FIGURE 9.17
Critical loads for elliptic cylinders (Ref. 9.46).

terized by a circumferential radius of curvature given as a function of the arc length

$$R_s = \frac{R_0}{1 - \xi(2s/R_0)} \qquad (9.40)$$

where $2\pi R_0$ is the length of the perimeter, s is the circumferential coordinate (arc length), and ξ is an eccentricity parameter. The essence of the results of Ref. 9.42 is that the load increases with increasing shortening after the first drop at the bifurcation point, so that shells with sufficient eccentricity may, after buckling, carry loads that are larger than the bifurcation-point load.

The first analysis of shells with an elliptic cross section was presented in Ref. 9.43. By use of the assumption that a linear membrane solution ($\sigma_x \equiv -\sigma_0$, $\sigma_y \equiv 0$) may be used to describe the prebuckling behavior, the bifurcation analysis is formulated in terms of a linear eigenvalue problem. Some of the results are shown in Fig. 9.17. Here R_m is the largest radius of circumferential curvature; i.e., if $A > B$, $R_m = A^2/B$ (Fig. 9.17). The critical axial stress for the elliptic cylinder in Fig. 9.17 is normalized with respect to the critical stress for an axially loaded circular cylinder with radius R_m. It may be seen that, for a thin elliptic cylinder, such definition of an equivalent circular cylinder ($a = R_m$) leads to an acceptable approximation. For instance, if $12(1 - v^2)R_0/h = 625$, the error will be less than 10 percent for any cylinder with $B/A > 0.1$.

The bifurcation analysis for the elliptic shell is not a rigorous solution, although, as we shall see below, it represents a reasonably close approximation to the buckling load. Lateral displacement due to the application of an axial load is not uniform in the circumferential direction. It seems that the only displacement modes orthogonal to the displacements on the primary path are those that are antisymmetric with respect to the two symmetry planes. Consequently, a rigorous analysis should consider both collapse in a symmetric mode at a limit point and bifurcation into an antisymmetric mode from a symmetric configuration obtained from nonlinear analysis. A solution to the nonlinear problem was presented in Ref. 9.44. Bifurcation buckling was not considered, but instead, a small antisymmetric initial imperfection in the geometry was included. In the presence of such imperfections, antisymmetric failure will also occur through the passing of a limit point rather than through bifurcation, and will be determined by the nonlinear analysis.

The solution in Ref. 9.44 is based on a finite difference energy approach (see Chap. 8). The resulting nonlinear algebraic equations are solved by use of the modified Newton-Raphson method (Sec. 9.3). Numerical results were presented for an elliptic cylinder with a length of 1.0 in., a thickness of 0.0144 in., and semiaxes of 1.75 and 1.0 in. Young's modulus was 10^7 psi, and Poisson's ratio was 0.3. The cylinder was subjected to a uniform shortening, with the edges free to rotate but restrained from moving in the radial and circumferential directions.

Since the buckling pattern was expected to be confined to the areas of least curvature, it appeared that antisymmetric behavior with respect to the plane through $\varphi = 0$ (Fig. 9.17) need not be considered. Hence the analysis was restricted to a 180° panel with symmetry conditions enforced at $\varphi = 0, \pi$. A uniform finite difference grid was chosen with 11 points in the axial and 29 points in the circumferential direction. Results obtained with finer grids indicated that use of the chosen grid led to accurate computations of the collapse load.

Due to the symmetry of the prebuckling deformation about the plane at midlength and about the normal plane through $\varphi = \pi/2$, it was necessary to excite nonsymmetric deformations by the use of small antisymmetric imperfections. Despite the presence of these imperfections, at collapse a deformation pattern developed that was symmetric about both of these planes. Therefore the continued analysis was restricted to panels covering half of the cylinder length and one quarter of the circumference. As the load is gradually increased, a very

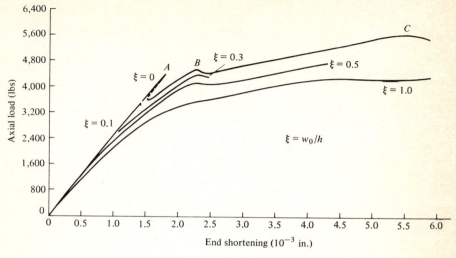

FIGURE 9.18
Equilibrium paths for elliptic cylinders.

sharp maximum is reached in the load-displacement curve, as shown in Fig. 9.18 (Ref. 9.44). Beyond this, maximum convergence cannot be obtained; hence the postbuckling curve cannot be directly determined.

The displacement mode that developed at collapse of a perfect shell was used as a guide in the choice of a suitable initial imperfection mode for the subsequent study of imperfect shells. It was assumed that

$$w = -w_0 \sin \frac{\pi x}{L} \cos 6\theta \qquad (9.41)$$

Load-displacement curves were computed for several different values of the imperfection amplitude w_0. From Fig. 9.18 it can be seen that, for a sufficiently large imperfection amplitude, the first sharp maximum does not exist. The curve is smooth, and it is possible to find equilibrium configurations in the postbuckling range. After such configurations have been found, they can be used as starting values for an analysis in which the imperfection amplitude is gradually decreased until a point is found on the postbuckling curve for perfect shells. After such a point is found, it is easy to establish postbuckling load-displacement curves for perfect shells (Fig. 9.18).

It was found that the first sharp maximum point is reasonably close (within 10 percent) to the critical load given by bifurcation theory. After the first

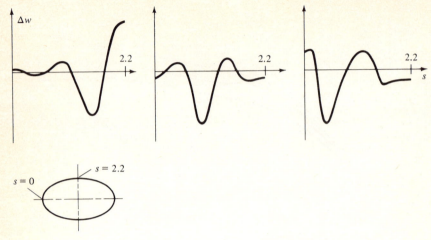

FIGURE 9.19
Collapse modes for elliptic cylinder.

maximum, the postbuckling curve exhibits two additional limit points (at *B* and *C* in Fig. 9.18) that correspond to secondary buckling.

In the neighborhood of a limit point, the developing collapse or buckle mode can be obtained as the difference between displacements for two neighboring solutions. Such collapse modes corresponding to each of the three points of maximum are shown in Fig. 9.19 (Ref. 9.44). It can be seen that the point of maximum deflection in these patterns moves toward the point of maximum curvature as the shortening increases. While the primary buckling load is rather sensitive to imperfections, it appears that the second maximum is not imperfection-sensitive. The results are similar to those obtained by Kempner and Chen for oval shells.

Elliptic or oval cylinders under other loading conditions have been considered in Refs. 9.45 to 9.47.

9.8 SHELLS WITH CUTOUTS

According to Ref. 9.41, one of the cases for which the bifurcation analysis yields a relatively poor estimate of the collapse load is a cylindrical shell with a cutout. Figure 9.20 (Ref. 9.41) shows how the displacement at the cutout edge (point *A* in the figure) varies with the applied axial load for a cylinder with two diametrically opposite rectangular cutouts. The critical load according to bifurcation

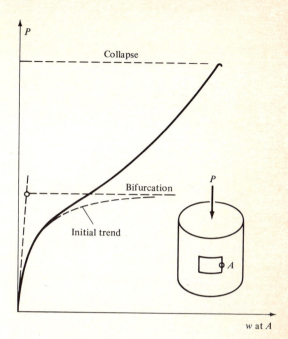

FIGURE 9.20
Equilibrium paths for cylinder with two cutouts.

analysis also is indicated in the figure. The initial trend of the lateral displacement at the cutout edge appears to be toward the horizontal asymptote, defined
by the bifurcation-point load. However, as the stresses are redistributed away
from the cutout edge, the slope of the load-displacement curve increases again,
and collapse does not occur until a load level is reached that is more than twice
the load indicated by bifurcation theory.

Numerical results from a computer program based on the finite difference
energy method are compared with experimental results in Ref. 9.48. The test
cylinders were machined from 6061-T6 aluminum tubes to the dimensions shown
in Fig. 9.21. Two diametrically opposite cutouts were made in each of the
cylinders. The width of the cutout and the shell thickness were varied in the test
series. Tests were made with cutouts covering 30° and 45° on cylinders with
either 0.014 or 0.009 in. thickness. For the thicker cylinders ($h = 0.014$ in.) and
with unreinforced cutouts, a relatively coarse grid can be used. It was found that
the use of the grid shown in Fig. 9.22 yields accurate values for the collapse load.

One of the test specimens with 45° cutouts and with $h = 0.014$ in. was
equipped with a large number of strain gages. Consequently, it was possible to

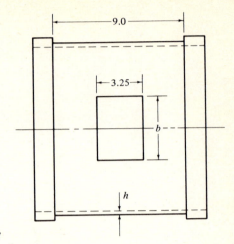

FIGURE 9.21
Test-specimen geometry

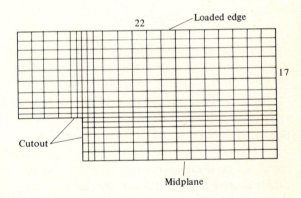

FIGURE 9.22
Finite difference grid for cylinder with cutout.

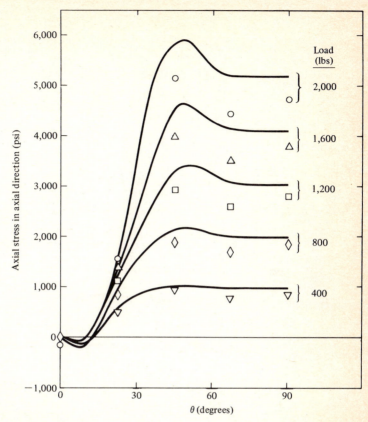

FIGURE 9.23
Axial stress 0.30 in. from end ring. (*Reproduced from Ref. 9.48, figs. 8 and 9, with the permission of Microform International Marketing Corporation, exclusive copyright licensee of Pergamon Press journal back files.*)

make comparisons between measured and computed stresses. Such comparisons are shown in Figs. 9.23 and 9.24 (Ref. 9.48). The solid lines represent computed stresses, and the points are the stress values determined by the strain gages. The deviations between measured and computed stresses is nowhere more than can be explained by variations in thickness on the test specimens. Measured and computed collapse loads for the cylinders with unreinforced cutouts are shown in Table 9.5.

The influence of reinforcements around the cutout edges is also studied in Ref. 9.48. As the reduction of the critical load due to the cutout is diminished with such reinforcement, the effect of random geometrical imperfections domi-

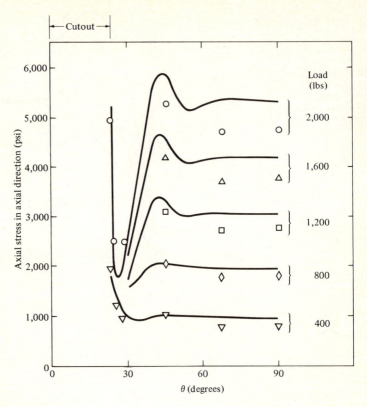

FIGURE 9.24
Axial stress at midlength (*Reproduced from Ref. 9.48, figs. 8 and 9, with the permission of Microform International Marketing Corporation, exclusive copyright licensee of Pergamon Press journal back files.*)

Table 9.5   MEASURED   AND   COMPUTED COLLAPSE LOADS

Shell thickness, in.	Cutout width, degrees	Critical load, lb	
		Computed	Measured
0.014	30	2,900	2,740
0.014	45	2,250	2,250
0.014	45	2,250	2,000

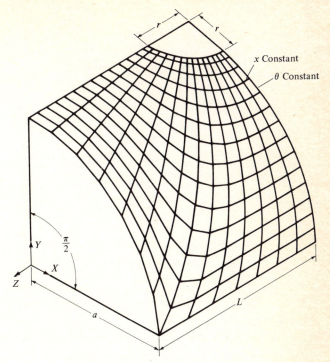

FIGURE 9.25
Finite difference grid for cylinder with circular cutout.

nates the effect of the cutout. Consequently, the agreement between tests and theory is not as close as for the cylinders with unreinforced cutouts.

The finite difference energy method also has been used for collapse analysis of cylinders with circular cutouts (Ref. 9.49). In this case the more general finite difference expressions defined by Eq. (8.70) were used. Figure 9.25 shows a segment of the shell with a finite difference grid. In contrast to the traditional finite difference approach, the in-plane displacements in Ref. 9.49 are not defined in the direction of the grid lines, but in the directions given by the natural cylindrical coordinates.

In Ref. 9.50 Starnes reports the results from a large number of tests on cylinders with circular cutouts. He finds from the experimental results that a parameter $\alpha = r/\sqrt{ah}$ determines the shell behavior, where r is the radius of the circular cutout. For cylinders with relatively small cutouts, this conclusion was verified by the analysis reported in Ref. 9.49. However, for cylinders with suf-

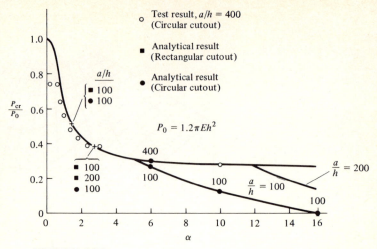

FIGURE 9.26
Critical loads for cylinders with cutouts, theoretical and experimental.

ficiently large cutouts, the value of α cannot alone determine the behavior of the shell. With two cutouts, for example, the critical load must be zero when $r/a = \pi/2$. This will occur for $a/h = 100$ at $\alpha = 5\pi$, and for $a/h = 400$ at $\alpha = 10\pi$. Therefore, for cylinders with relatively large cutouts, there must be separate curve branches for different a/h values in the P_{cr} versus α diagram. This was found to be the case when the critical loads were compared for two cylinders with $\alpha = 6$. The cylinder with $r/a = 0.3$ and $a/h = 400$ carries a significantly larger axial load than one with $r/a = 0.6$ and $a/h = 100$.

Results obtained from the analysis with uniform shortening are shown in Fig. 9.26 (Ref. 9.49), together with a curve based on these computed data and the obvious location of the end points of the curves. Test data for a cylinder with two cutouts were provided by Starnes. These are also plotted in the figure, and it appears that experimental and theoretical results are in very close agreement.

It may be noticed also that virtually the same critical load is obtained whether the cutout is circular or square. Figure 9.27 shows the outward displacement at the cutout edge (midlength) as a function of the applied axial load for two cylinders with $\alpha = 1.5$, one with a circular and one with a square cutout. The result that a cylinder with a square cutout can carry as high a load as a cylinder with a circular cutout is somewhat unexpected. The reason seems to be that larger displacements at the edge of the square cutout allow a more significant stress redistribution to occur.

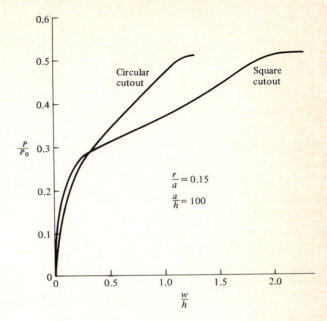

FIGURE 9.27
Equilibrium paths for cylinders with cutouts, $\alpha = 1.5$.

9.9 CONCLUSIONS

In the solution of the equations governing the behavior of the simple two-column structure in Sec. 9.2, we determined the primary path and a secondary path in the load-displacement diagram. The location of the bifurcation point then was found from plots of these equilibrium paths. This is an unusual situation; it generally is easier to determine the bifurcation-point load directly. In fact, as has been emphasized in the preceding chapters, linearized analyses that give only a bifurcation-point load frequently are used for reasons of economy, even though the information given by such analyses is limited. If the prebuckling behavior is linear, or if the effect of prebuckling nonlinearity can be neglected, the bifurcation-point analysis leads to a linear eigenvalue problem. In this case the bifurcation analysis is a relatively simple and eminently useful way to determine the stability limit.

A bifurcation analysis also is the simplest procedure for axisymmetrically loaded shells of revolution when nonlinearity in the prebuckling behavior is included. In such a case it is feasible to find points on the secondary path by use of imperfections that induce nonsymmetrical deformation. As in the finite difference analysis of an elliptic cylinder in Sec. 9.7, solutions obtained for imperfect shells can be used as starting values for analysis with stepwise decreasing imperfection amplitude. Once a point on the secondary path for the perfect shell is found, we can trace the secondary path. In addition to determining the bifurcation-point load, we then gain some information about postbuckling strength or possible imperfection sensitivity. However, the analysis is quite cumbersome, because the unknown displacement components are functions of two spatial coordinates. In using a nonlinear analysis for the prebuckling configuration and the linearized analysis to establish the location of the bifurcation point, we need only a one-dimensional analysis (see Sec. 9.5b). Therefore, if we can restrict ourselves to determination of the bifurcation-point load, which often is the case, considerable savings in computer time are possible.

In the general case (i.e., when the prebuckling deformation is not axisymmetric), a similar timesaving procedure that gives accurate although limited information is not available. In tracing the nonlinear primary path, we must solve the nonlinear partial (two-dimensional) differential equations for each step. If a bifurcation point exists along this path, its existence is indicated by a change of sign in the coefficient determinant of the equations for adjacent equilibrium. The value of this determinant is readily available if the regular Newton-Raphson method or one of the incremental methods is used (see Sec. 9.3). The bifurcation point may be found from a plot of the value of the determinant vs. applied load. If the modified Newton-Raphson method is used, the factored matrix is updated only occasionally, and the value of the determinant usually is available only for very few load steps.

We notice that bifurcation occurs only if there exists some mode of deformation that is orthogonal to the prebuckling displacement pattern. In the general case this is not likely to occur. It appears that if such a mode exists and is recognized, inclusion of a small imperfection that is approximately of the shape of the buckling mode will expedite the analysis. Since the analysis of points on the primary path must be computed from a nonlinear two-dimensional analysis anyhow, the presence of an imperfection does not change the computation time. An advantage of this approach is that the problem of possible bifurcation does not arise. In addition, some information is obtained about the secondary path.

Finally, if the structure contains a plane of symmetry with respect to geometry as well as loading, it is easy to recognize possible buckling patterns as those that are antisymmetric with respect to that plane. In the analysis of an elliptic cylinder in Sec. 9.7, an antisymmetric imperfection was used to induce antisymmetric displacements on the primary path. Another possible procedure in such a case is to take advantage of the symmetry plane in computation of points on the primary path and to consider possible bifurcation into antisymmetric modes by use of the linearized analysis. In that case, a determinant has to be evaluated at a number of load steps in the neighborhood of a bifurcation load. This determinant is not readily available, because it corresponds to the adjacent equilibrium equations with boundary conditions that are different from those corresponding to the primary branch. It is not clear in this case that much can be gained in computer economy by use of the linearized analysis.

To save on computer time for the analysis of shells of general shape, it is necessary to accept an approximation. It may be assumed that the primary path can be computed by use of a linear analysis. In that case loss of stability is due to bifurcation. This approach must be used with great caution. Clearly, if the shell geometry changes considerably in the prebuckling range, as is the case with long cylinders under bending (Sec. 9.5d), the critical load may be drastically overestimated. On the other hand, if the structure allows redistribution of stress, as is the case when cylinders with cutouts are subjected to an axial load (Sec. 9.8), the collapse load may be much *above* the load indicated by the linearized theory. Frequently, the experienced analyst can judge a priori whether the linearized analysis will lead to satisfactory results. For example, if a relatively short cylinder is subjected to bending, it seems rather obvious that neither a substantial change in cross section of the cylinder nor significant redistribution of stresses will occur in the prebuckling range.

REFERENCES

9.1 HUTCHINSON, J. W., and W. T. KOITER: Postbuckling Theory, *Appl. Mech. Rev.*, vol. 23, pp. 1353–1366, December 1970.

9.2 HOFMEISTER, L. D., G. A. GREENBAUM, and D. A. EVENSON: Large Strain, Elastoplastic Finite Element Analysis, *Proc. AIAA/ASME 11th Structures Struct. Dyn. Mat. Conf.*, 1970.

9.3 BALL, R. E.: A Geometrically Nonlinear Analysis of Arbitrarily Loaded Shells of Revolution, *NASA* CR-909, 1968.

9.4 RALL, L. R.: "Computational Solution of Nonlinear Operator Equations," Wiley, New York, 1969.

9.5 DANIEL, J. W.: "The Approximate Minimization of Functionals," Prentice-Hall, Englewood Cliffs, N.J., 1971.

9.6 TILLERSON, J. R., J. A. STRICKLIN, and W. C. HAISLER: Numerical Methods for the Solution of Nonlinear Problems in Structural Analysis, R. F. Hartung (ed.),

"Numerical Solution of Nonlinear Structural Problems," *AMD*, vol. 6, American Society of Mechanical Engineers, November 1973.

9.7 BROGAN, F. A., and B. O. ALMROTH: Practical Method for Elastic Collapse Analysis for Shell Structures, *AIAA J.*, vol. 9, pp. 2321–2325, December 1971.

9.8 VON KÁRMÁN, T., E. E. SECHLER, and L. H. DONNELL: The Strength of Thin Plates in Compression, *Trans. ASME*, vol. 54, pp. 53–58, 1932.

9.9 MARGUERRE, K.: Die mittragende Breite der gedruckten Platte, *Luftfahrtforsch.* vol. 14, pp. 121–128, 1937.

9.10 KOITER, W. T.: The Effective Width at Loads Far in Excess of the Critical Load for Various Boundary Conditions (in Dutch), *NLL Rep.* S287, 1943.

9.11 COX, H. L.: The Buckling of a Flat Plate under Axial Compression and Its Behavior after Buckling, *Aeronaut. Res. Council, R. and M.* 20201, 1945.

9.12 LEVY, S., and P. KRUPEN: Large-Deflection Theory for End Compression of Long Rectangular Plates Rigidly Clamped along Two Edges, *NACA* TN 884, 1943.

9.13 MAYERS, J., and E. NELSON: Maximum Strength Analysis of Post-buckled Rectangular Plates, *AIAA 6th Aerospace Sci. Meet.*, New York, 1968.

9.14 FÖPPL, L.: Achsensymmetrisches Ausknicken zylindrischer Schalen, *S.-B. Bayr. Akad. Wiss.* 1926, pp. 27–40.

9.15 FLÜGGE, W.: "Stresses in Shells," 2d ed. Springer, Berlin, 1973.

9.16 STEIN, M.: The Influence of Prebuckling Deformations and Stresses on the Buckling of Perfect Cylinders, *NASA* TR R-190, 1964.

9.17 FISCHER, G.: ÜBER den Einfluss der gelenkigen Lagerung auf die Stabilität dünnwandiger Kreiszylinderschalen unter Axiallast und Innendruck, *Z. Flugwiss.*, vol. 11, pp. 111–119, 1963.

9.18 ALMROTH, B. O.: Influence of Edge Conditions on the Stability of Axially Compressed Cylindrical Shells, *AIAA J.*, vol. 4, pp. 134–140, January 1966.

9.19 BUSHNELL, D.: Analysis of Buckling and Vibration of Ring-stiffened Segmented Shells of Revolution, *Intl. J. Solids Structures*, vol. 6, pp. 157–181, January 1970.

9.20 HOFF, N. J.: Low Buckling Stresses of Axially Compressed Circular Cylindrical Shells of Finite Length, *J. Appl. Mech.*, vol. 32, pp. 533–541, 1965.

9.21 ALMROTH, B. O., and D. BUSHNELL: Computer Analysis of Various Shells of Revolution, *AIAA Jour.*, vol. 6, pp. 1848–1856, October 1968.

9.22 FLÜGGE, W.: Die Stabilität der Kreiszylinderschale, *Ing. Arch.* vol. 3, pp. 463–506, 1932.

9.23 LUNDQUIST, E. E.: Strength Tests of Thin-walled Duraluminum Cylinders in Compression, *NACA Rep.* 473, 1933.

9.24 DONNELL, L. H.: A New Theory for the Buckling of Thin Cylinders under Axial Compression and Bending, *Trans. ASME*, vol. 56, p. 795, 1934.

9.25 ESSLINGER, M.: "Hochgeschwindigkeitsaufnahmen von Beulvorgang dünn-wandiger, axialbelasteter Zylinder, *Stahlbau*, vol. 39, pp. 73–76, March 1970.

9.26 HOFF, N. J.: The Perplexing Behavior of Thin Cylindrical Shells in Axial Compression, *Israel J. Technol.*, vol. 4, pp. 1–28, 1966.

9.27 VON KÁRMÁN, T., and H.-S. TSIEN: The Buckling of Thin Cylindrical Shells under Axial Compression, *J. Aeronaut. Sci.*, vol. 8, pp. 302–312, June 1941.

9.28 KEMPNER, J.: Postbuckling Behavior of Axially Compressed Circular Cylindrical Shells, *J. Aeronaut. Sci.*, vol. 21, pp. 329–335, May 1954.

9.29 HOFF, N. J., W. A. MADSEN, and J. MAYERS: Postbuckling Equilibrium of Axially Compressed Circular Cylindrical Shells, *AIAA J.*, vol. 4, pp. 126–133, January 1966.

9.30 TENNYSON, R. C.: A Note on the Classical Buckling Load of Circular Cylindrical Shells under Axial Compression, *AIAA J.*, vol. 1, pp. 475–476, February 1963.

9.31 BRAZIER, L. G.: On the Flexure of Thin Cylindrical Shells and Other "Thin" Sections, *Proc. Roy. Soc.*, ser. A, vol. 116, pp. 104–114, 1926.

9.32 AKSEL'RAD, E. L.: Refinement of the Upper Critical Loading of Pipe Bending, Taking Account of the Geometrical Nonlinearity (in Russian), *Izv. S.S.S.R., OTN Mekh.* no. 4, pp. 123–139, 1965.

9.33 CLARK, R. A., and E. REISSNER: Bending of Curved Tubes, *Advan. Appl. Mech.*, New York, vol. 2, pp. 93–133, 1951.

9.34 STEPHENS, W. E., J. H. STARNES, JR., and B. O. ALMROTH: "Collapse of Long Cylindrical Shells under Combined Bending and Pressure Loads," AIAA/ASME/SAE 15th Structures, Structural Dynamics and Materials Conf., Las Vegas, Nevada, 1974.

9.35 BUDIANSKY, B.: Buckling of Clamped Shallow Spherical Shells, *Proc. IUTAM Symp. Theory Thin Elastic Shells*, Delft, The Netherlands, 1959, pp. 64–94.

9.36 WEINITSCHKE, H. J.: On the Stability Problem of Shallow Spherical Shells, J. *Math. Phys.*, vol. 38, pp. 209–231, 1960.

9.37 WEINITSCHKE, H. J.: Asymmetric Buckling of Clamped Shallow Spherical Shells, *NASA* TN D-1510, pp. 481–490, 1962.

9.38 HUANG, N. C.: Unsymmetric Buckling of Thin Shallow Spherical Shells, *J. Appl. Mech.*, vol. 31, pp. 447–457, 1964.

9.39 KRENZKE, M. A., and I. J. KIERNAN: Elastic Stability of Near-perfect Shallow Spherical Shells, *AIAA J.*, vol. 1, pp. 2855–2857, December 1963.

9.40 FITCH, J. R., and B. BUDIANSKY: Buckling and Postbuckling Behavior of Spherical Caps under Axisymmetric Load, *AIAA J.*, vol. 8, pp. 686–693, April 1970.

9.41 ALMROTH, B. O., and F. A. BROGAN: Bifurcation Buckling as an Approximation of the Collapse Load for General Shells, *AIAA J.*, vol. 10, pp. 463–467, April 1972.

9.42 KEMPNER, J., and Y.-N. CHEN: Buckling and Postbuckling of an Axially Compressed Oval Cylindrical Shell, *Proc. Symp. Theory Thin Shells Honor Lloyd Hamilton Donnell*, University of Houston, Houston, Tex., 1967, pp. 141–183.

9.43   HUTCHINSON, J. W.: Buckling and Initial Postbuckling Behavior of Oval Cylindrical Shells under Axial Compression, *J. Appl. Mech.*, vol. 35, pp. 66–72, 1968.

9.44   ALMROTH, B. O., F. A. BROGAN, and M. B. MARLOWE: Collapse Analysis for Elliptic Cones, *AIAA* J., vol. 9, pp. 32–37, January 1971.

9.45   MARLOWE, M. B., and F. A. BROGAN: Collapse of Elliptic Cylinders under Uniform External Pressure, *AIAA J.*, vol. 9, pp. 2264–2266, November 1971.

9.46   ROMANO, R. J., and J. KEMPNER: Stress and Displacement Analysis of a Simply-supported Noncircular Cylindrical Shell under Lateral Pressure, *J. Appl. Mech.*, vol. 29, pp. 669–674, 1962.

9.47   KEMPNER, J., and Y.-N. CHEN: Buckling and Initial Postbuckling of Oval Cylindrical Shells under Combined Axial Compression and Bending, *Polytech. Inst. Brooklyn PIBAL Rep.* 72–23, 1972.

9.48   ALMROTH, B. O., and A. M. C. HOLMES: Buckling of Shells with Cutouts, Experiment and Analysis, *Intl. J. Solids Structures*, vol. 8, pp. 1057–1071, August 1972.

9.49   ALMROTH, B. O., F. A. BROGAN, and M. B. MARLOWE: Stability of Cylinders with Circular Cutouts, *AIAA J.*, vol. 11, pp. 1582–1584, November 1973.

9.50   STARNES, J. H., JR.: The Effect of a Circular Hole on the Buckling of Cylindrical Shells, Ph.D. thesis, California Institute of Technology, Pasadena, Calif. 1970.

VARIATIONAL METHODS

The energy criteria and the vectorial (summation of forces and moments) criteria for equilibrium and for stability are given approximately equal emphasis in this work. Energy methods often are less familiar, however. For this reason, a brief descriptive discussion of pertinent energy methods is presented in this Appendix. A systematic development of energy methods is given in the book by Langhaar (Ref. A.1), on which much of the following material is based. A critical discussion of the energy criterion for loss of stability is given by Koiter (Ref. A.2).

A static conservative system is in equilibrium if its potential energy is stationary, and the equilibrium is stable if the potential energy is a relative minimum. These familiar criteria from elementary mechanics are applied first to discrete and then to continuous structural systems, in the following sections.

A.1 SINGLE DEGREE OF FREEDOM SYSTEMS

For a structure whose potential energy V is a function of a single displacement parameter C, the character of the potential energy for a particular configuration $C = C_0$ may be determined by examination of the change in potential energy ΔV corresponding to an arbitrary virtual displacement of the structure from C_0 to $C_0 + C_1$, where C_1

is an infinitesimally small increment. In terms of a Taylor's series expansion in the vicinity of C_0, the change in potential energy is

$$\Delta V = V'(C_0)C_1 + \frac{1}{2!}V''(C_0)C_1{}^2 + \frac{1}{3!}V'''(C_0)C_1{}^3 + \cdots \qquad (A.1)$$

where $V'(C_0) \equiv dV/dC$ at $C = C_0$, etc. For stationary V, the first derivative must equal zero:

$$V'(C_0) = 0 \qquad (A.2)$$

For a relative minimum at $C = C_0$, the first derivative must equal zero and the second derivative must be nonnegative:

$$V''(C_0) \geq 0 \qquad (A.3)$$

Thus equilibrium depends on the term in the expansion that is linear in C_1, and stability on the term that is quadratic in C_1. (For $V'' = 0$, the sign of ΔV depends on the sign of the next higher ordered nonzero term in the expansion.)

A.2   MULTIPLE DEGREE OF FREEDOM SYSTEMS

These familiar criteria for equilibrium and stability are readily generalized for multiple degree of freedom systems. For a structure whose potential energy is a function of two displacement parameters B and C, for example, the change in potential energy resulting from arbitrary small virtual displacements B_1, C_1 may be written in a Taylor's series expansion as follows:

$$\begin{aligned}
\Delta V = {} & \frac{\partial V}{\partial B}(B_0,C_0)B_1 + \frac{\partial V}{\partial C}(B_0,C_0)C_1 + \frac{1}{2!}\left[\frac{\partial^2 V}{\partial B^2}(B_0,C_0)B_1{}^2 \right. \\
& \left. + 2\frac{\partial^2 V}{\partial B \partial C}(B_0,C_0)B_1 C_1 + \frac{\partial^2 V}{\partial C^2}(B_0,C_0)C_1{}^2\right] + \cdots
\end{aligned} \qquad (A.4)$$

For stationary V, the sum of the first-order terms must equal zero. Since B_1 and C_1 are independently arbitrary, this condition is expressed by two equations

$$\frac{\partial V}{\partial B}(B_0,C_0) = \frac{\partial V}{\partial C}(B_0,C_0) = 0 \qquad (A.5)$$

(that is, $\partial V/\partial B = \partial V/\partial C = 0$ at $B = B_0$, $C = C_0$). For the stationary V to be a relative minimum, the sum of the second-order terms must be positive for all possible values of B_1, C_1. A polynomial that is homogeneously quadratic, such as this sum of quadratic terms in B_1, C_1, is called a *quadratic form*. A quadratic form that is positive not only for certain values of the variables, but for all possible values, is said to be *positive definite*.

The sum of the first-order terms in the expression for ΔV is called the *first variation* of V and is denoted by the symbol δV. The second-order terms together are called the *second variation* of V and are denoted by $\delta^2 V/2$. Then the change in potential energy can be written

$$\Delta V = \delta V + \frac{1}{2!}\,\delta^2 V + \frac{1}{3!}\,\delta^3 V + \cdots$$

where, from Eq. (A.4),

$$\delta V = \frac{\partial V}{\partial B}\,(B_0,C_0)B_1 + \frac{\partial V}{\partial C}\,(B_0,C_0)C_1 \tag{A.6a}$$

$$\delta^2 V = \frac{\partial^2 V}{\partial B^2}\,(B_0,C_0)B_1{}^2 + 2\,\frac{\partial^2 V}{\partial B\,\partial C}\,(B_0,C_0)B_1 C_1 + \frac{\partial^2 V}{\partial C^2}\,(B_0,C_0)C_1{}^2 \tag{A.6b}$$

In this terminology, the criterion for equilibrium of the multiple degree of freedom system is that the first variation of V be equal to zero, and the criterion for stability is that, in addition, the second variation be positive definite.

The critical load for a structure is the limiting load at which the structure first loses its stability. For the multiple degree of freedom system, the critical load is the lowest load for which $\delta^2 V$ ceases to be positive definite as the load is increased from zero. In the algebraic theory of quadratic forms, it is shown that the appropriate condition for the limit of positive-definiteness of a quadratic form is that the determinant of the coefficients equal zero (Ref. A.1, p. 206). Thus, from Eq. (A.6b), the condition for initial loss of stability may be written

$$\begin{vmatrix} \dfrac{\partial^2 V}{\partial B^2}\,(B_0,C_0) & \dfrac{\partial^2 V}{\partial B\,\partial C}\,(B_0,C_0) \\[3ex] \dfrac{\partial^2 V}{\partial B\,\partial C}\,(B_0,C_0) & \dfrac{\partial^2 V}{\partial C^2}\,(B_0,C_0) \end{vmatrix} = 0 \tag{A.7}$$

Equation (A.7) is often developed without explicit reference to the theory of quadratic forms, as follows. For sufficiently small values of applied load P, the equilibrium of the structure is stable and $\delta^2 V$ is positive for all nonzero values of B_1 and C_1. As the load is increased from zero, a value is reached (call it $\bar{P}$) at which $\delta^2 V$ first equals zero for at least one B_1, C_1 pair. It is still positive for all other B_1, C_1. For $P = \bar{P}$, therefore, $\delta^2 V$ is a relative minimum for this one B_1, C_1 pair. Then it is also stationary for this B_1, C_1, and $\partial(\delta^2 V)/\partial B_1 = \partial(\delta^2 V)/\partial C_1 = 0$. Differentiation with respect to B_1 and C_1 in Eq. (A.6b) gives

$$\frac{\partial^2 V}{\partial B^2}\,(B_0,C_0)B_1 + \frac{\partial^2 V}{\partial B\,\partial C}\,(B_0,C_0)C_1 = 0$$

$$\frac{\partial^2 V}{\partial B\,\partial C}\,(B_0,C_0)B_1 + \frac{\partial^2 V}{\partial C^2}\,(B_0,C_0)C_1 = 0 \tag{A.8}$$

Equations (A.8) are homogeneous and linear in B_1, C_1. For a nontrivial solution, the determinant of the coefficients must equal zero. Consequently,

$$\begin{vmatrix} \dfrac{\partial^2 V}{\partial B^2}(B_0,C_0) & \dfrac{\partial^2 V}{\partial B\, \partial C}(B_0,C_0) \\[2ex] \dfrac{\partial^2 V}{\partial B\, \partial C}(B_0,C_0) & \dfrac{\partial^2 V}{\partial C^2}(B_0,C_0) \end{vmatrix} = 0$$

This equation is the same as Eq. (A.7).

A.3 CONTINUOUS SYSTEMS

Extension of these equilibrium and stability criteria from discrete to continuous systems requires use of the calculus of variations. Consider, as an example, a structure whose potential energy V is a function of a displacement variable $w(x)$. Then

$$V = \int_{x0}^{x1} F[w(x)]\, dx \qquad (A.9)$$

where the integrand F is a known function of w, but w is as yet an unknown function of x. A quantity such as V, whose value depends, not on a number of discrete variables, but on all the values of one or more continuous variables in a given interval, is called a *functional*. For such structural systems, the object of a stability analysis is to determine the form of the function $w(x)$ that minimizes the definite integral.

Let us first consider a more specific example, namely, an equilibrium analysis in terms of elementary linear beam theory of a cantilever beam subjected to uniformly distributed loading q, (in pounds per inch), as shown in Fig. A.1. Our objective will be the derivation of the differential equation of equilibrium and so-called natural boundary conditions by application of the energy criterion for equilibrium. For a conservative structural system, the total potential energy V of a loaded structure is defined as the sum of the strain energy U of the structure itself and the potential energy of the applied load Ω (Ref. A.1):

$$V = U + \Omega \qquad (A.10)$$

The potential energy of the applied loads is the negative of the work that would be done by the applied load if the structure were given a displacement caused, not by the applied load, but by an independent external agent. From Fig. A.1, therefore,

$$\Omega = \int_0^L qw\, dx \qquad (A.11)$$

For a beam subjected to transverse loading only,

$$U = \frac{EI}{2} \int_0^L (w'')^2\, dx \qquad (A.12)$$

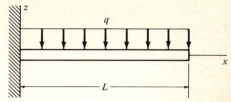

FIGURE A.1
Cantilever beam subjected to uniformly
distributed load.

where $w'' \equiv d^2w/dx^2$. Then

$$V = \int_0^L \left[\frac{EI}{2} (w'')^2 + qw \right] dx \qquad \text{(A.13)}$$

For the cantilever beam example, the integrand F in Eq. (A.9) is seen to have the form

$$F = \frac{EI}{2} (w'')^2 + qw \qquad \text{(A.14)}$$

Before making further use of the expression for the integrand, we must examine the boundary conditions. For the cantilever beam the boundary conditions are

$$w = w' = 0 \qquad \text{at } x = 0 \qquad \text{(A.15}a)$$

$$M = Q = 0 \qquad \text{at } x = L \qquad \text{(A.15}b)$$

These boundary conditions are seen to be of two kinds. Those at $x = 0$ refer to the kinematic quantities displacement and rotation, and are expressions of the physical constraints at the fixed end of the beam. The boundary conditions at $x = L$ refer to the kinetic quantities force and moment, and are conditions that must be met for equilibrium of the loaded beam. The two kinds of boundary conditions are termed *forced* and *natural* boundary conditions, respectively. The distinction is utilized in applications of the variational method, as will be seen in the following paragraph.

For the cantilever beam to be in equilibrium in a particular configuration $w_0(x)$, the potential energy must be stationary for that configuration. For the equilibrium to be stable, V at $w = w_0$ must also be a relative minimum. To determine the character of the potential energy at $w = w_0$, let

$$w \rightarrow w_0 + w_1$$

in Eq. (A.13), where the arrow is read " be replaced by," and w_1 is an arbitrary small virtual increment. Such an increment is called a *variation* in w. In the following discussion, both w_0 and w_1 must satisfy certain continuity and boundary conditions, namely, they must be continuous and twice differentiable in the interval $0 \le x \le L$, and they must satisfy the forced boundary conditions $w = w' = 0$ at $x = 0$ (Ref. A.1, p. 75). Functions that satisfy such requirements are called *admissible* functions.

For convenience, let $w_1(x) \equiv \varepsilon \zeta(x)$, where ε is an arbitrary small constant and $\zeta(x)$ is a function in the class of admissible functions and is otherwise arbitrary. Then

$$V + \Delta V = \int_0^L \left[\frac{EI}{2} (w_0'' + \varepsilon \zeta'')^2 + q(w_0 + \varepsilon \zeta) \right] dx \qquad \text{(A.16)}$$

The change in potential energy can be written

$$\Delta V = \varepsilon \int_0^L (EI w_0'' \zeta'' + q\zeta)\, dx + \varepsilon^2 \frac{EI}{2} \int_0^L (\zeta'')^2\, dx \qquad \text{(A.17)}$$

As with multiple degree of freedom systems, the first- and second-order terms in ΔV are called the first and second variations of V, respectively, and are denoted by δV and $\delta^2 V / 2$. Then

$$\Delta V = \delta V + \frac{1}{2!} \delta^2 V \qquad \text{(A.18)}$$

where

$$\delta V = \varepsilon \int_0^L (EI w_0'' \zeta'' + q\zeta)\, dx \qquad \text{(A.19)}$$

and

$$\tfrac{1}{2} \delta^2 V = \varepsilon^2 \frac{EI}{2} \int_0^L (\zeta'')^2\, dx \qquad \text{(A.20)}$$

For ε sufficiently small, $|\delta V| > |\delta^2 V|$. But δV changes sign if ε does, and therefore a necessary condition for V to be a relative minimum is that $\delta V = 0$, as before. For continuous systems this necessary condition for minimum V is the definition of stationary V; that is, the w_0 for which V is stationary is defined as the w_0 for which $\delta V = 0$.

For $\delta V = 0$, Eq. (A.19) gives, for arbitrary ε,

$$\int_0^L (EI w_0'' \zeta'' + q\zeta)\, dx = 0 \qquad \text{(A.21)}$$

Integration by parts and observation that $\zeta(x)$ is required to satisfy the forced boundary conditions gives, for a prismatic beam,

$$[EI w_0''(L)] \zeta'(L) - [EI w_0'''(L)] \zeta(L) + \int_0^L (EI w_0^{iv} + q)\, \zeta\, dx = 0 \qquad \text{(A.22)}$$

If this expression is to be satisfied for any function ζ, each of the three terms in Eq. (A.22) must equal zero. The first two, when set equal to zero, yield the relations

$$EI w_0'' = EI w_0''' = 0 \qquad \text{at } x = L \qquad \text{(A.23)}$$

For the third term in Eq. (A.22) to equal zero for arbitrary ζ, the factor $(EI_0^{iv} + q)$ must equal zero for every value of x in the interval $0 \le x \le L$, as may be seen, as follows. Suppose that, in any arbitrarily small region of the interval, $(EI w_0^{iv} + q)$ were positive (or alternatively, negative). Since ζ is an arbitrary admissible function, it might be positive in that region alone and zero elsewhere. Then the integrand would be positive (or negative) throughout the region and zero elsewhere, and the integral over the interval could not be zero. Consequently,

$$EIw_0^{iv} = -q \qquad 0 \le x \le L \qquad (A.24)$$

The three conditions for stationary V in Eqs. (A.23) and (A.24) are recognized as the natural boundary conditions [compare Eq. (A.15b)] and differential equation for equilibrium of the beam. Thus application of the energy criterion for equilibrium has led to both the governing differential equation and the natural boundary conditions in this example. The solution of the differential equation and forced and natural boundary conditions is readily found to be

$$w_0 = -\frac{q}{EI}\left(\frac{x^4}{24} - \frac{Lx^3}{6} + \frac{L^2x^2}{4}\right) \qquad (A.25)$$

For w equal to the w_0 in Eq. (A.25), V is stationary. Whether it is also a relative minimum depends on the sign of the second variation [Eq. (A.20)]. In this example $\delta^2 V$ is seen to be positive for any arbitrary function ζ, and therefore the equilibrium configurations given by Eq. (A.25) are stable. More will be said about the sign of the second variation later in the Appendix.

A.4 THE EULER EQUATIONS

Let us now consider as a more general example a structure for which the integrand F in Eq. (A.9) is a given function of one independent variable x, and one dependent variable w and its derivatives to second order, thus:

$$V = \int_{x_0}^{x_1} F(x,w,w',w'')\, dx \qquad (A.26)$$

To determine the character of V for a particular configuration $w = w_0$, we again let

$$w \to w_0 + w_1$$

where $w_1(x) = \varepsilon\zeta(x)$, ε is an arbitrary small constant, and $\zeta(x)$ is any admissible variation, i.e., any arbitrary function subject to the limitations that it satisfy the necessary continuity conditions (Ref. A.3, p. 190) and the forced boundary conditions (if any) at x_0 and x_1. Introduction into Eq. (A.26) gives

$$\Delta V = \int_{x_0}^{x_1} [F(x, w_0 + \varepsilon\zeta, w_0' + \varepsilon\zeta', w_0'' + \varepsilon\zeta'') - F(x,w_0,w_0',w_0'')]\, dx \qquad (A.27)$$

Expansion of the integrand in a Taylor's series gives, for the first variation, the expression

$$\delta V = \varepsilon \int_{x_0}^{x_1} \left(\frac{\partial F}{\partial w_0}\zeta + \frac{\partial F}{\partial w_0'}\zeta' + \frac{\partial F}{\partial w_0''}\zeta''\right) dx \qquad (A.28)$$

where $\partial F/\partial w_0$ represents $\partial F/\partial w$ for $w = w_0$, etc. Again, a necessary condition for V to be a relative minimum is that $\delta V = 0$. Because ε is arbitrary, this condition yields

$$\int_{x_0}^{x_1} \left(\frac{\partial F}{\partial w_0} \zeta + \frac{\partial F}{\partial w_0'} \zeta' + \frac{\partial F}{\partial w_0''} \zeta'' \right) dx = 0 \qquad (A.29)$$

By repeated integration by parts, Eq. (A.29) can be reduced to a set of boundary terms plus an integral in which $\zeta(x)$ appears as a factor, as in Eq. (A.22). The integral remaining after such integration by parts is found to be

$$\int_{x_0}^{x_1} \left(\frac{\partial F}{\partial w_0} - \frac{d}{dx} \frac{\partial F}{\partial w_0'} + \frac{d^2}{dx^2} \frac{\partial F}{\partial w_0''} \right) \zeta \, dx \qquad (A.30)$$

As previously noted, each of the boundary terms and the integral separately must equal zero for equilibrium. From the latter requirement the coefficient of ζ in the integrand must itself equal zero for all values of x in the interval, i.e.,

$$\frac{\partial F}{\partial w_0} - \frac{d}{dx} \frac{\partial F}{\partial w_0'} + \frac{d^2}{dx^2} \frac{\partial F}{\partial w_0''} = 0 \qquad x_0 \le x \le x_1 \qquad (A.31)$$

Equation (A.31) may be written without the subscript 0, with the understanding that solutions $w(x)$ represent configurations for which V is stationary. Thus

$$\frac{\partial F}{\partial w} - \frac{d}{dx} \frac{\partial F}{\partial w'} + \frac{d^2}{dx^2} \frac{\partial F}{\partial w''} = 0 \qquad (A.32)$$

Equation (A.32) is known as the *Euler equation* of the calculus of variations. It is sometimes called the *variational derivative* of F with respect to w. Its role for the continuous system is analogous to that of Eqs. (A.2) and (A.5) for single and multiple degree of freedom systems, respectively.

To illustrate use of the Euler equation in structural applications, we observe that the integrand in the potential energy expression for the cantilever beam in the preceding example is $F = EI(w'')^2/2 - qw$. Therefore $\partial F/\partial w = q$, $\partial F/\partial w' = 0$, and $\partial F/\partial w'' = EIw''$. Introduction into Eq. (A.32) and simplification gives, for constant EI,

$$EIw^{\text{iv}} = -q \qquad (A.33)$$

This is the differential equation for equilibrium of the beam [compare Eq. (A.24)]. Thus, for continuous systems, the criterion for equilibrium is represented by the Euler equation for the integrand in the total potential energy expression.

Equation (A.32) is the Euler equation for functionals of the form of Eq. (A.26) in which there is one dependent variable. When there are two dependent variables, such as $u(x)$ and $w(x)$ and, say, the highest-ordered derivatives in u and w are of first and second order, respectively, the Euler equations are found to be of the form

$$\frac{\partial F}{\partial u} - \frac{d}{dx}\frac{\partial F}{\partial u'} = 0$$

$$\frac{\partial F}{\partial w} - \frac{d}{dx}\frac{\partial F}{\partial w'} + \frac{d^2}{dx^2}\frac{\partial F}{\partial w''} = 0$$

(A.34)

For three dependent variables u, v, w and two independent variables x, y, if the highest-ordered derivatives are of first order in u and v and second order in w, the Euler equations are of the form (Ref. A.1, p. 92)

$$\frac{\partial F}{\partial u} - \frac{\partial}{\partial x}\frac{\partial F}{\partial u_{,x}} - \frac{\partial}{\partial y}\frac{\partial F}{\partial u_{,y}} = 0$$

$$\frac{\partial F}{\partial v} - \frac{\partial}{\partial x}\frac{\partial F}{\partial v_{,x}} - \frac{\partial}{\partial y}\frac{\partial F}{\partial v_{,y}} = 0$$

(A.35)

$$\frac{\partial F}{\partial w} - \frac{\partial}{\partial x}\frac{\partial F}{\partial w_{,x}} - \frac{\partial}{\partial y}\frac{\partial F}{\partial w_{,y}} + \frac{\partial^2}{\partial x^2}\frac{\partial F}{\partial w_{,xx}} + \frac{\partial^2}{\partial x \partial y}\frac{\partial F}{\partial w_{,xy}} + \frac{\partial^2}{\partial y^2}\frac{\partial F}{\partial w_{,yy}} = 0$$

Equations (A.34) and (A.35) are used extensively in Chaps. 3 to 6.

A.5 THE TREFFTZ CRITERION

The Euler equations represent the criterion for *stationary V*. Whether the V is a relative *minimum* depends on the sign of the second variation (except when $\delta^2 V = 0$, as noted). By definition, the integrand in the second-variation expression is a quadratic form in the variational displacements. The terminology "positive definite" is carried over from the algebraic theory of quadratic forms; the functional $\delta^2 V$ is said to be positive definite if it is positive for all possible variations. The critical load for a continuous structural system is the lowest load for which the definite integral $\delta^2 V$ is nonpositive for at least one possible variation. At this load the equilibrium changes from stable to unstable.

Let us again consider a specific example, namely, the determination of the critical load for a prismatic column subjected to compressive end loads P. The total potential energy may be written

$$V = U_m + U_b + \Omega \qquad \text{(A.36)}$$

where U_m and U_b are membrane and bending strain energy, and Ω is the potential energy of the applied loads. From elementary mechanics,

$$U_m = \frac{EA}{2}\int_0^L \varepsilon^2 \, dx$$

$$U_b = \frac{EI}{2}\int_0^L (w'')^2 \, dx$$

(A.37)

where ε is extensional strain on the centroidal axis. From Eq. (1.2),

$$\varepsilon = u' + \tfrac{1}{2}(w')^2$$

The potential energy of the axial load P is

$$\Omega = P[u(L) - u(0)] \qquad \text{(A.38)}$$

which may be written

$$\Omega = P \int_0^L u' \, dx \qquad \text{(A.39)}$$

Introduction into Eq. (A.36) gives

$$V = \int_0^L \left[\frac{EA}{2} (u' + \tfrac{1}{2}w'^2)^2 + \frac{EI}{2} (w'')^2 + Pu' \right] dx \qquad \text{(A.40)}$$

The expression for $\delta^2 V$ is obtained by letting

$$u \to u_0 + u_1$$
$$w \to w_0 + w_1$$

in Eq. (A.40), where u_0, w_0 denotes the configuration whose stability is under investigation, and the variations $u_1(x)$, $w_1(x)$ are infinitesimally small increments. For the undeflected form of the column,

$$u_0 = -\frac{P}{EA} x \qquad w_0 \equiv 0 \qquad \text{(A.41)}$$

Introduction into Eq. (A.40) and rearrangement gives, for the second variation, the equation

$$\tfrac{1}{2}\delta^2 V = \tfrac{1}{2} \int_0^L [EA(u_1')^2 + EI(w_1'')^2 - P(w_1')^2] \, dx \qquad \text{(A.42)}$$

The expression in Eq. (A.42) is seen to be positive definite for sufficiently small values of the applied load P. The critical value of P is the smallest load for which the definite integral is not positive definite.

The criterion for the limit of positive-definiteness for a continuous system is attributed to Trefftz (Ref. A.1, p. 211; see also the discussion in Ref. A.2, sec. 2). For small values of P, $\delta^2 V > 0$ for all nonzero variations $u_1(x)$, $w_1(x)$, in Eq. (A.42). For large values of P, $\delta^2 V < 0$ for some variations u_1, w_1. As P is increased from zero, a value is reached (call it $\bar{P}$) at which $\delta^2 V$ is for the first time zero for at least one variation u_1, w_1. It is still positive for all other variations. For $P = \bar{P}$, $\delta^2 V$ accordingly is stationary for this particular u_1, w_1. Then

$$\delta(\delta^2 V) = 0 \qquad \text{(A.43)}$$

i.e., the variational derivative with respect to u_1, w_1 of the integrand in $\delta^2 V$ equals zero.

For a functional of the form of Eq. (A.42), the Euler equations are [compare Eqs. (A.34)]:

$$\frac{\partial F}{\partial u_1} - \frac{d}{dx}\frac{\partial F}{\partial u_1'} = 0$$

$$\frac{\partial F}{\partial w_1} - \frac{d}{dx}\frac{\partial F}{\partial w_1'} + \frac{d^2}{dx^2}\frac{\partial F}{\partial w_1''} = 0$$

(A.44)

From Eq. (A.42),

$$F = EA(u_1')^2 + EI(w_1'')^2 - P(w_1')^2$$

Introduction of this expression for F into Eqs. (A.44) and simplification yields the final equations

$$u_1'' = 0 \qquad \text{(A.45a)}$$

$$EIw_1^{\text{iv}} + Pw_1'' = 0 \qquad \text{(A.45b)}$$

Equation (A.45b) is the differential equation for the critical load of the column [compare Eq. (1.23)]. Thus the criterion for loss of stability is represented by the Euler equations for the integrand in the second-variation expression.

In the general case, as noted, the potential energy increment may be written in the form

$$\Delta V = \delta V + \frac{1}{2!}\delta^2 V + \frac{1}{3!}\delta^3 V + \cdots$$

where terms on the right are linear, quadratic, etc., respectively, in the infinitesimally small variational displacements. Each nonzero term in the expansion is much larger than the sum of the succeeding terms. The first-order term vanishes identically for equilibrium configurations. Consequently, the sign of ΔV is governed by the sign of the second variation. For sufficiently small values of the applied load, the second variation is positive definite. The critical load is defined as the smallest load for which the second variation no longer is positive definite. On the basis of the Trefftz criterion, the equations for the critical load are given by the Euler equations for the integrand in the second-variation expression. Since that expression is a homogeneous quadratic functional, its variational derivatives necessarily are linear homogeneous differential equations. A discussion of the *sufficiency* of these criteria for the general case is presented in Ref. A.4.

It should be noted that the identification of the critical load with the vanishing of the second variation leaves open the question of whether the equilibrium is stable at the critical load itself, i.e., whether the equilibrium is stable for $P \leq P_{\text{cr}}$, or merely for $P < P_{\text{cr}}$. Since at $P = P_{\text{cr}}$ the second variation is equal to zero, the sign of ΔV is governed by the sign of the next higher ordered nonzero term in the expansion. This distinction is of no practical consequence in the determination of the critical load, but it is a central question in the discussion of initial-postbuckling behavior presented in Chap. 7.

REFERENCES

A.1   LANGHAAR, H. L.: "Energy Methods in Applied Mechanics." Wiley, New York, 1962.

A.2   KOITER, W. T.: Thermodynamics of Elastic Stability, *Proc. Third Canadian Congr. Appl. Mech.*, Calgary, 1971.

A.3   COURANT, R., and D. HILBERT: "Methods of Mathematical Physics," vol. I, Interscience, New York, 1953.

A.4   KOITER, W. T.: Purpose and Achievements of Research in Elastic Stability, *Recent Advanc. Eng. Sci.*, vol. 3, pp. 197–218, 1966.

INDEX

INDEX

Cylindrical shells:
    nonlinear equilibrium equations, 148, 149
    potential energy expression, 148, 149
    second-variation expression, 155
    stability equations, 154, 156, 160

Dead loading:
    circular rings, 120, 137
    cylindrical shells, 149, 157–159
    general shells, 194
Discontinuity zone, 162
Displacement method, 295
DMV equations (see Donnell-Mushtari-Vlasov equations)
Donnell approximations, 120, 137–139, 147
Donnell equations, 143
Donnell-Mushtari-Vlasov equations, 196

Effective length of columns, 28
Effective slenderness ratio, 66
Effective width of plates, 114
Eigenvalues, 9
Elastic foundation:
    circular ring, 133
    column, 35
Elastic lateral support:
    single spring, 31
    Winkler foundation, 35, 133
Elastic rotational restraint, 24
Elliptic cross sections, 340–344
Equilibrium paths:
    for axially compressed cylinder, 152, 182
    for cylinder subjected to bending, 335
    for cylinder with cutouts, 351
    for cylindrical panel, 3, 15, 352, 353
    for initially straight column, 11
    for noncircular cylinder, 343
    for rectangular flat plate, 89, 113
    for slightly crooked column, 15